Electrocatalytic Hydrogen Production: Catalysts and Applications

Xiang Peng

Published by **Materials Research Forum LLC**
Millersville, PA 17551, USA

Published as part of the book series
Materials Research Foundations
Volume 165 (2024)
ISSN 2471-8890 (Print)
ISSN 2471-8904 (Online)

Print ISBN 978-1-64490-306-3
ePDF ISBN 978-1-64490-307-0

Distributed worldwide by

Materials Research Forum LLC
105 Springdale Lane
Millersville, PA 17551
USA
https://www.mrforum.com

Printed in the United States of America
10 9 8 7 6 5 4 3 2 1

Table of Contents

Preface viii

Introduction 1

1. Introduction 1

2. Electrocatalytic Hydrogen Production Techniques 2

    2.1. Alkaline Water Electrolysis Cell (AWEC) 2

    2.2. Proton Exchange Membrane Electrolysis Cell (PEMEC) 3

    2.3. Solid Oxide Electrolysis Cell (SOEC) 4

    2.4. Anion Exchange Membrane Electrolysis Cell (AEMEC) 5

3. Application of Hydrogen 6

    3.1. Energy and Fuel 6

    3.2. Chemical Raw Material 7

    3.3. Reducing Agent in the Smelting Industry 7

4. Challenges in Electrocatalytic Hydrogen Production 8

    4.1. Electricity Consumption 8

    4.2. Catalysts 9

    4.3. Anodic Half-Cell Reaction 9

    4.4. Electrolyzer Components 11

References 11

Fundamentals of Electrocatalytic Hydrogen Production 16

1. Introduction 16

2. Reactions in Electrocatalytic Hydrogen Production 17

    2.1. Overview of HER 17

    2.2. Overview of OER 19

3. Performance Evaluation of the Electrocatalytic Hydrogen
Production 19

    3.1. Overpotential 19

    3.2. Tafel Slope 21

    3.3. Charge Transfer Properties 22

    3.4. Turnover Frequency 23

    3.5. Electrochemically Active Surface Area 24

    3.6. Stability 25

    3.7. Faraday's Efficiency 27

4.   Factors that Affect the HER Performance 28
        4.1. Type of the Catalyst 28
        4.2. Composition 30
        4.3. Structure 32
        4.4. Geometrical Morphology 33
        4.5. Electronic Structure 35
5.   Conclusion and Prospects 37
References 38

Precious Metal-based Catalysts for Electrocatalytic Hydrogen Production 43
1.   Introduction 43
2.   Pt-based Catalysts 44
        2.1. Morphology Optimization 47
        2.2. Structural Regulation 49
        2.3. Transition Metal Alloying 50
        2.4. Surface Modification 53
3.   Pd-based Catalysts 55
        3.1. Zero-Valent Pd Catalysts 55
        3.2. Pd-based Compounds 56
        3.3. Pd-based Alloys 57
4.   Other Precious Metal-based Catalysts 58
5.   Conclusion and Prospects 61
References 62

Single Atom Catalysts for Electrocatalytic Hydrogen Production 69
1.   Introduction 69
2.   Carrier Selection 70
        2.1. Carbon-based Carriers 70
        2.2. Metal-Organic Frameworks Carriers 74
        2.3. Transition Metal Compounds Carriers 76
3.   Active Center Metal Selection 79
4.   Coordination Structure Regulation 83
5.   Conclusion and Prospects 88
References 89

Transition Metal Compound Catalysts for HER 95
1.   Introduction 95

2. Transition Metal Carbides 96
   2.1. Structure and Properties of Transition Metal Carbides....................96
   2.2. Transition Metal Carbides for HER 97
3. Transition Metal Nitrides 100
   3.1. Structure and Physical Properties of Transition Metal
        Nitrides 100
   3.2. Transition Metal Nitrides for HER...101
4. Transition Metal Phosphides..103
   4.1. Structure and Physical Properties of Transition Metal
        Phosphides 103
   4.2. Transition Metal Phosphides for HER 104
5. Transition Metal Oxides...106
6. Transition Metal Sulfides 109
   6.1. Advantages of Transiton Metal Sulfides 109
   6.2. Transition Metal Sulfides for HER...109
7. Transition Metal Selenides...112
   7.1. Single Metal Selenide Catalysts 113
   7.2. Binary Metal Selenide Catalysts 114
8. Conclusion and Prospects..116
References 116

Composite Catalysts for Electrocatalytic Hydrogen Production........................126
1. Introduction 126
2. Metal/Metal Composites 127
   2.1. Intermetallic Compounds 127
   2.2. Solid Solution Alloys...129
   2.3. High-Entropy Alloys 131
3. Metal/Compound Composites..133
4. Compound/Compound Composites 138
   4.1. Crystal-Amorphous Composites 138
   4.2. Carbon-based Composites 139
   4.3. Mo-based Composites 139
5. Conclusion and Prospects..142
References 143

Surface Structure Modulation of the Catalysts 151

1. Introduction 151
2. Geometrical Morphology Regulation... 152
   2.1. Zero-Dimensional Structures... 152
   2.2. One-Dimensional Structure 153
   2.3. Two-Dimensional Structure 155
   2.4. Three-Dimensional Structures... 157
   2.5. Other Structures 159
3. Chemical State Modulation... 162
4. Electronic Structure Optimization... 165
   4.1. Heteroatom Doping 165
   4.2. Vacancy 169
   4.3. Phase Transition 172
   4.4. Strain.. 174
5. Conclusion and Prospects.. 178

   References 178

Heterostructure Catalysts for Electrocatalytic Hydrogen Production 186

1. Introduction 186
2. Heterostructure Catalysts for Electrocatalytic Hydrogen Production 187
   2.1. Metal/Metal Oxide Heterostructure Catalysts................................... 187
   2.2. Metal/Semiconductor Heterostructure Catalysts.............................. 196
   2.3. Mott-Schottky Heterostructure Catalysts 201
   2.4. Other Heterostructure Catalysts 208
3. Conclusion and Prospects.. 212

   References 213

Electrocatalytic Hydrogen Production Across a Wide pH Range 224

1. Introduction 224
2. Mechanisms and Challenges of HER Across a Wide pH Range.......... 225
   2.1. Catalytic Mechanisms of HER in Various Conditions.................. 225
   2.2. Challenges of HER Across a Wide pH Range 227
3. Construction Strategies for HER Catalysts Applied to a Wide pH Range.. 228
   3.1. Doping 228

    3.2. Constructing Heterogeneous Structures 232

    3.3. Alloying 240

  4. Conclusion and Prospects..249

  References 249

Catalysts for Oxygen Evolution Reaction **258**

  1. Introduction 258

  2. Mechanism of the Oxygen Evolution Reaction 259

  3. Catalysts for Oxygen Evolution Reaction...261

    3.1. Zero-Dimensional Nanoparticles..261

    3.2. One-Dimensional Nanomaterials 265

    3.3. Two-Dimensional Nanomaterials..272

    3.4. Three-Dimensional Nanomaterials..280

    3.5. Others..284

  4. Conclusion and Prospects..286

  References 287

Electrolyzer for Electrocatalytic Hydrogen Production **297**

  1. Introduction 297

  2. Alkaline Electrolyzers 298

    2.1. Alkaline Electrolyzer..298

    2.2. Anion Exchange Membrane Electrolyzer 300

  3. Acidic Electrolyzers 305

  4. Solid Oxide Electrolyzers..308

  5. Intermediate-Temperature Electrolyzers..310

  6. Conclusion and Prospects..312

  References 313

About the Author..**320**

Preface

Electrocatalytic hydrogen production has emerged as a promising avenue in the quest for sustainable energy solutions. As the world seeks to transition towards a cleaner and more efficient energy landscape, the development of efficient and cost-effective electrocatalysts for hydrogen evolution reaction (HER) holds great significance.

This book, "Electrocatalytic Hydrogen Production: Catalysts and Applications" provides a comprehensive exploration of the fundamental principles, materials, and strategies involved in the design and development of catalysts for electrocatalytic hydrogen production. It delves into various aspects of electrocatalytic hydrogen production, encompassing the fundamentals of reaction mechanisms, performance evaluation, and the design and characterization of advanced electrocatalysts.

This book was organized and written by Dr. Xiang Peng, with significant contributions from outstanding authors such as Song Xie, Baochai Xu, Shuai Feng, Yunfan Qiu, Rong Li, Xuerong Xu, and Hao Dong.

In Chapter 1, we lay the foundation by introducing the field of electrocatalytic hydrogen production, outlining its importance, and providing an overview of the content covered in subsequent chapters. Chapter 2 delves into the fundamental reactions involved in electrocatalytic hydrogen production, performance evaluation, and measurement techniques. It also discusses the factors that influence the performance of hydrogen production.

Chapters 3 to 6 focus on different types of catalysts for HER. Chapter 3 explores precious metal-based catalysts, while Chapter 4 delves into the realm of single atomic catalysts. Chapter 5 delves into the synthesis and properties of various compound electrocatalysts, including carbides, nitrides, phosphides, oxides, sulfides, selenides, and others. Chapter 6 investigates the integration of different materials to form composite electrocatalysts, including metal/metal composites, metal/compound composites, and compound/compound composites, all aimed at enhancing HER performance.

Chapter 7 delves into the importance of surface structure regulation in electrocatalysts, covering topics such as geometrical morphology regulation, chemical state regulation, and electronic structure regulation. The design, construction, preparation, characterization, and regulation strategies of heterostructured electrocatalysts are highlighted in Chapter 8.

Expanding the scope, Chapter 9 addresses the construction of HER electrocatalysts for wide pH range applications, enabling the utilization of electrocatalytic hydrogen production under different operating conditions. Chapter 10 explores electrocatalysts for the oxygen evolution reaction (OER), providing insights into reaction kinetics, precious metal-based electrocatalysts, and non-precious metal-based electrocatalysts.

Finally, Chapter 11 focuses on electrolyzers and electrolytes for electrocatalytic hydrogen production, examining their design and optimization for efficient and sustainable hydrogen generation.

Throughout this book, we aim to present the latest advancements in electrocatalytic hydrogen production, offering researchers, scientists, and engineers a comprehensive resource to understand the principles and strategies involved in developing efficient and sustainable electrocatalysts for HER. It is our hope that this book will inspire further innovation and collaboration in the pursuit of clean and abundant hydrogen energy.

Dr. Xiang Peng

Wuhan Institute of Technology

Electrocatalytic Hydrogen Production: Catalysts and Applications   Materials Research Forum LLC
Materials Research Foundations **165** (2024)   https://doi.org/10.21741/9781644903070

CHAPTER 1

Introduction

Xiang Peng*

Hubei Key Laboratory of Plasma Chemistry and Advanced Materials, Engineering Research
Center of Phosphorus Resources Development and Utilization of Ministry of Education, School
of Materials Science and Engineering, Wuhan Institute of Technology, Wuhan 430205, China

xpeng@wit.edu.cn

Abstract

Hydrogen has long been recognized as a clean and carbon-zero energy carrier in the renewable energy sector. Its potential as an efficient supplement to the future power grid and its significant roles in transportation, diverse industries, and construction, replacing carbon-related sources. Electrocatalytic hydrogen production driven by renewable energy-derived electricity emerges as one of the most promising techniques for green hydrogen production in the future. Technological advancements and large-scale production aimed at reducing the expense of green hydrogen production have become the shared goals of hydrogen energy enterprises worldwide.

Keywords

Hydrogen Recycling, Alkaline Water Electrolysis Cell, Proton Exchange Membrane Electrolysis Cell, Solid Oxide Electrolysis Cell, Applications of Hydrogen, Challenges in Electrocatalytic Hydrogen Production

1.   Introduction

Hydrogen, known for its high energy density (142 kJ g^{-1}) surpassing that of fossil fuels, chemical fuels, and biofuels, has long been recognized as a clean and carbon-zero energy carrier in the renewable energy sector [1-3]. Its potential as an efficient supplement to the future power grid and its significant roles in transportation, diverse industries, and construction, replacing carbon-related sources, are illustrated in **Figure 1** [4-6]. Moreover, hydrogen combustion results in the production of water, making it a sustainable and environmentally friendly option [7, 8]. Over 30 countries or coalitions have formulated national hydrogen strategies, emphasizing its importance.

However, the current industrial production method involves pyrolyzing fossil fuels such as coal, oil, and natural gas, which not only wastes energy but also has adverse environmental impacts [9, 10]. Consequently, hydrogen production driven by renewable energy-derived electricity emerges as one of the most promising techniques for green hydrogen production in the future [11, 12]. However, the expense of green hydrogen production remains higher than that of fossil fuel hydrogen production, with electricity consumption accounting for 60% ~ 70% of the total cost [3, 13, 14]. Therefore, technological advancements and large-scale production aimed at reducing

Materials Research Forum LLC
https://doi.org/10.21741/9781644903070

the expense of green hydrogen production have become the shared goals of hydrogen energy enterprises worldwide. In the context of sustainable development and addressing climate change, hydrogen, as a potential energy solution, holds tremendous development potential. Through further research and innovation, we can expect to see hydrogen plays a more significant role in the energy sector in the future, contributing to the achievement of sustainable energy transition.

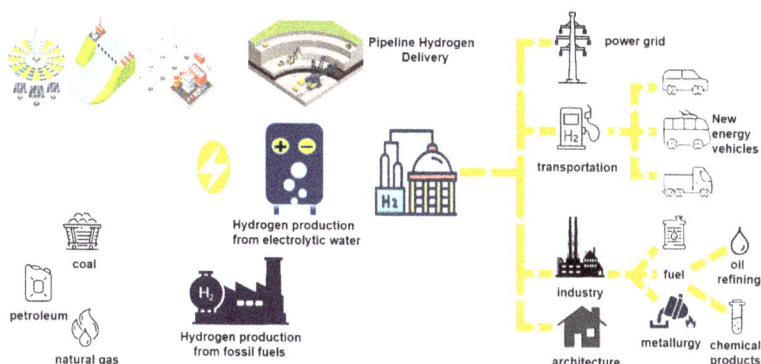

Figure 1. Hydrogen recycling.

2.   Electrocatalytic Hydrogen Production Techniques

In the field of hydrogen production by water electrolysis, there are generally four technologies available, namely alkaline water electrolysis cell (AWEC), proton exchange membrane electrolysis cell (PEMEC), anion exchange membrane electrolysis cell (AEMEC), and solid oxide electrolysis cell (SOEC) [15]. Among these technologies, the AWEC stands as the most extensively employed technique in the industry.

2.1.   Alkaline Water Electrolysis Cell (AWEC)

The AWEC comprises an anode, cathode, and membrane, which work together to separate hydrogen and oxygen gases, as depicted in **Figure 2**. Typically, alloys are used as electrodes to facilitate the splitting of water molecules into hydrogen and oxygen products through the application of electric current, with an energy conversion efficiency ranging from 59% to 70% [16]. In this system, a 20% to 30% concentration of KOH aqueous solution is commonly employed as the electrolyte [17]. The utilization of non-precious metal-based electrocatalysts and the prolonged lifespan of the electrolyzer in alkaline environments contribute to the cost-effectiveness of this hydrogen production technique [18].

The alkaline water electrolysis technology has several advantages. Firstly, it is relatively mature and technologically reliable and has been widely used in industrial applications. Secondly, compared to other water electrolysis technologies, alkaline water electrolysis has lower costs, partially due to the use of non-precious metal electrocatalysts, which reduces material expenses.

Additionally, AWEC operates under alkaline conditions, and thus the electrolyzer has a longer lifespan, reducing maintenance and replacement costs. However, AWEC also faces some challenges. Firstly, its energy conversion efficiency is relatively lower. Secondly, alkaline water electrolysis has higher requirements for electrolyte concentration and purity, which may increase the complexity of operations and handling. Additionally, precautions need to be taken to prevent solution leakage and corrosion issues since alkaline solutions are used in the electrolyzer.

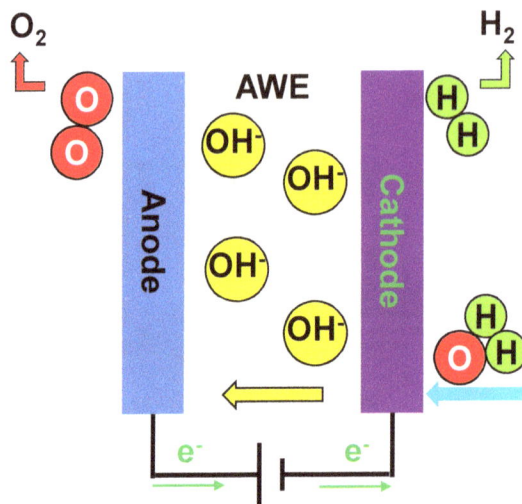

Figure 2. Schematic diagram of the AWEC for hydrogen production.

2.2.   Proton Exchange Membrane Electrolysis Cell (PEMEC)

The membrane electrode in the PEMEC used for hydrogen production consists of an anode layer, a cathode layer, and a proton exchange membrane, as depicted in **Figure 3**. This system operates in acid media at a temperature ranging from 50-80 °C [19]. Both the anode and cathode layers are supported by the membrane. The proton exchange membrane is an ionic conductive membrane that exhibits excellent physical and chemical stability, along with efficient proton transfer capabilities [20, 21]. Within the proton exchange membrane, the hydrophobic polymer electrolyte skeleton and hydrophilic groups are separated to form hydrophilic channels, enabling water molecule proton transfer through the membrane [22]. The protons generated at the anode traverse to the cathode, where they combine to form hydrogen gas through the HER, establishing a pathway for ion transport within the electrolytic cell. Moreover, the proton exchange membrane facilitates the separation of hydrogen and oxygen gases to prevent interpenetration. The performance of PEMEC greatly relies on the proton conductivity of the membrane, ion exchange capacity, mechanical strength, water uptake, and swelling rate.

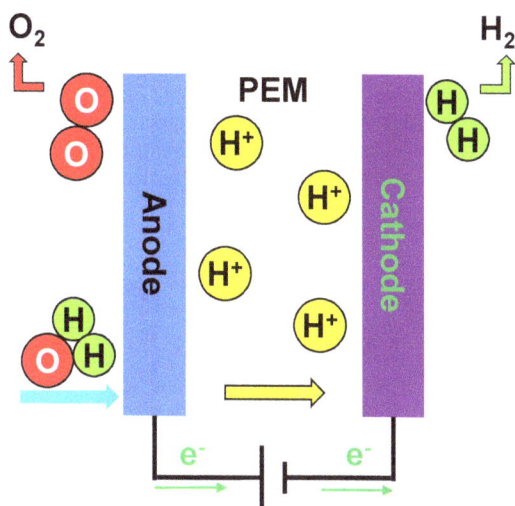

Figure 3. Schematic diagram of the PEMEC for hydrogen production.

A PEMEC typically operates in an acid medium, utilizing a voltage range of 1.4 to 2.0 V. This voltage range enables the generation of high current densities and hydrogen products at elevated pressures. One of the key advantages of PEMEC is its ability to start up within milliseconds, thereby accommodating the volatile nature of renewable energy power generation systems. This characteristic makes it highly compatible with renewable energy systems, positioning it as an ideal solution for the production of green hydrogen through water electrolysis. However, it's important to note that the anode and cathode catalysts used in PEMEC are noble metals, such as Pt (platinum), Ir (iridium), and Ru (ruthenium), which are loaded onto a porous substrate. The anode employs a porous Ti substrate as the gas diffusion layer, while the cathode utilizes a porous carbon substrate for the same purpose. The incorporation of these expensive membrane materials and noble catalysts contributes to the overall high cost of the system.

2.3   Solid Oxide Electrolysis Cell (SOEC)

Figure 4 illustrates the structure and working mechanism of a SOEC. SOEC employs solid ceramics as electrolytes and operates at high temperatures ranging from 500 to 1000 °C. These solid electrolytes exhibit excellent ion transport characteristics and high chemical stability at high temperatures. This enables SOEC to operate under extreme conditions, such as high temperatures and corrosive environments. The elevated temperature reduces the Gibbs free energy of the reaction, resulting in a low equilibrium voltage requirement for water electrolysis. For instance, at 800 °C and 0.1 MPa, the equilibrium voltage for electrolyzing water vapor is only 0.85 V, leading to significantly reduced electricity consumption [23]. The high operational temperature of SOEC facilitates excellent reaction kinetics, allowing for conversion efficiencies reaching approximately 100%.

One notable advantage of SOEC is that it does not rely on noble metal-based electrocatalysts for hydrogen production. Commonly used anode materials in SOEC include strontium-doped lanthanum manganite (LSM), yttrium-stabilized zirconia (YSZ), etc. [24, 25]. The cathode can be composed of nickel-yttrium-stabilized zirconia (Ni-YSZ) [26]. The use of SOEC presents promising opportunities for efficient and sustainable hydrogen production. The high operating temperature enables favorable reaction kinetics and eliminates the need for expensive noble metal catalysts, contributing to cost reduction and overall system efficiency.

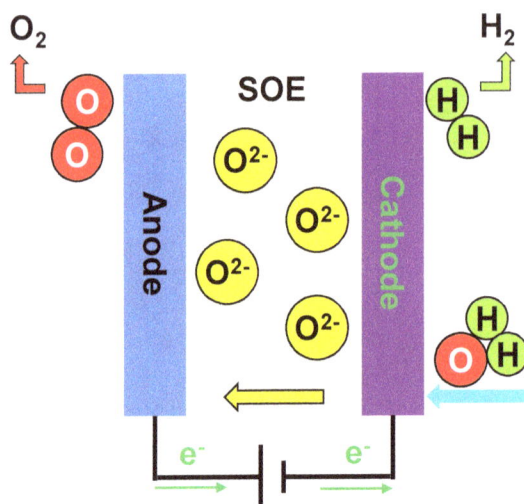

Figure 4. Schematic diagram of the SOEC for hydrogen production.

2.4   Anion Exchange Membrane Electrolysis Cell (AEMEC)

The AEMEC, as shown in **Figure 5**, differs fundamentally from PEMEC as it involves the exchange of ions from protons to hydroxide ions within the membrane. Hydroxide ions have a relative molecular mass 17 times larger than protons, resulting in slower migration. AEMEC can operate at relatively low temperatures of 40 to 60 °C, achieving energy conversion efficiencies of 60 to 79% [27]. The absence of metal cations or carbonate precipitation prevents clogging in the hydrogen production system.

AEMEC utilizes non-precious metal catalysts such as nickel, cobalt, and iron, enabling the production of highly pure hydrogen. The non-precious metal catalysts exhibit good activity and stability, effectively promoting hydrogen generation at lower temperatures while reducing system costs. Furthermore, AEMEC can operate at relatively low temperatures, enhancing system safety and reliability, and making it more suitable for certain special applications. The system exhibits rapid response characteristics, aligning well with the requirements of renewable energy power generation systems. However, before its industrial application, certain aspects of

AEMEC, such as the low permeability of hydroxide ions through the membrane, the membrane's inferior mechanical stability, and the sluggish catalytic kinetics, require optimization.

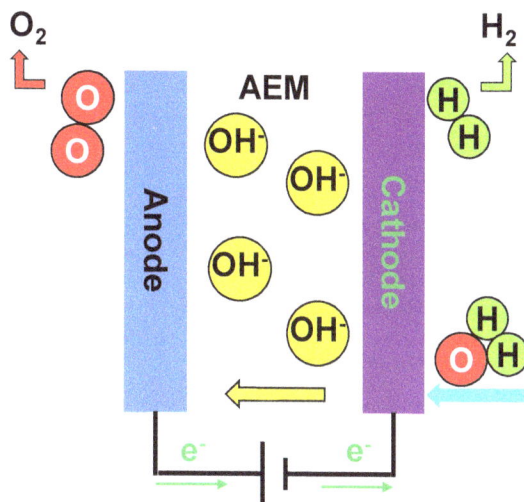

Figure 5. *Schematic diagram of the AEMEC for hydrogen production.*

3.   Application of Hydrogen

Developing and implementing hydrogen on a large scale is crucial for reducing carbon emissions and achieving carbon neutrality. Hydrogen serves not only as a fuel within energy systems but also as a valuable chemical feedstock in the chemical industry and as a reducing agent in the smelting industry.

3.1.   Energy and Fuel

The successful conversion of sustainable wind and solar energy into electricity has been achieved. However, the intermittent and unpredictable nature of wind and solar power generation poses challenges to the continuity and stability of grid-connected power supply. To address this issue, utilizing electricity generated from wind and solar power to electrolyze water for green hydrogen production can be a solution. This approach not only effectively harnesses wind and solar energy but also reduces the cost of producing green hydrogen. Implementing a hydrogen energy system can enhance the safety and stability of a renewable energy-driven power generation system. Hydrogen energy serves as a bridge, connecting renewable energy sources with the power grid, thereby accelerating the transformation of the energy system. This integration allows for a more reliable and resilient energy supply.

Hydrogen, as a representative form of chemical energy, can be converted back into electricity using fuel cells such as proton exchange membrane fuel cells (PEMFCs) and solid oxide fuel cells (SOFCs). Hydrogen fuel cells offer numerous advantages, including high energy density, high energy conversion efficiency, and zero carbon emissions. These characteristics have attracted increasing attention in the field of automotive power supply, as they provide a clean and efficient alternative to traditional combustion engines. By leveraging the potential of hydrogen fuel cells, we can further enhance the utilization of hydrogen as an energy carrier, promoting the adoption of clean and sustainable energy systems while reducing greenhouse gas emissions.

3.2. Chemical Raw Material

Hydrogen serves as a crucial raw material for the production of various chemical products in the industry, including ammonia, urea, methanol, and more. Ammonia, primarily synthesized through the Haber-Bosch process, possesses a higher energy density than hydrogen. It can be utilized for energy storage and power generation, offering a carbon dioxide (CO_2) emission-free alternative [28]. Additionally, the combination of ammonia and CO_2 can yield urea, a vital nitrogen fertilizer and sustainable hydrogen carrier [29].

The technique of hydrogenation proves effective in producing clean oil products and enhancing product quality. In the petrochemical industry, hydrogenation techniques encompass heavy oil hydrocracking for aromatics and ethylene production, residuum hydrodesulfurization for ultralow sulfur fuel and inferior catalytic diesel, and gasoline hydroconversion for high-octane gasoline, among others. However, it is worth noting that hydrogenation techniques are associated with high costs and energy consumption.

Hydrogen can react with CO_2 to generate carbon-based fuels like methanol, methane, formic acid, and formaldehyde, among others, which can be conveniently stored and transported in their liquefied states. Particularly, methanol exhibits a high hydrogen content of 12.6% (mass fraction) and a significant energy density of 5.53 kWh kg^{-1}, making it an important liquid fuel and hydrogen energy carrier. Methanol can be converted back into hydrogen and carbon monoxide for use as fuel in PEMFCs or employed directly in methanol fuel cells. This conversion of hydrogen into methanol, methane, and hydrocarbons enables efficient storage and transportation of hydrogen produced from renewable energy sources, thus addressing safety and cost concerns associated with hydrogen production, storage, and transportation in the hydrogen industry. Consequently, it provides solutions for highly efficient hydrogen utilization.

3.3. Reducing Agent in the Smelting Industry

Using coke as a reducing agent in traditional iron smelting processes results in significant carbon emissions and the release of harmful gases. However, substituting coke with hydrogen as a reducing agent can substantially reduce carbon emissions and promote cleaner metallurgical transformations, as the byproduct of hydrogen reaction is water. Green hydrogen, in particular, holds great promise as a replacement for coke in actual production processes. In addition, hydrogen can serve as an energy carrier in metallurgical processes including iron and steel production, as well as the production of other metals. The use of hydrogen as a reducing agent in metallurgical processes results in the production of cleaner metals. This contributes to improving the quality of the produced metals and reducing the environmental impact.

Moreover, hydrogen is commercially utilized for extracting tungsten from ores like wolframite, scheelite, and ferberite. The same principle can be applied to copper production from ores such as copper oxide, black copper, and cone black copper. While the use of hydrogen in the metallurgical industry offers numerous benefits, there are also challenges to overcome. One of the main challenges is reducing the production costs of green hydrogen to make it more competitive compared to traditional fuels and reducing agents. The cost of green hydrogen is relatively high, which could increase smelting costs by 20% to 30%. Therefore, reducing the production cost of green hydrogen is crucial for the widespread adoption of hydrogen in the smelting industry.

4.   Challenges in Electrocatalytic Hydrogen Production

4.1.   Electricity Consumption

Electrolytic water technology for hydrogen production holds great promise for achieving large-scale production of green hydrogen. However, the high cost associated with this technology, primarily due to electricity consumption, currently presents a barrier to replacing fossil fuel-based hydrogen production with high carbon emissions. Approximately 70% of the cost of hydrogen production through water electrolysis is attributed to electricity consumption [30]. Therefore, reducing the expense of electricity is crucial to improving the economic viability of hydrogen production and enabling its widespread application in various fields.

There are two main approaches to address this challenge. Firstly, reducing the voltage requirement of the water electrolysis process can directly decrease electricity consumption. This can be achieved by using highly efficient catalysts at both electrodes and optimizing the structure of the electrolyzer and electrolyte. By enhancing the efficiency of the electrolysis process, the overall electricity consumption can be significantly reduced.

Secondly, lowering the price of electricity generated from sustainable energy sources plays a key role in reducing the cost of hydrogen production. This can be accomplished by developing advanced systems for wind power and solar cells that offer high energy conversion efficiency. The advancement of these technologies will enable the production of electricity at a lower cost, making green hydrogen more economically competitive.

By focusing on both improving the efficiency of water electrolysis and reducing the cost of sustainable electricity generation, the overall cost of green hydrogen production can be significantly reduced. This will facilitate the widespread adoption of green hydrogen across various sectors.

4.2.   Catalysts

The theoretical voltage required for water electrolysis is 1.23 V (H_2O (l) → H_2 (g) + 1/2O_2 (g), ΔG = +237.2 kJ mol^{-1}, ΔE^o = 1.23 V). However, in practice, the voltage required is commonly higher than 1.23 V due to overpotentials at both electrodes, as shown in **Figure 6** [31]. Catalysts are necessary to reduce the overpotentials on both electrodes. Typically, Pt group metals are used to facilitate the hydrogen evolution reaction (HER) at the cathode, while IrO_2 and RuO_2 are employed to accelerate the oxygen evolution reaction (OER) at the anode. The use of these precious metals and compounds contributes to the high cost of this hydrogen production technique, limiting its large-scale industrial application.

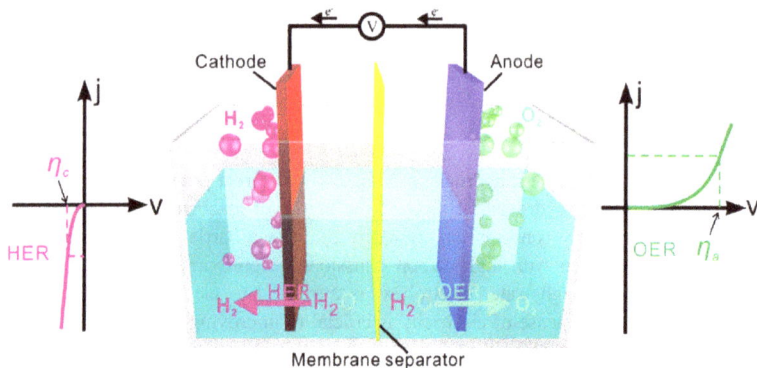

Figure 6. Half-cell reactions in electrocatalytic water splitting. Reproduced with permission from ref [31]. Copyright 2020, American Chemical Society.

To address this issue, there have been proposals and developments in the field of noble-metal-based catalysts with ultra-low mass loading. The aim is to reduce the cost of the catalysts used in electrochemical hydrogen production systems. Furthermore, there is growing interest in the design and construction of non-noble metal-based catalysts as alternatives to expensive noble metal-based materials for hydrogen production. For instance, developing non-noble metal-based materials with specific morphology, chemical state, and electronic structure on the surface can mimic the surface structure of noble metal-based materials. This optimization of the catalyst surface facilitates improved adsorption behavior of reactants/intermediates, resulting in the presentation of high-performance and low-cost catalysts for hydrogen production.

4.3.   Anodic Half-Cell Reaction

The reaction potential for the anodic OER, (2H_2O (l) → O_2 (g) + 4H^+ (aq) + 4e^-) is 1.23 V *vs.* reversible hydrogen electrode (RHE) in the overall water electrolysis. The high reaction potential leads to significant electrical consumption at the anode. Additionally, the OER is a 4-electron transfer process that exhibits slow kinetics. The combination of a high-energy barrier and slow kinetics in the anodic half-cell reaction leads to unsatisfactory energy conversion efficiency and high costs associated with hydrogen production.

Materials Research Forum LLC
https://doi.org/10.21741/9781644903070

One effective approach to mitigate these challenges is to replace the OER with thermodynamically favorable anodic oxidation reactions. Implementing alternative anodic oxidation reactions presents a promising avenue for advancing the field of hydrogen production, facilitating the development of efficient and cost-effective electrolytic systems. For example, alternative reactions such as urea oxidation reaction [32-34], methanol oxidation reaction [35-38], hydrazine oxidation reaction [39-42], glucose oxidation reaction [43-46], iron oxidation reaction [47], and others can be employed to couple with the HER and lower the required reaction potential at the anode. By adopting these thermodynamically easier anodic oxidation reactions, the voltage for overall reaction can be decreased. For example, Peng et al. proposed a novel strategy to integrate the HER with the waste Fe upgrading reaction (FUR), since iron corrodes easily and even self-corrodes to form magnetic iron oxide species and generate corrosion currents [47]. The linear sweep voltammetry (LSV) curve in **Figure 7a** presented that only 0.11 V *vs*. RHE is essential to afford a current density of 10 mA cm^{-2} in FUR in the neutral electrolyte (0.5 M Na_2SO_4), which is 2.01 V smaller than that in OER and even 1.71 V smaller than that of the commercial IrO_2 electrocatalyst. The results suggest that by replacing OER with FUR, the potential required to achieve a current density of 10 mA cm^{-2} can decrease by 95%. As a result, the overall reaction demonstrated extremely highly energy-efficient hydrogen production in neutral media, as illustrated in **Figure 7b**. The constructed HER/FUR overall reaction couple required an ultralow voltage of 0.68 V for the current density of 10 mA cm^{-2} with a small power equivalent of 2.69 kWh per m^3 H_2, which is even less than the theoretical limit of conventional overall water splitting of 2.94 kWh per m^3 H_2.

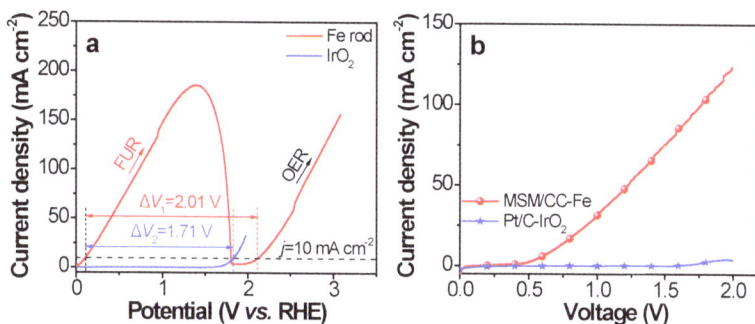

Figure 7. (a) LSV curves of FUR with a waste steel rod as the working electrode. (b) LSV curves of MSM/CC∥Fe and Pt/C∥IrO$_2$ couples. Reproduced with permission from ref [47]. Copyright 2023, Elservier.

The alternative anodic oxidation reaction strategy offers several advantages, including improved energy efficiency and reduced costs in hydrogen production. By exploring alternative reactions and optimizing the conditions for their implementation, researchers aim to enhance the performance and economic viability of water electrolysis for hydrogen production.

4.4. Electrolyzer Components

The production of electrolytic hydrogen involves complex and costly materials and components. One key component in PEMEC is the proton exchange membrane, which accounts for approximately 10% of the cell's cost. To achieve industrial-scale PEMEC, the proton exchange membrane must possess certain characteristics. Firstly, the membrane should exhibit good proton conductivity to reduce transfer impedance. This property facilitates efficient proton transport within the electrolysis cell. Secondly, excellent mechanical properties are necessary to provide support for the catalyst layer and maintain the structural integrity of the membrane. Thirdly, the membrane should possess strong chemical and thermal stability to ensure the long-term stability and durability of the electrolysis cell. Fourthly, the membrane should have low permeability to effectively separate the oxygen and hydrogen generated during the electrolysis process at the electrodes.

Furthermore, PEMEC operates in a highly acidic and oxidizing environment, which places additional demands on the equipment. This often requires the use of expensive metal materials such as iridium, platinum, and titanium, resulting in high costs associated with this technique [48]. The complexity and cost of materials and components contribute to the high expenses associated with electrolytic hydrogen production. Efforts are being made to develop alternative materials and improve their efficiency and cost-effectiveness to overcome these challenges and enable widespread adoption of this technology.

References

[1] L. Schlapbach, A. Züttel, Hydrogen-storage materials for mobile applications, Nature, 414 (2001) 353-358. https://doi.org/10.1038/35104634

[2] X.M. Guo, E. Trably, E. Latrille, H. Carrère, J.-P. Steyer, Hydrogen production from agricultural waste by dark fermentation: a review, Int. J. Hydrogen Energ., 35 (2010) 10660-10673. https://doi.org/10.1016/j.ijhydene.2010.03.008

[3] S.E. Hosseini, M.A. Wahid, Hydrogen production from renewable and sustainable energy resources: Promising green energy carrier for clean development, Renew. Sustain. Energy Rev., 57 (2016) 850-866. https://doi.org/10.1016/j.rser.2015.12.112

[4] P.M. Grant, C. Starr, T.J. Overbye, A power grid for the hydrogen economy, Sci. Am., 295 (2006) 76-83. https://doi.org/10.1038/scientificamerican0706-76

[5] M.Z. Oskouei, H. Mehrjerdi, Optimal allocation of power-to-hydrogen units in regional power grids for green hydrogen trading: Opportunities and barriers, J. Clean. Prod., 358 (2022) 131937. https://doi.org/10.1016/j.jclepro.2022.131937

[6] Y. Teng, Z. Wang, Y. Li, Q. Ma, Q. Hui, S. Li, Multi-energy storage system model based on electricity heat and hydrogen coordinated optimization for power grid flexibility, CSEE J. Power Energy, 5 (2019) 266-274. http://doi.org/10.17775/CSEEJPES.2019.00190

[7] A.A. Vostrikov, A.V. Shishkin, O.N. Fedyaeva, Conjugated processes of bulk aluminum

and hydrogen combustion in water-oxygen mixtures, Int. J. Hydrogen Energ., 45 (2020) 1061-1071. https://doi.org/10.1016/j.ijhydene.2019.10.152

[8] G. Chisholm, T. Zhao, L. Cronin, Hydrogen from water electrolysis, in: T.M. Letcher (Ed.) Storing Energy, Elsevier, Amsterdam, 2022, pp. 559-591. https://doi.org/10.1016/B978-0-12-824510-1.00015-5

[9] P.J. Megía, A.J. Vizcaíno, J.A. Calles, A. Carrero, Hydrogen production technologies: From fossil fuels toward renewable sources. A mini review, Energy & Fuels, 35 (2021) 16403-16415. https://doi.org/10.1021/acs.energyfuels.1c02501

[10] M. Balat, Possible methods for hydrogen production, Energy Sources Part A 31 (2008) 39-50. https://doi.org/10.1080/15567030701468068

[11] N. Burton, R. Padilla, A. Rose, H. Habibullah, Increasing the efficiency of hydrogen production from solar powered water electrolysis, Renew. Sustain. Energy Rev., 135 (2021) 110255. https://doi.org/10.1016/j.rser.2020.110255

[12] J.A. Turner, Sustainable hydrogen production, Science, 305 (2004) 972-974. https://doi.org/10.1126/science.1103197

[13] M. Younas, S. Shafique, A. Hafeez, F. Javed, F. Rehman, An overview of hydrogen production: current status, potential, and challenges, Fuel, 316 (2022) 123317. https://doi.org/10.1016/j.fuel.2022.123317

[14] B.C. Tashie-Lewis, S.G. Nnabuife, Hydrogen production, distribution, storage and power conversion in a hydrogen economy-a technology review, Chem. Eng. J. Adv., 8 (2021) 100172. https://doi.org/10.1016/j.ceja.2021.100172

[15] M. Martino, C. Ruocco, E. Meloni, P. Pullumbi, V. Palma, Main hydrogen production processes: An overview, Catalysts, 11 (2021) 547. https://doi.org/10.3390/catal11050547

[16] C. Bin, X. Heping, L. Tao, L. Cheng, L. Kuiwu, Z. Yuan, Principles and progress of advanced hydrogen production technologies in the context of carbon neutrality, Adv. Eng. Sci., 54 (2022) 106-116. https://doi.org/10.15961/j.jsuese.202100686

[17] S. Marini, P. Salvi, P. Nelli, R. Pesenti, M. Villa, M. Berrettoni, G. Zangari, Y. Kiros, Advanced alkaline water electrolysis, Electrochim. Acta, 82 (2012) 384-391. https://doi.org/10.1016/j.electacta.2012.05.011

[18] J. Brauns, T. Turek, Alkaline water electrolysis powered by renewable energy: A review, Processes, 8 (2020) 248. https://doi.org/10.3390/pr8020248

[19] A. Buttler, H. Spliethoff, Current status of water electrolysis for energy storage, grid balancing and sector coupling via power-to-gas and power-to-liquids: A review, Renew. Sustain. Energy Rev., 82 (2018) 2440-2454. https://doi.org/10.1016/j.rser.2017.09.003

[20] K. Ayers, The potential of proton exchange membrane-based electrolysis technology, Curr. Opin. Electroche., 18 (2019) 9-15. https://doi.org/10.1016/j.coelec.2019.08.008

[21] J.M. Spurgeon, N.S. Lewis, Proton exchange membrane electrolysis sustained by water

vapor, Energy Environ. Sci., 4 (2011) 2993-2998. https://doi.org/10.1039/c1ee01203g

[22] Y. Sui, Y. Du, H. Hu, J. Qian, X. Zhang, Do acid-base interactions really improve the ion conduction in a proton exchange membrane?–A study on the effect of basic groups, J. Mater. Chem. A, 7 (2019) 19820-19830. https://doi.org/10.1039/C9TA06508C

[23] A. Brisse, J. Schefold, High temperature electrolysis at EIFER, main achievements at cell and stack level, Energy Procedia, 29 (2012) 53-63. https://doi.org/10.1016/j.egypro.2012.09.008

[24] P. Hjalmarsson, X. Sun, Y.-L. Liu, M. Chen, Durability of high performance Ni-yttria stabilized zirconia supported solid oxide electrolysis cells at high current density, J. Power Sources, 262 (2014) 316-322. https://doi.org/10.1016/j.jpowsour.2014.03.133

[25] A. Ploner, A. Hauch, S. Pylypko, S. Di Iorio, G. Cubizolles, J. Mougin, Optimization of solid oxide cells and stacks for reversible operation, ECS Trans., 91 (2019) 2517. https://doi.org/10.1149/09101.2517ecst

[26] D. Grondin, J. Deseure, A. Brisse, M. Zahid, P. Ozil, Simulation of a high temperature electrolyzer, J. Appl. Electrochem., 40 (2010) 933-941. https://doi.org/10.1007/s10800-009-0030-0

[27] S.S. Kumar, V. Himabindu, Hydrogen production by PEM water electrolysis-A review, Mater. Sci. Energy Technol., 2 (2019) 442-454. https://doi.org/10.1016/j.mset.2019.03.002

[28] A. Valera-Medina, H. Xiao, M. Owen-Jones, W.I. David, P. Bowen, Ammonia for power, Prog. Energy Combust. Sci., 69 (2018) 63-102. https://doi.org/10.1016/j.pecs.2018.07.001

[29] A.N. Rollinson, J. Jones, V. Dupont, M.V. Twigg, Urea as a hydrogen carrier: a perspective on its potential for safe, sustainable and long-term energy supply, Energy Environ. Sci., 4 (2011) 1216-1224. https://doi.org/10.1039/c0ee00705f

[30] G. Cipriani, V. Di Dio, F. Genduso, D. La Cascia, R. Liga, R. Miceli, G.R. Galluzzo, Perspective on hydrogen energy carrier and its automotive applications, Int. J. Hydrogen Energ., 39 (2014) 8482-8494. https://doi.org/10.1016/j.ijhydene.2014.03.174

[31] J. Zhu, L. Hu, P. Zhao, L.Y.S. Lee, K.-Y. Wong, Recent advances in electrocatalytic hydrogen evolution using nanoparticles, Chem. Rev., 120 (2020) 851-918. https://doi.org/10.1021/acs.chemrev.9b00248

[32] L. Wang, Y. Zhu, Y. Wen, S. Li, C. Cui, F. Ni, Y. Liu, H. Lin, Y. Li, H. Peng, B. Zhang, Regulating the local charge distribution of Ni active sites for the urea oxidation reaction, Angew. Chem. Int. Ed., 60 (2021) 10577-10582. https://doi.org/10.1002/anie.202100610

[33] S. Geng, Y. Zheng, S. Li, H. Su, X. Zhao, J. Hu, H. Shu, M. Jaroniec, P. Chen, Q. Liu, S. Qiao, Nickel ferrocyanide as a high-performance urea oxidation electrocatalyst, Nat. Energy, 6 (2021) 904-912. https://doi.org/10.1038/s41560-021-00899-2

[34] Q. Xu, T. Yu, J. Chen, G. Qian, H. Song, L. Luo, Y. Chen, T. Liu, Y. Wang, S. Yin,

Coupling interface constructions of FeNi$_3$-MoO$_2$ heterostructures for efficient urea oxidation and hydrogen evolution reaction, ACS Appl. Mater. Interfaces, 13 (2021) 16355-16363. https://doi.org/10.1021/acsami.1c01188

[35] X. Peng, S. Xie, X. Wang, C. Pi, Z. Liu, B. Gao, L. Hu, W. Xiao, P.K. Chu, Energy-saving hydrogen production by methanol oxidation reaction coupled hydrogen evolution reaction co-catalyzed by phase separation induced heterostructure, J. Mater. Chem. A, 10 (2022) 20761-20769. https://doi.org/10.1039/D2TA02955C

[36] C. Wan, J. Jin, X. Wei, S. Chen, Y. Zhang, T. Zhu, H. Qu, Inducing the SnO$_2$-based electron transport layer into NiFe LDH/NF as efficient catalyst for OER and methanol oxidation reaction, J. Mater. Sci. Technol., 124 (2022) 102-108. https://doi.org/10.1016/j.jmst.2022.01.022

[37] Y. Guo, X. Yang, X. Liu, X. Tong, N. Yang, Coupling methanol oxidation with hydrogen evolution on bifunctional Co-doped Rh electrocatalyst for efficient hydrogen generation, Adv. Funct. Mater., 33 (2023) 2209134. https://doi.org/10.1002/adfm.202209134

[38] S.L. Candelaria, N.M. Bedford, T.J. Woehl, N.S. Rentz, A.R. Showalter, S. Pylypenko, B.A. Bunker, S. Lee, B. Reinhart, Y. Ren, Multi-component Fe-Ni hydroxide nanocatalyst for oxygen evolution and methanol oxidation reactions under alkaline conditions, ACS Catal., 7 (2017) 365-379. https://doi.org/10.1021/acscatal.6b02552

[39] Y. Liu, J. Zhang, Y. Li, Q. Qian, Z. Li, Y. Zhu, G. Zhang, Manipulating dehydrogenation kinetics through dual-doping Co$_3$N electrode enables highly efficient hydrazine oxidation assisting self-powered H$_2$ production, Nat. Commun., 11 (2020) 1853. https://doi.org/10.1038/s41467-020-15563-8

[40] Y. Li, J. Zhang, Y. Liu, Q. Qian, Z. Li, Y. Zhu, G. Zhang, Partially exposed RuP$_2$ surface in hybrid structure endows its bifunctionality for hydrazine oxidation and hydrogen evolution catalysis, Sci. Adv., 6 (2020) cabb4197. https://doi.org/10.1126/sciadv.abb4197

[41] Q. Qian, J. Zhang, J. Li, Y. Li, X. Jin, Y. Zhu, Y. Liu, Z. Li, A. El-Harairy, C. Xiao, G. Zhang, Y. Xie, Artificial heterointerfaces achieve delicate reaction kinetics towards hydrogen evolution and hydrazine oxidation catalysis, Angew. Chem. Int. Ed., 60 (2021) 5984-5993. https://doi.org/10.1002/anie.202014362

[42] S. Zhou, Y. Zhao, R. Shi, Y. Wang, A. Ashok, F. Héraly, T. Zhang, J. Yuan, Vacancy-rich MXene-immobilized Ni single atoms as a high-performance electrocatalyst for the hydrazine oxidation reaction, Adv. Mater., 34 (2022) 2204388. https://doi.org/10.1002/adma.202204388

[43] Y. Zhang, Y. Qiu, Z. Ma, Y. Wang, Y. Zhang, Y. Ying, Y. Jiang, Y. Zhu, S. Liu, Core-corona Co/CoP clusters strung on carbon nanotubes as a Schottky catalyst for glucose oxidation assisted H$_2$ production, J. Mater. Chem. A, 9 (2021) 10893-10908. https://doi.org/10.1039/D0TA11850H

[44] Y. Zhang, B. Zhou, Z. Wei, W. Zhou, D. Wang, J. Tian, T. Wang, S. Zhao, J. Liu, L. Tao, Coupling glucose-assisted Cu(I)/Cu(II) redox with electrochemical hydrogen production,

Adv. Mater., 33 (2021) 2104791. https://doi.org/10.1002/adma.202104791

[45]   Y. Xin, F. Wang, L. Chen, Y. Li, K. Shen, Superior bifunctional cobalt/nitrogen-codoped carbon nanosheet arrays on copper foam enable stable energy-saving hydrogen production accompanied with glucose upgrading, Green Chem., 24 (2022) 6544-6555. https://doi.org/10.1039/D2GC02426H

[46]   X. Lin, H. Zhong, W. Hu, J. Du, Nickel-cobalt selenide electrocatalytic electrode toward glucose oxidation coupling with alkaline hydrogen production, Inorg. Chem., 62 (2023) 10513-10521. https://doi.org/10.1021/acs.inorgchem.3c01679

[47]   X. Peng, S. Xie, S. Xiong, R. Li, P. Wang, X. Zhang, Z. Liu, L. Hu, B. Gao, P. Kelly, P.K. Chu, Ultralow-voltage hydrogen production and simultaneous Rhodamine B beneficiation in neutral wastewater, J. Energy Chem., 81 (2023) 574-582. https://doi.org/10.1016/j.jechem.2023.03.022

[48]   H. Dau, C. Limberg, T. Reier, M. Risch, S. Roggan, P. Strasser, The mechanism of water oxidation: From electrolysis via homogeneous to biological catalysis, ChemCatChem, 2 (2010) 724-761. https://doi.org/10.1002/cctc.201000126

Electrocatalytic Hydrogen Production: Catalysts and Applications
Materials Research Foundations **165** (2024)

Materials Research Forum LLC
https://doi.org/10.21741/9781644903070

CHAPTER 2

Fundamentals of Electrocatalytic Hydrogen Production

Xiang Peng*

Hubei Key Laboratory of Plasma Chemistry and Advanced Materials, Engineering Research Center of Phosphorus Resources Development and Utilization of Ministry of Education, School of Materials Science and Engineering, Wuhan Institute of Technology, Wuhan 430205, China

xpeng@wit.edu.cn

Abstract

When a voltage is applied to the electrodes, H_2O molecules undergo decomposition into H_2 and O_2 gases at the cathode and anode, respectively. Generally, the electrochemical water splitting reaction can be divided into two half-cell reactions, the hydrogen evolution reaction (HER) at the cathode to generate H_2 and the oxygen evolution reaction (OER) at the anode to produce O_2. The efficiency of mass conversion in the electrochemical water splitting process is highly dependent on the choice of electrolytes used.

Keywords

Half-Cell Reactions, Reaction Mechanism, Hydrogen Production, Performance Evaluation, Effect Factors

1. Introduction

A typical electrochemical water splitting system is composed cathode, anode, electrolyte and electrolyzer, as shown in **Figure 1**. When a voltage is applied to the electrodes, H_2O molecules undergo decomposition into H_2 and O_2 gases at the cathode and anode, respectively. Generally, the electrochemical water splitting reaction can be divided into two half-cell reactions, the hydrogen evolution reaction (HER) at the cathode to generate H_2 and the oxygen evolution reaction (OER) at the anode to produce O_2.

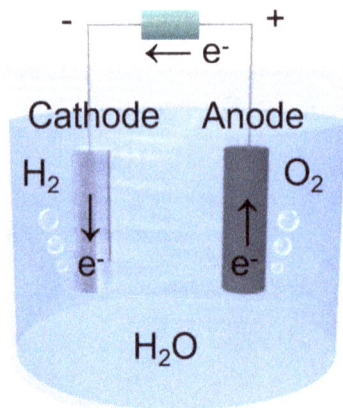

Figure 1. Schematic diagram of hydrogen production system by electrolysis of water.

2. Reactions in Electrocatalytic Hydrogen Production

The half-cell reactions of HER at the cathode and OER at the anode together constitute the overall reaction of electrocatalytic hydrogen production. This overall reaction can be expressed as follows:

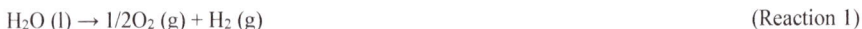

$$H_2O \ (l) \rightarrow 1/2O_2 \ (g) + H_2 \ (g) \hspace{4cm} \text{(Reaction 1)}$$

2.1. Overview of HER

HER is a half-cell reaction that occurs at the cathode in the electrolytic water reaction. Under the influence of an applied voltage, two electrons are transferred at its cathode, resulting in the production of H_2 gas.

In acidic conditions, the presence of hydrogen ions in the solution itself leads to a relatively low overpotential for the hydrogen evolution catalyst. The reaction pathway for the HER can be described through distinct steps. Reaction 3 illustrates the Volmer pathway, where hydrogen ions in the solution combine with electrons on the electrode, resulting in the formation of a hydrogen atom on the electrode [1]. This reaction can be further divided into two directions: the combination of hydrogen ions in the solution with electrons on the electrode, known as the Heyrovsky pathway [2], and the combination of two existing hydrogen atoms on the electrode to form a hydrogen molecule, referred to as the Tafel pathway [3]. It is through these pathways that hydrogen is generated via electrocatalysis in an acidic environment.

The reaction steps involved in the HER process can vary depending on the pH of the electrolyte. Below are the specific reaction equations for the HER:

The reaction in acidic solutions:

$$2H^+ + 2e^- \rightarrow H_2 \text{ (g)}$$ (Reaction 2)

Comprises the following steps:

Volmer reaction:

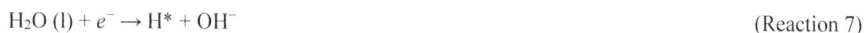

$$H^+ + e^- \rightarrow H^*$$ (Reaction 3)

Heyrovsky reaction:

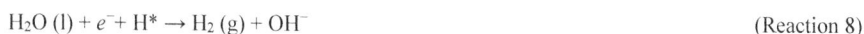

$$H^* + H^+ + e^- \rightarrow H_2 \text{ (g)}$$ (Reaction 4)

Or Tafel reaction:

$$H^* + H^* \rightarrow H_2 \text{ (g)}$$ (Reaction 5)

The reaction in alkaline solutions:

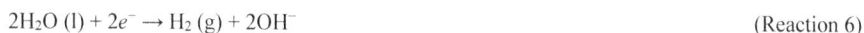

$$2H_2O \text{ (l)} + 2e^- \rightarrow H_2 \text{ (g)} + 2OH^-$$ (Reaction 6)

Comprises the following steps:

Volmer reaction:

$$H_2O \text{ (l)} + e^- \rightarrow H^* + OH^-$$ (Reaction 7)

Heyrovsky reaction:

$$H_2O \text{ (l)} + e^- + H^* \rightarrow H_2 \text{ (g)} + OH^-$$ (Reaction 8)

Or Tafel reaction:

$$H^* + H^* \rightarrow H_2 \text{ (g)}$$ (Reaction 9)

In simple terms, the HER in acidic solutions typically involves two steps.

First, the Volmer reaction occurs, where a H^+ from the electrolyte attaches itself to the catalyst's surface.

Next, the Heyrovsky reaction or the Tafel reaction takes place. In Heyrovsky reaction step, the adsorbed hydrogen atom (H^*) on the catalyst's surface combines with another H^+ from the electrolyte and an electron (e^-) to form a hydrogen molecule (H_2).

While in Tafel reaction step, two H^* that are already on the catalyst's surface combine with each other to form an H_2 molecule.

These two steps, Volmer, Heyrovsky or Tafel, together make up the HER process. Understanding and optimizing these steps are essential for designing efficient catalysts and advancing the development of electrocatalytic hydrogen production systems.

2.2.   Overview of OER

The OER is the complementary half-cell reaction that takes place at the anode during water electrolysis. It is a more complex process than the HER, involving the transfer of four electrons and multiple reaction intermediates being adsorbed and desorbed during the process [4]. The OER requires a higher driving voltage and exhibits relatively slower reaction kinetics, which can significantly impact the efficiency of water electrolysis.

In an alkaline electrolyte, the OER process can be described as follows:

Alkaline solution:

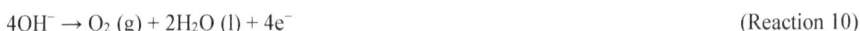

$$4OH^- \rightarrow O_2 \,(g) + 2H_2O \,(l) + 4e^- \qquad\qquad \text{(Reaction 10)}$$

Reaction steps:

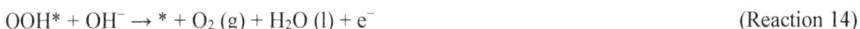

$$* + OH^- \rightarrow OH^* + e^- \qquad\qquad \text{(Reaction 11)}$$

$$OH^* + OH^- \rightarrow O^* + H_2O + e^- \qquad\qquad \text{(Reaction 12)}$$

$$O^* + OH^- \rightarrow OOH^* + e^- \qquad\qquad \text{(Reaction 13)}$$

$$OOH^* + OH^- \rightarrow * + O_2 \,(g) + H_2O \,(l) + e^- \qquad\qquad \text{(Reaction 14)}$$

The oxygen evolution process encompasses the formation of OH*, O*, and OOH* intermediates. This reaction presents a greater challenge than the HER, requiring a theoretical potential exceeding 1.23 V *vs.* RHE. The reaction mechanism involves the oxidation of active sites on the catalyst's surface, as well as the adsorption and desorption of active intermediates and products. Optimizing catalysts for the OER primarily entails adjusting the type and electronic structure of active centers to augment the binding strength of intermediates, thereby improving reaction kinetics and enhancing activity. Ongoing research on the OER mechanism utilizes advanced in-situ characterization techniques and widely employed density functional theory calculations (DFT). These investigations contribute to the continuous enhancement and optimization of catalyst performance by deepening our understanding of the OER mechanism.

3.   Performance Evaluation of the Electrocatalytic Hydrogen Production

3.1.   Overpotential

Overpotential (η) refers to the additional voltage required to drive an electrochemical reaction at a specific current density compared to the thermodynamic potential. It is a measure of the energy barrier or inefficiency in the electrochemical process. Typically, overpotential is determined by subtracting the thermodynamic potential from the applied potential. An accurate assessment of overpotential is crucial for evaluating the thermodynamic performance of an electrocatalyst. A smaller overpotential indicates higher catalytic activity, as it signifies the catalyst's ability to reduce the energy required for the reaction. The overpotential is influenced by various factors, including diffusion resistance, electrolyte resistance, and the kinetics of the reaction [5, 6]. To determine the overpotential, different methods can be employed depending on the specific electrochemical reaction. For example, in the case of the HER, the theoretical potential is considered to be 0 V *vs.* RHE. The overpotential for HER (η_c) can be calculated by subtracting the measured potential from the RHE potential. Similarly, for the OER (η_a), the theoretical potential is 1.23 V *vs.* RHE. The overpotential for the OER is obtained by subtracting 1.23 V from the measured potential after calibration using the *iR* compensation method. The calculation formulas are given by Equations 1 and 2.

$$\eta_c = E_{RHE} - E_{iR} \qquad\qquad\qquad\qquad \text{(Equation 1)}$$

$$\eta_a = E_{RHE} - E_{iR} - 1.23\,V \qquad\qquad\qquad \text{(Equation 2)}$$

Linear sweep voltammetry (LSV) is an electrochemical analysis technique which involves electrolyzing a dilute solution of the analyte and analyzing the resulting current-potential curve. In modern practice, the three-electrode system, consisting of a working electrode, counter electrode, and reference electrode, is widely employed in voltammetry methods. LSV is a widely utilized technique for qualitative and quantitative analysis of substances, as well as mechanism research. LSV applies a linearly varying voltage to the working electrode. It serves as an electrochemical scanning analysis method, where the working electrode acts as a probe, the linearly changing potential signal serves as the scanning signal, and the collected current signal provides feedback for qualitative and quantitative substance detection through scanning. The potential scanning rate commonly ranges from 0.001 to 0.1 V s^{-1} in LSV, and it can be performed once or multiple times, as depicted in **Figure 2 [7]**. During HER and OER tests, the potential data is converted to the RHE using Equation 3 to account for the reference electrode (such as Hg/Hg$_2$Cl$_2$) potential.

$$E_{RHE} = E_{Hg/Hg2Cl2} + 0.242 + 0.059 \times pH \qquad\qquad \text{(Equation 3)}$$

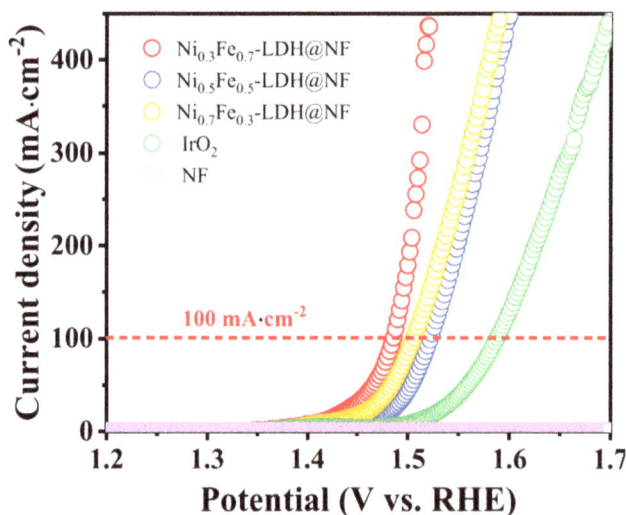

Figure 2. *LSV diagram of $Ni_{0.3}Fe_{0.7}$-LDH@NF, $Ni_{0.5}Fe_{0.5}$-LDH@NF, $Ni_{0.7}Fe_{0.3}$-LDH@NF and IrO_2 in 1 M KOH. Reproduced with permission from ref [7]. Copyright 2023, Elsevier.*

3.2. Tafel Slope

The Tafel slope is a parameter used to quantify the electrochemical kinetics of a reaction. In electrocatalysis, the Tafel slope provides insights into the reaction mechanism and the rate-determining steps. It is determined by analyzing the current-potential curve obtained through techniques such as LSV. The Tafel slope is calculated by fitting a linear regression to the Tafel region of the curve, which corresponds to moderate overpotentials. It can be expressed using the following equation:

$$\eta = a + b \times \log|j| \qquad \text{(Equation 4)}$$

Here, b represents the Tafel slope. According to the Tafel formula, the current density at zero overpotential is calculated as exchange current density (j_0), which reflects the inherent activity of the catalyst at equilibrium potential. The Tafel slope is typically expressed in units of volts per decade or millivolts per decade. For the HER, a smaller Tafel slope indicates faster reaction kinetics and higher catalytic activity. It represents the kinetics at which protons are reduced to produce hydrogen gas. On the other hand, for the OER, a larger Tafel slope suggests slower reaction kinetics and lower catalytic activity. It reflects the rate at which oxygen gas is evolved by the oxidation of water molecules. The Tafel slope is an important parameter in electrocatalysis research as it provides quantitative information about the reaction mechanism, the efficiency of catalysts, and the potential for improving catalytic performance. **Figure 3** illustrates the typical Tafel plots observed for Mo-based catalysts in HER, as extracted from the

LSV curves [8]. It demonstrates that the MoON-2 electrocatalyst shares the comparable Tafel slope with the commercial Pt/C catalyst, indicating the fast reaction kinetics. Therefore, when designing a catalyst, careful consideration should be given to achieving a small Tafel slope.

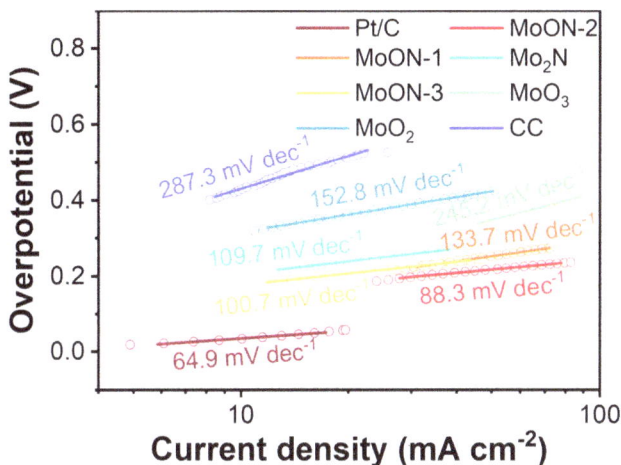

Figure 3. Tafel plots of the molybdenum-based catalysts for HER measured in 0.5 M H_2SO_4. Reproduced with permission from ref [8]. Copyright 2024, Elsevier.

3.3. Charge Transfer Properties

Charge transfer resistance (R_{ct}) is a parameter that characterizes the resistance encountered by charge carriers during the transfer process at the electrode-electrolyte interface in an electrochemical system. It represents the hindrance or impedance to the flow of electrons or ions between the electrode and the electrolyte. The charge transfer resistance arises due to several factors, including the kinetics of the reaction, the properties of the electrode surface, and the composition and conductivity of the electrolyte. The R_{ct} can be determined experimentally by techniques such as electrochemical impedance spectroscopy (EIS). EIS involves applying an alternating current (AC) signal to the electrochemical system and measuring the resulting impedance response. By analyzing the impedance spectrum, including the real and imaginary components, the charge transfer resistance can be extracted.

A higher R_{ct} indicates a slower charge transfer process and can be associated with factors like sluggish reaction kinetics, electrode fouling or passivation, or poor electrical conductivity of the electrolyte. Conversely, a smaller R_{ct} signifies more efficient charge transfer and better electrochemical performance in terms of reaction rates and overall system efficiency. Understanding and minimizing R_{ct} is crucial for improving the performance of electrochemical devices and optimizing the design and selection of catalysts and electrolytes. **Figure 4** illustrates the Nyquist plots of the Mo-based catalysts compared to commercial Pt/C catalyst in HER [8]. A smaller R_{ct} value indicates a faster catalytic reaction rate.

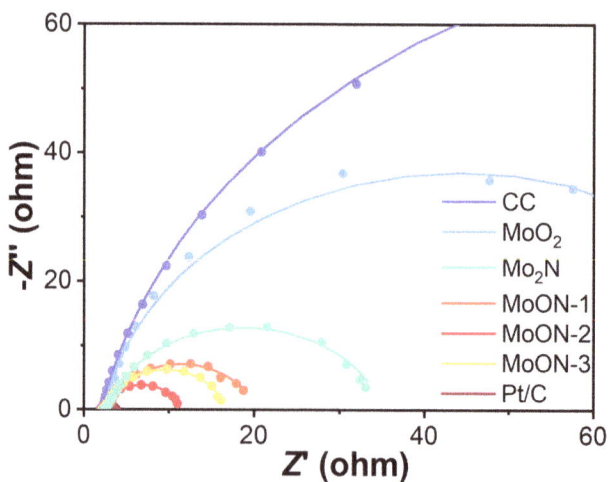

Figure 4. *Nyquist plots of the molybdenum-based catalysts obtained in 0.5 M H_2SO_4.*
Reproduced with permission from ref [8]. Copyright 2024, Elsevier.

3.4. Turnover Frequency

Turnover frequency (TOF) is a term commonly used in catalysis to describe the catalytic activity of a catalyst. It represents the number of catalytic reactions occurring per active site or per unit time. TOF is usually expressed in units of moles of reactant converted per second or hour. TOF provides a measure of the efficiency and effectiveness of a catalyst in facilitating a specific chemical reaction. A higher TOF indicates a catalyst that can convert reactants into products at a faster rate, reflecting its higher catalytic activity. In contrast, a lower TOF suggests a slower catalytic rate or reduced efficiency.

TOF is calculated by dividing the rate of the catalytic reaction by the number of active sites on the catalyst. The rate of the reaction can be determined experimentally by measuring the conversion of reactants over a given period, typically through techniques such as spectroscopy or chromatography. The number of active sites can be estimated or determined through various analytical methods, such as surface area measurements or microscopic imaging. The formula for TOF, as shown in Equation 5, is as follows:

$$TOF = \frac{I}{\alpha \times F \times N} \qquad \text{(Equation 5)}$$

Here, I is the electric current, α is the number of transfer electrons involved in the formation of the product of interest, F is the Faraday constant, and N is the molar amount of the active substance.

TOF is influenced by several factors, including the nature of the catalyst, its surface properties, the reaction conditions (temperature, pressure, etc.), and the reactant concentrations. Comparing

Materials Research Forum LLC
https://doi.org/10.21741/9781644903070

the TOF of different catalysts can provide insights into their relative catalytic efficiencies and help in the selection and design of catalysts for specific reactions. TOF is a valuable metric in catalysis research and industrial applications, as it allows for the quantitative evaluation and comparison of catalyst performance, as shown in **Figure 5** [9].

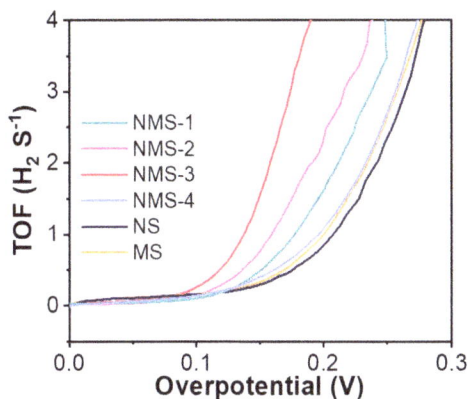

Figure 5. TOF of the electrocatalysts measured in 1.0 M KOH electrolyte. Reproduced with permission from ref [9]. Copyright 2024, Elsevier.

3.5.   Electrochemically Active Surface Area

Electrochemically active surface area (ECSA) refers to the portion of the electrode surface that is accessible and actively involved in electrochemical reactions. It represents the area available for electrochemical processes to occur, such as redox reactions. In the electrocatalysis systems, the reactions take place at the catalyst-electrolyte interface. However, not all of the catalyst surface area may be effectively participating in the desired electrochemical reactions. This can be due to factors such as surface contamination, passivation, or insufficient contact between the catalyst and the electrolyte. ECSA directly influences the rate of electrochemical reactions. A higher ECSA implies a larger active surface area available for reactant adsorption and charge transfer, leading to increased reaction rates and higher catalytic activity. Conversely, a lower ECSA can limit the reaction rate and reduce the overall efficiency of the electrochemical process.

ECSA is commonly measured using techniques such as cyclic voltammetry or electrochemical impedance spectroscopy. These methods involve characterizing the electrochemical behavior of the catalyst in a controlled electrochemical cell and analyzing the resulting current-potential or impedance response. By comparing the experimental data with theoretical models or reference electrodes, the ECSA can be estimated, as demonstrated in **Figure 6** [10]. Accurate determination of ECSA is crucial for evaluating the performance of the catalysts and understanding the electrochemical behavior of the catalysts. It provides insights into the efficiency of catalysts, the kinetics of electrochemical reactions, and the effect of different surface modifications or treatments on catalyst.

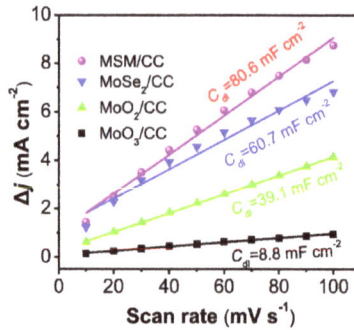

Figure 6. Double layer capacitance of the electrocatalysts measured in 1.0 M KOH electrolyte. Reproduced with permission from ref [10]. Copyright 2023, Elsevier.

3.6.   Stability

The stability of a catalyst for electrocatalytic hydrogen production is a critical factor in ensuring the long-term efficiency and durability of the catalyst. Hydrogen production catalysts undergo continuous electrochemical reactions, and their stability refers to their ability to maintain their catalytic activity and structural integrity over extended periods of operation. Several factors contribute to the stability of a catalyst for hydrogen production:

(i) Chemical Stability: The catalyst should exhibit high chemical stability in the electrolyte environment to prevent degradation or dissolution of the catalyst material. Corrosion or chemical reactions with the electrolyte can lead to loss of catalytic activity and deterioration of the catalyst.

(ii) Electrochemical Stability: The catalyst must withstand the electrochemical conditions during hydrogen production without undergoing degradation or decomposition. It should resist processes such as oxidation, reduction, or side reactions that can impair its catalytic performance.

(iii) Structural Stability: The catalyst should maintain its physical structure and morphology during prolonged operation. Structural stability is crucial in preventing catalyst agglomeration, particle growth, or surface area loss, which can reduce the active sites available for catalytic reactions.

(iv) Catalyst Support Stability: Many catalysts are supported on conductive materials, such as carbon or metal oxides. The stability of the catalyst-support interface is important to maintain the integrity and dispersion of the catalyst particles on the support material.

The electrochemical chronopotentiometry (*V-t*) curve is a graphical representation of the potential response of an electrochemical system as a function of time under a constant current. It is obtained by monitoring the potential of an electrochemical cell while applying a constant current, as demonstrated in **Figure 7** [11]. In addition, the electrochemical chronoamperometry (*i-t*) curve is also used to study the dynamic behavior and characteristics of an electrochemical reaction. By analyzing the changes in current over time, valuable insights can be gained regarding the reaction kinetics, mechanisms, and factors influencing the reaction rate. The shape of the electrochemical *i-t* curve can vary depending on the specific experimental conditions and the nature of the electrochemical system under investigation. However, it is common to divide the curve into three stages, each representing different phases of the electrochemical process:

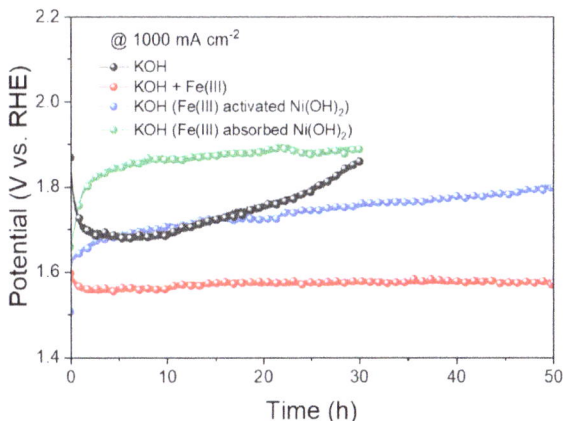

Figure 7. Chronopotentiometry curves plots of the Ni(OH)₂, Fe(III)-activated, and Fe(III)-absorbed Ni(OH)₂ obtained from KOH as well as the Ni(OH)₂ obtained from KOH + Fe(III). Reproduced from ref [11]. Copyright 2023, Elsevier.

The first stage is transient. At the beginning of the experiment, there is typically a transient stage characterized by a rapid change in current. This initial response is related to factors such as the charging of the double layer at the electrode-electrolyte interface or the redistribution of species near the electrode.

The second stage is the steady-state stage. Following the transient stage, the current reaches a steady state, where it remains relatively constant over time. This stage represents a stable electrochemical regime where the reaction rate is balanced by the mass transport of reactants and products to and from the electrode surface.

The third stage is the decay stage. In some cases, a decay stage may be observed after the steady state. The current gradually decreases over time, indicating a decrease in reaction rate or the occurrence of side reactions or catalyst deactivation.

3.7.   Faraday's Efficiency

Faraday's efficiency is a measure of the efficiency with which an electrochemical system converts electrical charge into the desired electrochemical reaction or product. It quantifies the ratio of the actual amount of a specific product generated or consumed during an electrochemical reaction to the theoretically expected amount based on Faraday's law of electrolysis. Faraday's law states that the amount of substance produced or consumed during an electrochemical reaction is directly proportional to the quantity of electric charge passed through the system. The efficiency is determined by calculating the ratio of the generated hydrogen to the applied charge. The Faraday's efficiency calculation follows these steps:

(1) Theoretical total molar hydrogen:

$$n(H_2)_T = \frac{i \times t}{2 \times F} \qquad\qquad \text{(Equation 6)}$$

Where $n(H_2)_T$ represents the total molar mass of theoretically generated hydrogen, i is the current (A), t is time (s), and F is the Faraday constant ($F = 96485$ C mol^{-1}).

(2) Under the standard experimental conditions ($P = 101.325$ kPa, $T = 298.15$ K), the actual molar total hydrogen produced is:

$$n(H_2)_M = \frac{V}{V_m} \qquad\qquad \text{(Equation 7)}$$

In this equation, $n(H_2)_M$ is the molar mass of H$_2$, V is the volume of hydrogen produced, and V_m is the molar volume of gas at 298.15 K.

(3) Faraday's efficiency is expressed as a percentage and is calculated by dividing the actual amount of the desired product produced or consumed by the theoretical amount and multiplying by 100. The overall formula for calculating Faraday's efficiency is:

$$FE = \frac{n(H_2)_M}{n(H_2)_T} \times 100\% \qquad\qquad \text{(Equation 8)}$$

Here, FE denotes the Faraday efficiency, which is expressed as a percentage.

A Faraday's efficiency of 100% indicates that all the electrical charge passed through the system was utilized to generate the desired product. However, in practical electrochemical systems, Faraday's efficiency is often less than 100% due to various factors such as side reactions, incomplete reaction kinetics, competing reactions, mass transport limitations, and electrode or catalyst inefficiencies, as demonstrated in **Figure 8** [10].

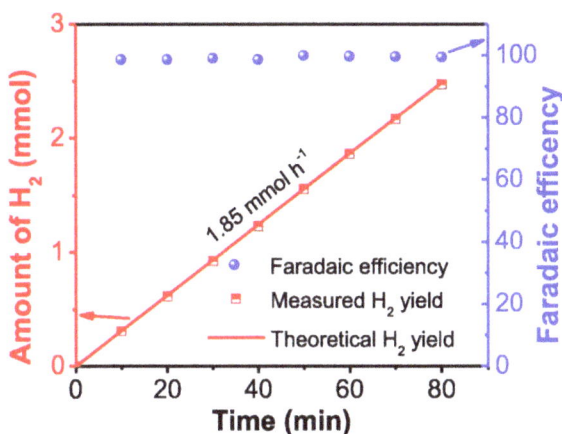

Figure 8. Experimental and theoretical amounts of H_2 produced by the MSM/CC electrocatalyst and Faraday's efficiency at 100 mA. Reproduced with permission from ref [10]. Copyright 2023, Elsevier.

4.   Factors that Affect the HER Performance

4.1.   Type of the Catalyst

The construction of catalysts for the HER can be broadly categorized into three types based on their distinct physical and chemical properties: precious metal-based catalysts, low-cost transition metal-based catalysts, and non-metallic base catalysts. Evaluating the catalytic activity of HER catalysts involves a significant parameter functional relationship between the metal-H (M-H) adsorption energy (E_{M-H}) and the exchange current density (j_0) of different catalysts. This relationship is often depicted as a volcano curve, as shown in **Figure 9** [12]. A catalyst's HER performance is typically better when possessing an appropriate E_{M-H}. Pt, a precious metal catalyst, stands at the peak of the volcano curve with ΔG_H very close to zero, setting the performance benchmark for HER catalysts. However, when exploring alternatives to precious metal Pt, it is crucial to compare the E_{M-H}. This comparison reveals that transition metal-based materials (such as Co, Ni, Mo, MoS_2) and non-metallic carbon-based materials (such as nitrogen-doped graphene, C_3N_4) closely resemble precious metal Pt-based materials in terms of performance [13, 14]. Consequently, these transition metal-based and non-metallic carbon-based materials hold significant potential as high-performance HER catalysts, providing alternatives to replace precious metal Pt.

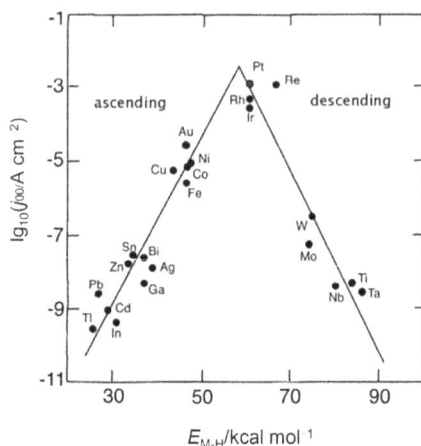

Figure 9. Volcano curve for the HER in acid solutions. Reproduced with permission from ref [12]. Copyright 2015, Elsevier.

In electrocatalytic reactions, the electronic structure effect plays a crucial role, emphasizing the importance of selecting an appropriate catalyst with the right adsorption-free energy to achieve energy-efficient reactions. In the context of the HER, the ΔG_H is a significant factor in describing its activity [15]. The volcano curve, observed in both theoretical and experimental results, provides a semi-quantitative relationship between the HER reaction rate and the ΔG_{H*} of M-H$_{ads}$, which is closely associated with the electronic structure of the metal surface.

The interaction between the H$_{ads}$ orbital and the metal d orbital gives rise to a bonding state (low energy and filled) and an anti-bonding state (high energy and partially filled) located just above the metal d orbital. The strength of the M-H bond depends on the occupancy of the anti-bonding orbital, with higher occupancy indicating weaker bond strength. When the anti-bonding orbital is positioned above the Fermi level, it exhibits strong adsorption capacity for H$_{ads}$, resulting in $\Delta G_{H*} < 0$ [16]. The classic volcano diagram, as shown in **Figure 10**, illustrates that an excellent catalyst should possess a moderate binding strength with the adsorption intermediate. If the binding energy is too weak, desorption becomes easy, while excessive binding strength makes desorption challenging [17]. Therefore, achieving a balance of binding strength is crucial for optimal catalyst performance.

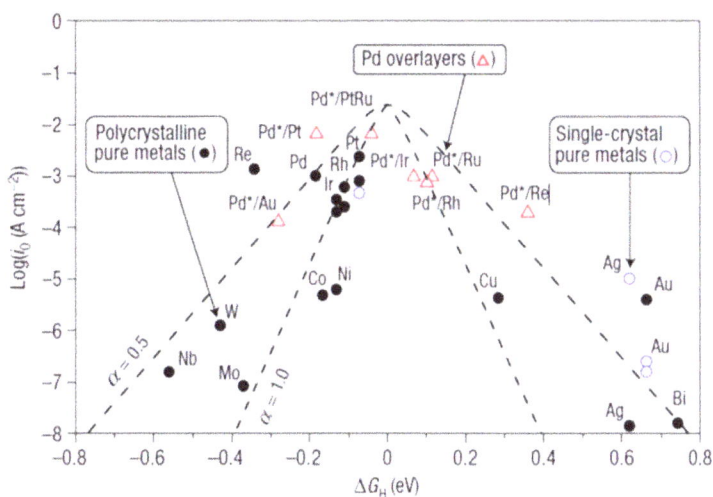

Figure 10. *Volcano plots for HER. Reproduced with permission from ref [17]. Copyright 2006, Springer Nature.*

In industrial production, catalysts for electrolytic water electrodes often rely on precious metals like platinum, palladium, and rhodium. These metals possess a ΔG_H close to 0 eV and demonstrate low overpotential, making them highly efficient for the HER. However, their high cost and limited availability present obstacles to widespread market adoption. Consequently, researchers are investigating alternative catalysts based on transition metals, which offer abundant reserves and lower costs while maintaining excellent performance [5].

4.2.   Composition

Studies have consistently shown that bimetallic and multi-metallic materials often exhibit superior catalytic activity compared to single-metal materials. This enhanced performance is attributed to the synergistic effect that arises from the electronic interaction at the interface of different phases within the composite [18, 19]. In addition, combining two or more materials can significantly enhance the hydrogen evolution activity of a single transition metal compound through this synergistic effect. As a result, extensive research is currently focused on improving the hydrogen evolution performance of electrolytic water by combining carbon materials or transition metal compounds with transition metal compounds.

Carbon materials, such as carbon nanotubes (CNT), graphene, and mesoporous carbon materials, possess good conductivity and diverse shapes, making them suitable not only as standalone catalyst materials but also for combination with other transition metal materials. This combination offers the advantages of high conductivity and high catalytic activity [20]. Carbon materials are widely employed in electrocatalytic HER due to their unique properties. CNT, for instance, as an allotrope of sp^2 hybrid carbon, exhibits exceptional characteristics in the realm of carbon-based nanomaterials. Combining CNT with transition metal compounds results in a strong synergistic coupling between the two components. The functional groups and mesoporous morphology of CNT surfaces facilitate the incorporation and stable dispersion of active nanoparticles, leading to improved utilization of active sites [21]. For example, Begum et al. [22] synthesized a crust-like Fe-Co_3O_4/CNT composite nanomaterial, the X-Ray Diffraction (XRD) pattern is shown in **Figure 11**. The as-prepared catalyst demonstrated excellent catalytic activity and durability due to the synergistic chemical coupling between Fe-Co_3O_4 and CNT. Particularly, it shows superior electrocatalytic activity with lowering overpotentials of 120 mV for HER and of 300 mV for OER to afford the current density of 10 mA cm^{-2} in 1. M KOH solution.

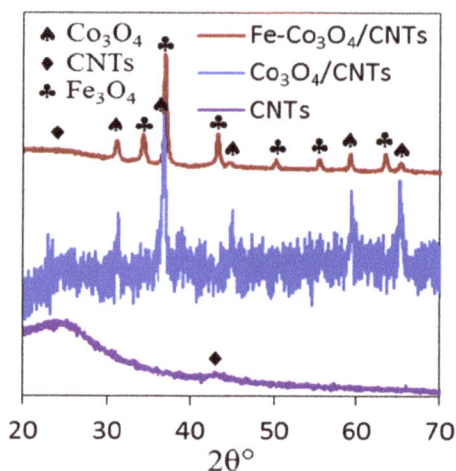

Figure 11. *The XRD patterns of Fe-Co_3O_4/CNT. Reproduced with permission from ref [22]. Copyright 2018, Elsevier.*

Transition metal compound composites have also been extensively studied to enhance the efficiency of hydrogen evolution. These composites combine the advantages of various transition metal compounds and exhibit excellent electrochemical properties. For example, Guo et al. [23] utilized $NiCo_2O_4$ with a hierarchical nanopore (HNP) structure as a precursor and obtained NiO/NiCoP composites through a simple phosphating treatment. The strong synergistic coupling effect between TMO and TMP, along with the unique HNP structure characteristics and increased active centers, contributed to the significant electrocatalytic HER activity of the composite. Similarly, Tao et al. [24] prepared a Co-W precursor using an anion exchange method and obtained Co_2P/$CoWO_4$/NF composites with a nano-layered morphology. The anion exchange method endowed the catalyst with unique characteristics including strong electron interaction, improved electronic structure, and optimized catalytic activity. Additionally, the metallic properties of Co_2P and $CoWO_4$, along with their interfacial interactions, facilitated electrocatalytic electron and charge transfer in HER. The composite exhibited excellent electrocatalytic performance with low overpotential and Tafel slope. In 1.0 mol L^{-1} KOH, an overpotential of only 81 mV was required to produce a current density of 10 mA cm^{-2}, with a small Tafel slope of 47 mV dec^{-1}.

4.3.   Structure

The design of nanostructures plays a crucial role in determining the specific surface area, exposed active site, and stability of a catalyst, thereby significantly influencing its electrocatalytic hydrogen evolution. For instance, Pan et al. [25] conducted a study where they synthesized a floral-like organic frame material (Co-MOF-800) through calcination, as illustrated in **Figure 12a**. This catalyst exhibited an overpotential of 200 mV at a current density of 10 mA cm^{-2} and a Tafel slope of 77 mV dec^{-1} in an acidic medium. In another study, Duan et al. [26] successfully synthesized a novel core-shell structure Co-MOF-800 by carbonizing ZIF-8 coated with glucosamine at high temperature, as illustrated in **Figure 12b**. The electrocatalytic HER of Co-MOF-800 in an alkaline medium displayed an impressively low overpotential of only 175 mV at a current density of 10 mA cm^{-2}. Furthermore, the catalyst maintained its performance for 85 hours without significant changes, indicating excellent stability. The core-shell structure exhibited superior catalytic activity stability compared to the flower-like structure of Co-MOF-800, exposing more active sites and demonstrating better HER performance.

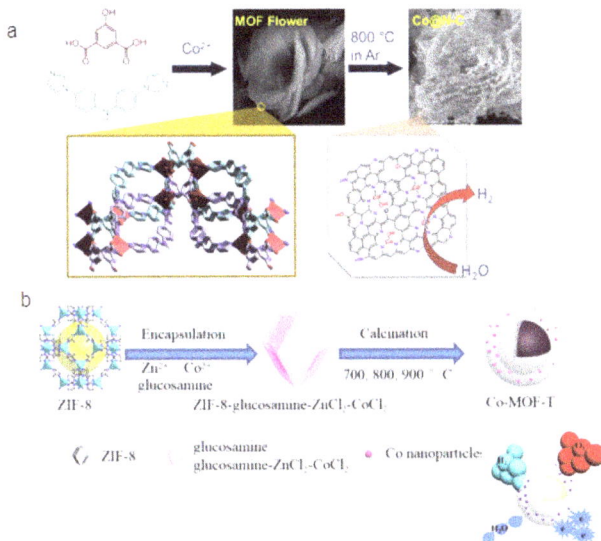

Figure 12. (a) Schematic diagram of the synthesis of the floral structure Co-MOF-T. reproduced with permission from ref [25]. Copyright 2019, Elsevier. (b) Synthesis diagram of porous core-shell structure Co-MOF-T. Reproduced with permission from ref [26]. Copyright 2023, Elsevier.

Song et al. [27] conducted a study where they employed a micro-nano structure control strategy to synthesize a three-dimensional (3D) MoS_2/carbon nanoflower composite catalyst (MoS_2/CF) using a continuous "in situ hydrothermal calcination" synthesis method. This approach effectively addresses the challenges associated with interlayer accumulation and poor conductivity in MoS_2. By expanding the layer spacing to approximately 9.5 Å and reducing the number of layers to 1-2, the catalyst exhibits abundant active sites and efficient charge transport performance. These improved physical and chemical properties enable the MoS_2/CF-750 catalyst to demonstrate remarkable electrocatalytic performance for the HER in an acidic solution (0.5 M H_2SO_4). An overpotential of 125 mV is required to achieve the current density of 10 mA cm^{-2}, and it shows long-term stability for continuous work for more than 16 hours.

To further enhance the electrocatalytic performance of MoS_2-based catalysts, a heterostructure construction strategy was utilized to synthesize MoS_2-based nanoarray catalysts grown on carbon aerogel with uniaxial orientation. By adjusting the stoichiometric ratio of Mo and S sources, a series of Mo_xS_y@GCA vertical nanoarray dual-function catalysts enriched with S edges were fabricated. This unique structure effectively prevents stacking between Mo_xS_y layers and exposes a greater number of edge active sites. Additionally, the incorporation of GCA facilitates improved longitudinal conductivity by establishing charge transfer paths between layers. Leveraging these advantages, the best-performing Mo_4S_{16}@GCA catalyst demonstrates excellent performance for both HER and OER. It achieves low overpotentials of 54.13 mV and 370 mV, respectively, at a current density of 10 mA cm^{-2}, which is comparable to the performance of precious metal Pt/C catalyst. Moreover, it exhibits a low starting potential of 24.28 mV vs. RHE for HER and 1.53 V vs. RHE for OER.

4.4.   Geometrical Morphology

Electrocatalysis is widely recognized as a surface interface reaction process, and researchers have been focusing on optimizing catalyst morphology to enhance surface atom utilization. Various strategies have been employed to increase the specific surface area of catalysts, including constructing morphologies of different sizes and dimensions (one dimension, two dimensions, and three dimensions). These strategies aim to improve the non-intrinsic catalytic activity of the materials. Significant advancements have been made through methods such as constructing porous materials, preparing ultra-thin two-dimensional materials, synthesizing single atoms or nanoclusters, and constructing heterogeneous structures [28-32].

Porous electrodes, in particular, provide a common and cost-effective means of constructing morphologies. They greatly increase the specific surface area of the electrode, resulting in higher current density during electrolysis and reduced overpotential for hydrogen evolution. For example, a team led by Duan [33] prepared a Ni-P/$LaNi_5$ catalyst through composite electrodeposition and impurity removal with lye. This catalyst exhibited numerous holes of varying sizes on its surface, significantly enhancing the specific surface area. Moreover, the calculated apparent activation energy of the Ni-P/$LaNi_5$ catalyst (35.44 kJ mol^{-1}) was much lower than that of a traditional Ni-P catalyst, indicating its superior electrocatalytic hydrogen evolution capability compared to a regular Ni-P catalyst. This highlights the advantages of constructing porous catalysts.

Two-dimensional nanoarray materials offer the exposure of more active sites during the reaction process and exhibit excellent stability, effectively preventing agglomeration and stacking [13, 30]. Compared to bulk materials, two-dimensional layered materials exhibit strong constrained charge transfer between layers. Therefore, reducing the layer thickness becomes a direct approach to adjusting the electronic structure of layered materials. Density functional theory (DFT) calculations have shown that ultra-thin nanosheets near the Fermi level exhibit significantly higher electron density compared to bulk materials. This improvement enhances the electrical conductivity of solid materials and facilitates faster electron transfer rates during water electrolysis [34]. Novoselov et al. [35] successfully isolated monolayer MoS_2 from bulk MoS_2 using a micromechanical method, leading to progress in the preparation of single or few-layer MoS_2 nanosheets. Additionally, the indirect band gap present in bulk MoS_2 transforms into a direct band gap in single or few-layer MoS_2 nanosheets, significantly enhancing the intrinsic activity of MoS_2 and improving its HER activity [36].

Cui et al. [37] prepared two-dimensional porous MoS_2 nanosheets through KOH treatment, as shown in **Figure 13**, which resulted in a porous nanosheet structure providing abundant edge active sites. This geometrical morphology modification technique improved the HER performance of MoS_2 in acidic solutions. It was concluded that the sample treated with 37.5 wt% KOH achieved the highest density of bore edge electrochemically active sites, the smallest overpotential of $\eta = 240.7$ mV for the current density of 10 mA cm^{-2} in hydrogen production and a Tafel slope of 88.4 mV dec^{-1}.

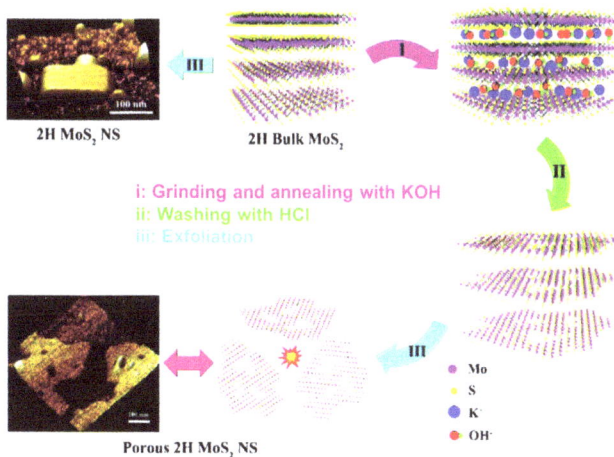

Figure 13. *Schematic diagram of the process for preparing porous MoS_2 nanosheets from bulk MoS_2 nanosheets. Reproduced with permission from ref [37]. Copyright 2018, the Royal Society of Chemistry.*

The influence of heterogeneous structures on catalysts is of paramount importance in designing efficient catalysts for HER [38]. In this regard, Tian et al. [39] successfully synthesized a dandelion-like $NiMoP_2$-Ni_2P catalyst with a heterogeneous structure comprising nanowires and nanoparticles. The synthesis process involved a combination of hydrothermal and phosphating methods. Scanning electron microscopy (SEM) was employed to observe the morphology of the synthesized product, revealing a dandelion array morphology on the surface of carbon fiber. Each dandelion structure consisted of smooth-surfaced $NiMo(OH)_x$ nanowires with a diameter of 100 nm, as depicted in **Figure 14**. The catalyst exhibited a uniformly grown hierarchical heterostructure, which facilitates efficient electron transport and collection, leading to enhanced electrocatalytic performance for the HER. The catalyst exhibited a low overpotential of 53 mV to achieve a current density of 10 mA cm^{-2}, along with a small Tafel slope of 58 mV dec^{-1}. These results indicate the excellent HER activity of the catalyst. Importantly, the catalyst possesses the advantages of low cost and good catalytic performance, thereby providing valuable insights for the application of heterogeneous structures in the field of electrocatalytic hydrogen production.

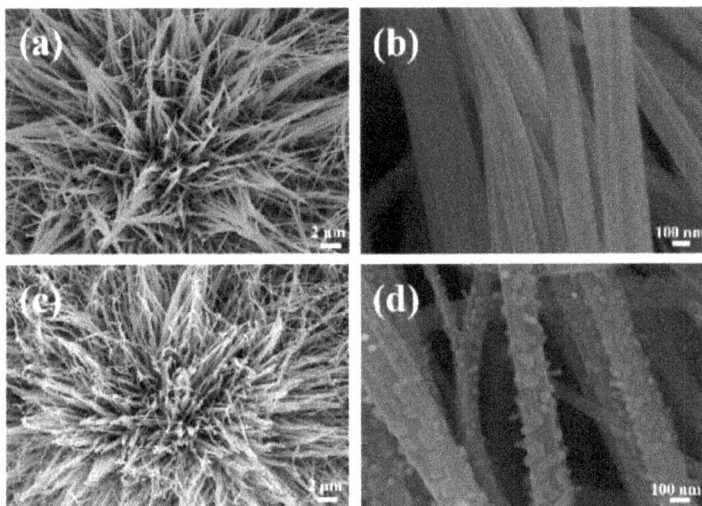

Figure 14. SEM images of (a, b) NiMo(OH)x/CC and (c, d) NiMoP$_2$-Ni$_2$P/CC. Reproduced with permission from ref [39]. Copyright 2021, Elsevier.

4.5.   Electronic Structure

The electronic structure of a catalyst plays a crucial role in controlling its activity [40]. It directly influences the electrical conductivity and adsorption strength of intermediate products, which are equivalent to the reaction barriers, thereby determining the reaction kinetics. Consequently, the electronic structure serves as a significant descriptive parameter for explaining catalytic behavior [41]. Element doping has been widely employed to enhance catalyst performance by regulating the physical and chemical properties of materials. Heteroatom doping can modify the chemical state and electronic structure of the catalyst surface, optimize the ΔG_H

of hydrogen intermediates, improve the activity of electrocatalytic hydrogen evolution, and enhance catalyst stability [40].

The Engel-Brewer valence bond theory [42] states that metals on the left side of the transition metal series (Fe, Co, Ni, etc.), which possess partially filled d orbitals and unpaired electrons unsuitable for synthesizing chemical bonds in pure metals, can exhibit good electrocatalytic synergy when combined with metals on the right side of the transition metal series (W, Mo, La, Ta, Zr, etc.) during HER. Building upon this discovery, researchers have combined various transition metal elements to prepare catalysts, leveraging their synergistic effects to further enhance electrochemical hydrogen evolution capabilities. For example, molybdenum (Mo), a transition metal, has been found to enhance the catalytic activity of water electrolysis when combined with nickel. For instance, Ma et al. [42] demonstrated that Ni-Mo alloy coatings exhibit excellent hydrogen evolution performance due to the synergistic effect between nickel and molybdenum. Nickel forms weak Ni-H bonds, while molybdenum forms strong Mo-H bonds [43]. Under alkaline conditions, the Ni-Mo alloy exhibits a synergistic effect that enhances the HER activity. Additionally, highly electronegative phosphorus (P) atoms can attract electrons from metal surfaces, and negatively charged P sites serve as effective adsorbents for protons and oxygen-containing intermediates in both HER and OER processes. This reduces the energy barrier for H_2 and O_2 desorption [44]. Therefore, the combination of molybdenum, nickel, and phosphorus can yield excellent hydrogen evolution performance in binary metal phosphides.

Traditionally, researchers have primarily focused on metal cation doping, such as Co-doped Ni-P, Ru-doped Ni-P, Ru-Fe-P-doped Ni, and various multimetallic compounds [45-47]. However, non-metallic elements (O, C, N, P, S, etc.) doping in transition metal compounds has also been proven effective in adjusting the electronic state of the transition metal and modifying its electronic structure to improve catalytic performance. Nitrogen (N) and carbon (C) are commonly used non-metallic elements in catalyst preparation due to their ability to regulate electronic structure and provide good stability. Liu et al. [48] demonstrated that doping phosphorus (P) into the lattice of MoS_2 can adjust its electronic structure, reduce the hydrogen adsorption free energy from 2.2 eV to 0.04 eV, enhance the intrinsic activity of MoS_2, and improve HER performance. Xiao et al. [49] constructed N-doped MoS_2 and found that the introduction of N further activated the edge active site of MoS_2 and enhanced the conductivity of the base plane during the HER process. DFT calculation also showed that nitrogen addition increased the hydrogen adsorption free energy of the edge site (**Figure 15a-b**) as well as enhanced charge transfer efficiency on the basal surface. Gao et al. [50] introduced nitrogen into MoS_2 while inducing defects to further improve the conductivity of the basal surface and increase the catalytic activity, as shown in **Figure 16**. Consequently, the N-MoS_2/CN catalyst possesses superior HER activity with an overpotential of 114 mV for the current density of 10 mA cm^{-2} and excellent stability over 10 hours, delivering one of the best MoS_2-based HER electrocatalysts.

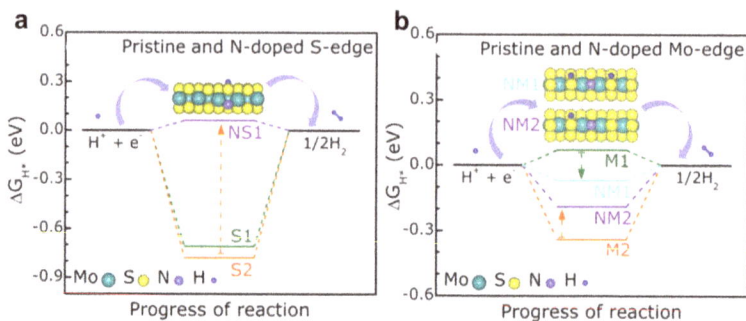

Figure 15. *Hydrogen adsorption free energy at the (a) S edge site and (b) Mo edge site. Reproduced with permission from ref [49]. Copyright 2017, Wiley-VCH.*

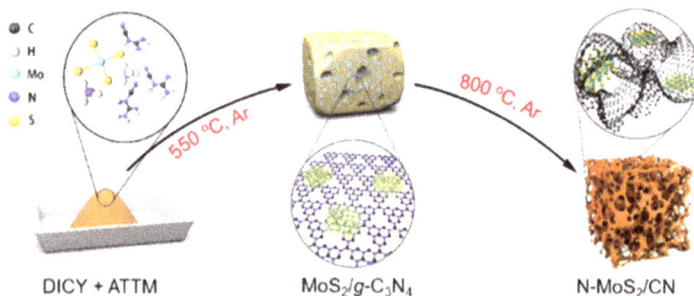

Figure 16. *Preparation diagram of N-doped MoS₂ on porous carbon mesh. Reproduced with permission from ref [50]. Copyright 2019, American Chemical Society.*

5.   Conclusion and Prospects

In conclusion, this chapter provides a comprehensive foundation for understanding the fundamentals of electrocatalytic hydrogen production. It covers the essential reactions involved in the process and introduces the performance parameters used to evaluate catalysts. Moreover, the chapter extensively discusses the key factors that influence the performance of catalysts in electrocatalytic hydrogen production. These factors include the type of catalyst, its composition, structure, geometrical morphology, and electronic structure, all of which play critical roles in determining catalytic activity.

To further advance the field of electrocatalytic hydrogen production, further research should focus on deepening our understanding of the electronic structure and its impact on catalytic behavior. Exploring novel catalyst compositions, employing heteroatom doping strategies, and investigating non-metallic doping techniques hold significant promise for designing more efficient and cost-effective catalysts.   Additionally, the development of advanced characterization techniques and theoretical modelling approaches will provide valuable insights into the intricate mechanisms underlying catalytic reactions.

Furthermore, integrating electrocatalytic hydrogen production systems with renewable energy sources, such as solar and wind power, offers exciting prospects for sustainable and environmentally friendly hydrogen production. Optimizing catalysts and electrode designs, as well as developing scalable and economically viable electrolysis systems, will be pivotal for the practical implementation of electrocatalytic hydrogen production on a large scale.

References

[1]   W. Yang, S. Chen, Recent progress in electrode fabrication for electrocatalytic hydrogen evolution reaction: A mini review, Chem. Eng. J., 393 (2020) 124726. https://doi.org/10.1016/j.cej.2020.124726

[2]   Z. Zhou, Z. Pei, L. Wei, S. Zhao, X. Jian, Y. Chen, Electrocatalytic hydrogen evolution under neutral pH conditions: Current understandings, recent advances, and future prospects, Energy Environ. Sci., 13 (2020) 3185-3206. https://doi.org/10.1039/D0EE01856B

[3]   Z. Chen, H. Qing, K. Zhou, D. Sun, R. Wu, Metal-organic framework-derived nanocomposites for electrocatalytic hydrogen evolution reaction, Prog. Mater. Sci., 108 (2020) 100618. https://doi.org/10.1016/j.pmatsci.2019.100618

[4]   D.M. Morales, M.A. Kazakova, S. Dieckhöfer, A.G. Selyutin, G.V. Golubtsov, W. Schuhmann, J. Masa, Trimetallic Mn-Fe-Ni oxide nanoparticles supported on multi-walled carbon nanotubes as high-performance bifunctional ORR/OER electrocatalyst in alkaline media, Adv. Funct. Mater., 30 (2020) 1905992. https://doi.org/10.1002/adfm.201905992

[5]   S. Anantharaj, S.R. Ede, K. Sakthikumar, K. Karthick, S. Mishra, S. Kundu, Recent trends and perspectives in electrochemical water splitting with an emphasis on sulfide, selenide, and phosphide catalysts of Fe, Co, and Ni: a review, ACS Catal., 6 (2016) 8069-8097. https://doi.org/10.1021/acscatal.6b02479

[6]   W. Feng, W. Pang, Y. Xu, A. Guo, X. Gao, X. Qiu, W. Chen, Transition metal selenides for electrocatalytic hydrogen evolution reaction, ChemElectroChem, 7 (2020) 31-54. https://doi.org/10.1002/celc.201901623

[7]   Y. Zhai, X. Ren, Y. Sun, D. Li, B. Wang, S. Liu, Synergistic effect of multiple vacancies to induce lattice oxygen redox in NiFe-layered double hydroxide OER catalysts, Appl. Catal. B-Environ., 323 (2023) 122091. https://doi.org/10.1016/j.apcatb.2022.122091

[8]   S. Feng, D. Li, H. Dong, S. Xie, Y. Miao, X. Zhang, B. Gao, P.K. Chu, X. Peng, Tailoring the Mo-N/Mo-O configuration in MoO_2/Mo_2N heterostructure for ampere-level current density hydrogen production, Appl. Catal. B-Environ., 342 (2024) 123451. https://doi.org/10.1016/j.apcatb.2023.123451

[9]   R. Li, S. Xie, Y. Zeng, Q. Zhao, M. Mao, Z. Liu, P.K. Chu, X. Peng, Synergistic dual-

regulating the electronic structure of NiMo selenides composite for highly efficient hydrogen evolution reaction, Fuel, 358 (2024) 130203. https://doi.org/10.1016/j.fuel.2023.130203

[10]   X. Peng, S. Xie, S. Xiong, R. Li, P. Wang, X. Zhang, Z. Liu, L. Hu, B. Gao, P. Kelly, Ultralow-voltage hydrogen production and simultaneous Rhodamine B beneficiation in neutral wastewater, J. Energy Chem., 81 (2023) 574-582. https://doi.org/10.1016/j.jechem.2023.03.022

[11]   Z. Huang, A. Reda Woldu, X. Peng, P.K. Chu, Q.-X. Tong, L. Hu, Remarkably boosted water oxidation activity and dynamic stability at large-current–density of Ni(OH)$_2$ nanosheet arrays by Fe ion association and underlying mechanism, Chem. Eng. J., 477 (2023) 147155. https://doi.org/10.1016/j.cej.2023.147155

[12]   Y. Mao, J. Chen, H. Wang, P. Hu, Catalyst screening: Refinement of the origin of the volcano curve and its implication in heterogeneous catalysis, Chin. J. Catal., 36 (2015) 1596-1605. https://doi.org/10.1016/S1872-2067(15)60875-0

[13]   J. Hou, Y. Wu, B. Zhang, S. Cao, Z. Li, L. Sun, Rational design of nanoarray architectures for electrocatalytic water splitting, Adv. Funct. Mater., 29 (2019) 1808367. https://doi.org/10.1002/adfm.201808367

[14]   R. Paul, L. Zhu, H. Chen, J. Qu, L. Dai, Recent Advances in Carbon-Based Metal-Free Electrocatalysts, Adv. Mater., 31 (2019) 1806403. https://doi.org/10.1002/adma.201806403

[15]   V. Vij, S. Sultan, A.M. Harzandi, A. Meena, J.N. Tiwari, W.-G. Lee, T. Yoon, K.S. Kim, Nickel-based electrocatalysts for energy-related applications: Oxygen reduction, oxygen evolution, and hydrogen evolution reactions, ACS Catal., 7 (2017) 7196-7225. https://doi.org/10.1021/acscatal.7b01800

[16]   W. Nabgan, B. Nabgan, T.A. Tuan Abdullah, H. Alqaraghuli, N. Ngadi, A.A. Jalil, B.M. Othman, A.M. Ibrahim, T.J. Siang, Ni-Pt/Al nano-sized catalyst supported on TNPs for hydrogen and valuable fuel production from the steam reforming of plastic waste dissolved in phenol, Int. J. Hydrogen Energ., 45 (2020) 22817-22832. https://doi.org/10.1016/j.ijhydene.2020.06.128

[17]   J. Greeley, T.F. Jaramillo, J. Bonde, I. Chorkendorff, J.K. Nørskov, Computational high-throughput screening of electrocatalytic materials for hydrogen evolution, Nat. Mater., 5 (2006) 909-913. https://doi.org/10.1038/nmat1752

[18]   Y. Li, X. Bao, D. Chen, Z. Wang, N. Dewangan, M. Li, Z. Xu, J. Wang, S. Kawi, Q. Zhong, A minireview on nickel-based heterogeneous electrocatalysts for water splitting, ChemCatChem, 11 (2019) 5913-5928. https://doi.org/10.1002/cctc.201901682

[19]   H. Zhang, A.W. Maijenburg, X. Li, S.L. Schweizer, R.B. Wehrspohn, Bifunctional heterostructured transition metal phosphides for efficient electrochemical water splitting, Adv. Funct. Mater., 30 (2020) 2003261. https://doi.org/10.1002/adfm.202003261

[20]   J. Xiang, Z. Peng, J. Zhao, J. Liang, Preparation of sorbitol by hydrogenation of glucose over the forming Ru/PAC catalyst, J. Nanjing Tech. University (Nat. Sci. Ed.), 44 (2022) 464-472.

[21]   A. Zhang, J. Wu, L. Xue, S. Yan, S. Zeng, Probing heteroatomic dopant-activity synergy over Co_3O_4/doped carbon nanotube electrocatalysts for oxygen reduction reaction, Inorg. Chem., 59 (2019) 403-414. https://doi.org/10.1021/acs.inorgchem.9b02663

[22]   H. Begum, S. Jeon, Highly efficient and stable bifunctional electrocatalyst for water splitting on Fe-Co_3O_4/carbon nanotubes, Int. J. Hydrogen Energ., 43 (2018) 5522-5529. https://doi.org/10.1016/j.ijhydene.2018.01.053

[23]   M. Guo, Y. Qu, F. Zeng, C. Yuan, Synthetic strategy and evaluation of hierarchical nanoporous NiO/NiCoP microspheres as efficient electrocatalysts for hydrogen evolution reaction, Electrochim. Acta, 292 (2018) 88-97. https://doi.org/10.1016/j.electacta.2018.09.159

[24]   K. Tao, H. Dan, Y. Hai, L. Liu, Y. Gong, Ultrafine Co_2P anchored on porous $CoWO_4$ nanofiber matrix for hydrogen evolution: Anion-induced compositional/morphological transformation and interfacial electron transfer, Electrochim. Acta, 328 (2019) 135123. https://doi.org/10.1016/j.electacta.2019.135123

[25]   Z. Pan, N. Pan, L. Chen, J. He, M. Zhang, Flower-like MOF-derived Co-N-doped carbon composite with remarkable activity and durability for electrochemical hydrogen evolution reaction, Int. J. Hydrogen Energ., 44 (2019) 30075-30083. https://doi.org/10.1016/j.ijhydene.2019.09.117

[26]   X. Duan, N. Pan, C. Sun, K. Zhang, X. Zhu, M. Zhang, L. Song, H. Zheng, MOF-derived Co-MOF, O-doped carbon as trifunctional electrocatalysts to enable highly efficient Zn–air batteries and water-splitting, J. Energy Chem., 56 (2021) 290-298. https://doi.org/10.1016/j.jechem.2020.08.007

[27]   Y. Cheng, K. Pang, X. Wu, Z. Zhang, X. Xu, J. Ren, W. Huang, R. Song, In situ hydrothermal synthesis MoS_2/Guar gum carbon nanoflowers as advanced electrocatalysts for electrocatalytic hydrogen evolution, ACS Sustain. Chem. Eng., 6 (2018) 8688-8696. https://doi.org/10.1021/acssuschemeng.8b00994

[28]   D. Song, D. Hong, Y. Kwon, H. Kim, J. Shin, H.M. Lee, E. Cho, Highly porous Ni-P electrode synthesized by an ultrafast electrodeposition process for efficient overall water electrolysis, J. Mater. Chem. A, 8 (2020) 12069-12079. https://doi.org/10.1039/D0TA03739G

[29]   Y. Wang, H. Arandiyan, X. Chen, T. Zhao, X. Bo, Z. Su, C. Zhao, Microwave-induced plasma synthesis of defect-rich, highly ordered porous phosphorus-doped cobalt oxides for overall water electrolysis, J. Phys. Chem. C, 124 (2020) 9971-9978. https://doi.org/10.1021/acs.jpcc.0c01135

[30]   D. Pletcher, Electrocatalysis: present and future, J. Appl. Electrochem., 14 (1984) 403-415. https://doi.org/10.1007/BF00610805

[31] M. Khalid, P.A. Bhardwaj, A.M. Honorato, H. Varela, Metallic single-atoms confined in carbon nanomaterials for the electrocatalysis of oxygen reduction, oxygen evolution, and hydrogen evolution reactions, Catal. Sci. Technol., 10 (2020) 6420-6448. https://doi.org/10.1039/D0CY01408G

[32] S. Niu, J. Yang, H. Qi, Y. Su, Z. Wang, J. Qiu, A. Wang, T. Zhang, Single-atom Pt promoted Mo_2C for electrochemical hydrogen evolution reaction, J. Energy Chem., 57 (2021) 371-377. https://doi.org/10.1016/j.jechem.2020.08.028

[33] Q. Duan, S. Wang, L. Wang, Electro-deposition of the porous composite $Ni-P/LaNi_5$ electrode and its electro-catalytic performance toward hydrogen evolution reaction, Acta Phys.-Chim. Sin., 29 (2013) 123-130. https://doi.org/10.3866/PKU.WHXB201210095

[34] Y. Sun, S. Gao, F. Lei, C. Xiao, Y. Xie, Ultrathin two-dimensional inorganic materials: new opportunities for solid state nanochemistry, Acc. Chem. Res., 48 (2015) 3-12. https://doi.org/10.1021/ar500164g

[35] K.S. Novoselov, D. Jiang, F. Schedin, T.J. Booth, V.V. Khotkevich, S.V. Morozov, A.K. Geim, Two-dimensional atomic crystals, PNAS, 102 (2005) 10451-10453. https://doi.org/10.1073/pnas.0502848102

[36] T. Li, G. Galli, Electronic properties of MoS_2 nanoparticles, J. Phys. Chem. C, 111 (2007) 16192-16196. https://doi.org/10.1021/jp075424v

[37] Z. Cui, H. Chu, S. Gao, Y. Pei, J. Ji, Y. Ge, P. Dong, P.M. Ajayan, J. Shen, M. Ye, Large-scale controlled synthesis of porous two-dimensional nanosheets for the hydrogen evolution reaction through a chemical pathway, Nanoscale, 10 (2018) 6168-6176. https://doi.org/10.1039/C8NR01182F

[38] H. Zhao, J. Liang, Q. Zheng, Construction of core-shell heterostructure $NiO@Co_{0.5}Fe_{0.5}P$ as efficient bifunctional electrocatalysts for overall water splitting, J. Alloys Compd., 905 (2022) 164264. https://doi.org/10.1016/j.jallcom.2022.164264

[39] G. Tian, S. Wei, Z. Guo, S. Wu, Z. Chen, F. Xu, Y. Cao, Z. Liu, J. Wang, L. Ding, J. Tu, H. Zeng, Hierarchical $NiMoP_2-Ni_2P$ with amorphous interface as superior bifunctional electrocatalysts for overall water splitting, J. Mater. Sci. Technol., 77 (2021) 108-116. https://doi.org/10.1016/j.jmst.2020.09.046

[40] X. Du, J. Huang, J. Zhang, Y. Yan, C. Wu, Y. Hu, C. Yan, T. Lei, W. Chen, C. Fan, Modulating electronic structures of inorganic nanomaterials for efficient electrocatalytic water splitting, Angew. Chem. Int. Ed., 58 (2019) 4484-4502. https://doi.org/10.1002/anie.201810104

[41] M. Blasco-Ahicart, J. Soriano-López, J.J. Carbó, J.M. Poblet, J.-R. Galan-Mascaros, Polyoxometalate electrocatalysts based on earth-abundant metals for efficient water oxidation in acidic media, Nat. Chem., 10 (2018) 24-30. https://doi.org/10.1038/nchem.2874

[42] M.M. Jakšić, Advances in electrocatalysis for hydrogen evolution in the light of the

Brewer-Engel valence-bond theory, J. Mol. Catal., 38 (1986) 161-202.
https://doi.org/10.1016/0304-5102(86)87056-0

[43]   S. Trasatti, Work function, electronegativity, and electrochemical behaviour of metals: II.
       Potentials of zero charge and "electrochemical" work functions, J. Electroanal. Chem.
       Interfacial Electrochem., 33 (1971) 351-378. https://doi.org/10.1016/0368-
       1874(71)80045-X

[44]   Y. Li, Z. Dong, L. Jiao, Multifunctional transition metal-based phosphides in energy-
       related electrocatalysis, Adv. Energy Mater., 10 (2020) 1902104.
       https://doi.org/10.1002/aenm.201902104

[45]   B.-Y. Guo, Y.-W. Wang, B. Dong, Controllable leaching of Ag_2WO_4 template for
       production of Ni-Co-P nanosheets as electrocatalyst for hydrogen evolution, Int. J.
       Electrochem. Sci., 17 (2022) 22091. https://doi.org/10.20964/2022.09.03

[46]   K. Wu, K. Sun, S. Liu, W.-C. Cheong, Z. Chen, C. Zhang, Y. Pan, Y. Cheng, Z. Zhuang,
       X. Wei, Atomically dispersed Ni-Ru-P interface sites for high-efficiency pH-universal
       electrocatalysis of hydrogen evolution, Nano Energy, 80 (2021) 105467.
       https://doi.org/10.1016/j.nanoen.2020.105467

[47]   Y. Wang, Z. Chen, Q. Li, X. Wang, W. Xiao, Y. Fu, G. Xu, B. Li, Z. Li, Z. Wu, Porous
       needle-like Fe-Ni-P doped with Ru as efficient electrocatalyst for hydrogen generation
       powered by sustainable energies, Nano Res., 16 (2023) 2428-2435.
       https://doi.org/10.1007/s12274-022-4980-4

[48]   T. Ling, D.Y. Yan, H. Wang, Y. Jiao, Z. Hu, Y. Zheng, L. Zheng, J. Mao, H. Liu, X.W.
       Du, M. Jaroniec, S.Z. Qiao, Activating cobalt(II) oxide nanorods for efficient
       electrocatalysis by strain engineering, Nat. Commun., 8 (2017) 1509.
       https://doi.org/10.1038/s41467-017-01872-y

[49]   W. Xiao, P. Liu, J. Zhang, W. Song, Y.P. Feng, D. Gao, J. Ding, Dual-functional N
       dopants in edges and basal plane of MoS_2 nanosheets toward efficient and durable
       hydrogen evolution, Adv. Energy Mater., 7 (2017) 1602086.
       https://doi.org/10.1002/aenm.201602086

[50]   H. Wang, X. Xiao, S. Liu, C.-L. Chiang, X. Kuai, C.-K. Peng, Y.-C. Lin, X. Meng, J.
       Zhao, J. Choi, Y.-G. Lin, J.-M. Lee, L. Gao, Structural and electronic optimization of
       MoS_2 edges for hydrogen evolution, J. Am. Chem. Soc., 141 (2019) 18578-18584.
       https://doi.org/10.1021/jacs.9b09932

Electrocatalytic Hydrogen Production: Catalysts and Applications
Materials Research Foundations **165** (2024)

Materials Research Forum LLC
https://doi.org/10.21741/9781644903070

CHAPTER 3

Precious Metal-based Catalysts for Electrocatalytic Hydrogen Production

Xiang Peng*

Hubei Key Laboratory of Plasma Chemistry and Advanced Materials, Engineering Research Center of Phosphorus Resources Development and Utilization of Ministry of Education, School of Materials Science and Engineering, Wuhan Institute of Technology, Wuhan 430205, China

xpeng@wit.edu.cn

Abstract

Electrocatalysts play a crucial role in water electrolysis technology by reducing the activation energy of the reactions, accelerating the reaction kinetics, enhancing the electrolytic efficiency, and improving the product selectivity. It is evident that precious metals, such as platinum (Pt), palladium (Pd), ruthenium (Ru), and rhodium (Rh), occupy the top positions on the volcano curve, indicating the excellent hydrogen evolution reaction performance of these precious metal catalysts.

Keywords

Hydrogen Evolution Reaction, Precious Metal Catalyst, Pt-Based Catalyst, Pd-Based Catalyst, Surface Regulation

1. Introduction

Electrolytic water is the process of splitting water into hydrogen and oxygen gases, which usually requires the application of an external potential to drive the reactions [1-3]. Electrocatalysts play a crucial role in water electrolysis technology by reducing the activation energy of the reactions, accelerating the reaction kinetics, enhancing the electrolytic efficiency, and improving the product selectivity [4-6].

Using density functional theory (DFT), Nørskov et al. calculated the Gibbs free energy of hydrogen adsorption (ΔG_{H*}) for different types of catalysts and combined it with the exchange current density (j_0) to construct a "volcano curve", as depicted in **Figure 1**. The volcano curve serves as an important parameter for evaluating the catalytic activity of the hydrogen evolution reaction (HER) electrocatalyst [7]. The closer the ΔG_{H*} value is to zero, the better the HER performance of the catalyst. It is evident that precious metals, such as platinum (Pt), palladium (Pd), ruthenium (Ru), and rhodium (Rh), occupy the top positions on the volcano curve, indicating the excellent HER performance of these precious metal catalysts.

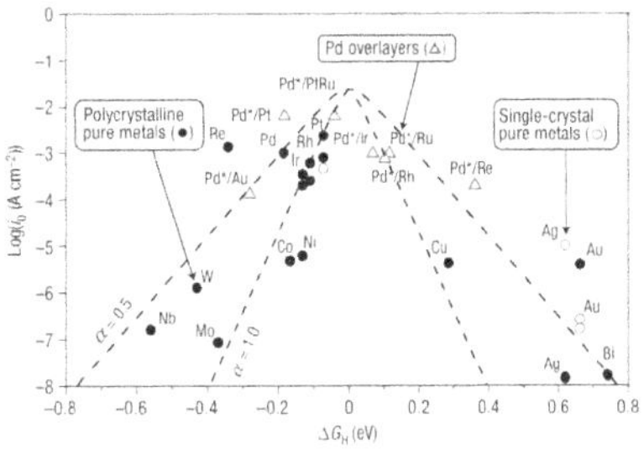

Figure 1. *The volcano curves of various metal materials for HER. Reproduced with permission from ref [7]. Copyright 2006, Springer Nature.*

2.   Pt-based Catalysts

According to the volcano curve for HER, the ΔG_{H^*} value of Pt is very close to zero, making it the best electrocatalyst for HER. Due to its excellent innate HER activity, Pt-based catalysts have garnered significant attention [8, 9]. Pt exhibits different crystal face exposures, such as (111), (100), and (110). Pt-based catalysts with different crystal face exposures exhibit distinct characteristics and find applications in catalytic activity, surface energy, and electrochemical properties.

Markovic et. al [10] employed their developed rotating disk technique to obtain single-crystal Pt rotating disk electrodes (RD$_{Pt(hkl)}$E) and systematically studied the HER capabilities of different crystal faces in a sulfuric acid solution over a temperature range of 274-333 K. The Pt$_{(hkl)}$ single crystal was pretreated and assembled in an RDPt(hkl)E configuration with an area of 0.283 cm^2. It was then flame annealed before mounting it into the disk position of an insertable rotating disk electrodes (RDE) assembly. Subsequently, the modified RDE was transferred into a standard electrochemical cell and immersed into 0.05 M H$_2$SO$_4$. Upon immersion, the electrolyte was equilibrated for 5 min with the hydrogen gas.

All experimental measurements were performed using a standard three-compartment electrochemical cell equipped with a water bath to maintain a constant electrolyte temperature within a ± 0.5 °C range through circulation. The measurements were conducted nonisothermally, meaning that the temperature of the reference electrode was kept at 298 K while the temperature of the working electrode varied from 274 to 333 K. The reference electrode used was a saturated calomel electrode (SCE), which was separated by a bridge to prevent Cl⁻ contamination of electrolytes. Meanwhile, all the potentials were corrected to a "constant-temperature" scale and referenced to the reversible hydrogen electrode (RHE) under specific conditions of one standard atmosphere, 298 K, and 0.05 M H$_2$SO$_4$ electrolyte. Current-potential curves were obtained potentiodynamically with a sweep rate of 10 mV s⁻¹ and recorded simultaneously on a chart recorder and digitally on a computer.

The polarization curve in the micropolarization region (± 10 mV) demonstrates the linear relationship between current density and overpotential, as shown in **Figure 2**. From the Figure, the exchange current density (j_0) can be calculated. It was observed that j_0 increased in the order of Pt (111) < Pt (100) < Pt (110) at a fixed temperature of 274 K, as shown in **Table 1**. The j_0 value on the Pt (110) surface was found to be three times larger than that on the Pt (111) surface. Furthermore, it was discovered that each crystal face exhibits a unique, temperature-dependent Tafel slope. The activation energies for the HER followed the sequence $\Delta H^{\#}_{111} > \Delta H^{\#}_{100} > \Delta H^{\#}_{110}$, which corresponds to the order of activity.

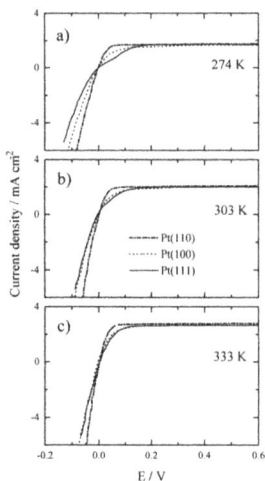

Figure 2. Polarization curves for the HER on $Pt_{(hkl)}$ in 0.05 M H_2SO_4 in the micro-polarization potential region at (a) 274 K, (b) 303 K, and (c) 333 K. Straight lines are the slopes used to obtain the exchange current densities. Reproduced with permission from ref [10]. Copyright 1997, American Chemical Society.

Table 1. *Kinetic Parameters for the HER on $Pt_{(hkl)}$ in 0.05 M H_2SO_4 at different temperatures.*

$Pt_{(hkl)}$	Tafel slope (mV dec^{-1})	exchange current density (mA cm^{-2})			activation energy (kJ mol^{-1})	mechanism
		274 K	303 K	333 K		
$Pt_{(110)}$	2.3 RT/2F	0.65	0.98	1.35	9.5	Tafel - Volmer
$Pt_{(100)}$	2.3 RT/3F^b 2(2.3RT/F)^c	0.36	0.60	0.76	12	Heyrovsky - Volmer
$Pt_{(111)}$	≈2.3 RT/F	0.21	0.45	0.83	18	-

^{a} Obtained from a linear polarization method. ^{b} Low current density. ^{c} High current density.

The difference in activation energies among crystal faces can be attributed to the structure-sensitive properties of the adsorption behavior of the active intermediate (H_{ad}). On the Pt (110) crystal face, the reaction follows the Tafel-Volmer mechanism, with the Tafel step serving as the rate-determining step (RDS). On the Pt (100) crystal face, the reactions follow the Heyrovsky-Volmer sequence, with the Heyrovsky (ion-atom) reaction being the RDS. However, the specific reaction mechanism for the Pt(111) crystal face could not be determined with this experimental design.

Further result indicates that the electrocatalytic activity of Pt-based catalysts with different crystal faces follows the order of Pt (111) < Pt (100) < Pt (110) for HER in alkaline media [11], as shown in **Figure 3**. However, the activation energy of Pt-based catalysts in alkaline media is typically higher than that in acidic media [12]. This is mainly due to the presence of a strong metal-OH_{ad} interaction in the alkaline media and the high water ionization energy barrier [13].

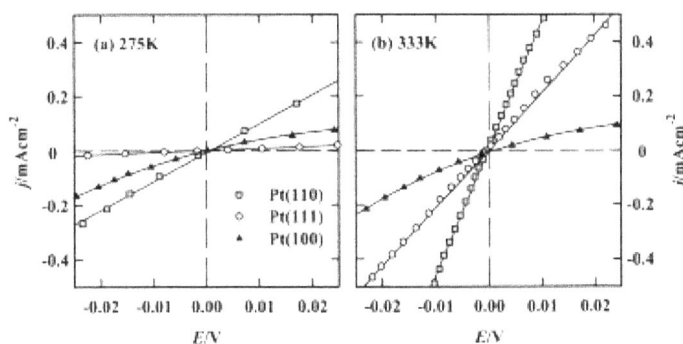

Figure 3. *The micro-polarization region ± 20 mV of the reversible potential on $Pt_{(hkl)}$ at: (a) 275 and (b) at 333 K in 0.1 M KOH. Reproduced with permission from ref [11]. Copyright 2002, Elsevier.*

So far, the commercial carbon-supported platinum (Pt/C) electrocatalyst has shown the most optimal adsorption and recombination of hydrogen intermediates (e.g., H*) compared to other carbon-supported metals. This ensures low overpotential and fast kinetics for HER in acidic media. However, the agglomeration of the nanoparticles (NPs) and dissolution of the metal atoms in an acidic environment during the long-term operation result in the loss of active sites. This rapid degradation of catalytic activity and stability in acidic electrolytes is a major concern. In contrast, the industry favors hydrogen production through water-alkali electrolysis due to its unlimited availability of reactants, good manufacturing safety, stable output, and high product purity. However, initiating HER in alkaline media requires significant dissociation of water molecules because of the low abundance of H^+. As a result, the reaction kinetics is two to three orders of magnitude slower compared to acidic media. Furthermore, the commercial Pt/C catalyst has a Pt loading of approximately 20%, making it expensive and unsuitable for large-scale industrial applications, which has drawn considerable attention.

Therefore, the solution lies in enhancing the catalytic capability of Pt-based catalysts and reducing the amount of Pt loading to develop highly efficient catalysts with exceptional activity, fast kinetics, long-term stability, and high tolerance in both acidic and alkaline environments. To overcome the challenges associated with Pt-based catalysts, various strategies have been developed, including morphology optimization, structural regulation, transition metal alloying, and surface modification.

2.1.   Morphology Optimization

The electrocatalytic HER typically takes place on the surface of the catalyst material. Optimizing the geometrical morphology of Pt-based catalysts is beneficial for modifying the atomic exposure and porous structure, thereby improving their specific surface area and extrinsical catalytic activity. For example, Liu et al. [14] designed a highly curved onion-like carbon nanosphere (OLC) support to anchor stable, atomically dispersed Pt (assigned as Pt_1OLC), which has been synthesized and applied as a catalyst exhibiting high catalytic activity. The schematic illustration of the raw materials is shown in **Figure 4**. The synthesis of the catalyst involves a two-step process. First, the starting material of the surface-oxidized detonation nanodiamonds (DND) are thermally deoxidized at different temperatures to precisely adjust the type and distribution density of oxygen, during which the DND is automatically converted into OLC support at high temperatures. Then the Pt atoms are deposited on the support using a single cycle atomic layer deposition method.

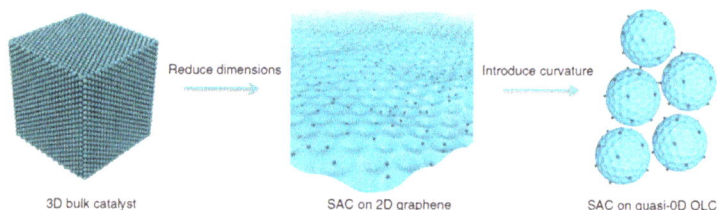

3D bulk catalyst   SAC on 2D graphene   SAC on quasi-0D OLC

Figure 4. The schematic illustration of the preparation of Pt_1/OLC. Reproduced with permission from ref [14]. Copyright 2019, Springer Nature.

The high-angle annular dark-field scanning transmission electron microscopy (HAADF-STEM) shown in **Figure 5a** provides insights into the morphology of Pt atoms distributed on OLC scaffolds. The HAADF-STEM analysis reveals that isolated Pt atoms are uniformly dispersed within the support, without the presence of significant nanoparticles (NPs) or clusters. Simultaneously, the OLC particles, following oxidative annealing, exhibit typical multi-shell fullerene structures, with an approximate diameter of 5 nm and an interlayer distance of 0.35 nm, as shown in **Figure 5b**.

Based on the geometric details obtained from the aforementioned experiments, an atomic model for first principles testing can be proposed using DFT. The PtO_2C_{295} model extracted from X-ray absorption fine structure (XAFS) results are constructed by encapsulating C_{60} fullerenes within the defective fullerenes of C_{235}, as shown in **Figure 6a**. The optimized Pt-O and Pt-C bond lengths, obtained by combining the Pt atom with the C atom and two O atoms, are in good agreement with the extended X-ray absorption fine structure (EXAFS) data. Notably, the fixed Pt atoms protrude from the surface of the spherical fullerene structure. The diffusion of Pt on the surface of PtO_2C_{295} requires overcoming a high energy barrier of 3.20 eV, indicating excellent structural stability, as indicated in **Figure 6b**. In sharp contrast, the Pt system loaded with the C_{300} without Pt-O binding exhibits a lower diffusion barrier of 0.75 eV and poor stability. As anticipated, the Pt metal-based catalysts exhibit negligible onset potentials near the thermodynamic potential for HER. The Pt_1/OLC catalyst, with only 0.27 wt% of Pt, achieves a very low overpotential of approximately 38 mV at a current density of 10 mA cm^{-2}, which is comparable to that of a commercial Pt/C catalyst with a 20 wt% Pt loading and significantly better than that of a Pt/C catalyst loaded with 5 wt% Pt.

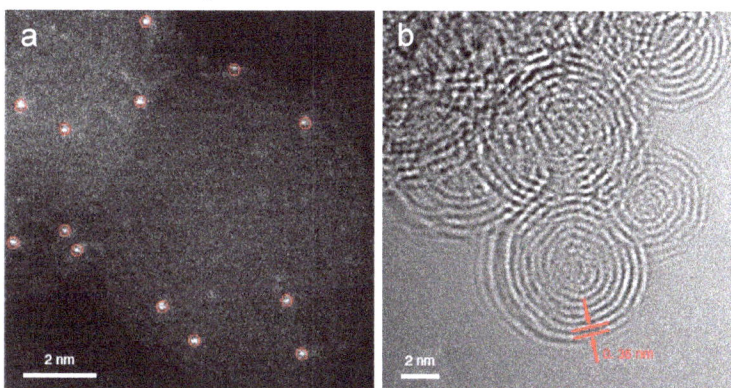

Figure 5. (a) The HAADF-STEM image of Pt_1/OLC with the Pt single atoms (highlighted by red circles). (b) TEM image of Pt_1/OLC. Reproduced with permission from ref [14]. Copyright 2019, Springer Nature.

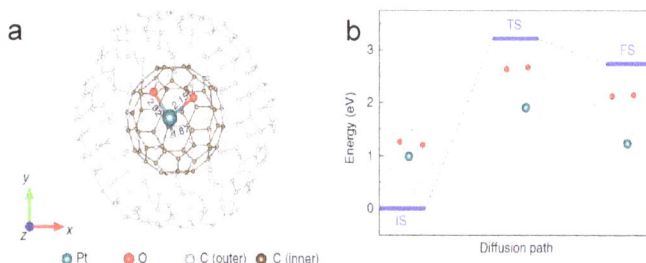

Figure 6. (a) the optimized atomic model of PtO_2C_{295} which exhibits Pt-O bonding following the experiments. (b) The lowest diffusion barrier (most probable diffusion path) of a Pt atom on PtO_2C_{295}. Reproduced with permission from ref [14]. Copyright 2019, Springer Nature.

2.2.   Structural Regulation.

Through the regulating of the crystal structure, defect sites, and crystal surface of Pt-based catalysts, the surface topology and electronic structure can be modified, thereby controlling the activity and selectivity of the catalyst [15, 16]. It has been observed that the activity of Pt is influenced by its electronic structure, which is sensitive to lattice strain [17, 18]. This dependence can be leveraged in catalyst design. The implementation of core-shell structures and elastic substrates has led to significant improvements in the electrocatalytic performance of strain-engineered Pt catalysts.

He et al. [19] employed a strain strategy to deposit ultra-thin Pt shells on Pd-based nanocubes. The strain was induced in the Pt (100) lattice through the expansion and contraction of phosphorylated and dephosphorized nanocubes, allowing for the regulation of strain ranging from -5.1% to 5.9%. This approach of strain control was successfully utilized to adjust the electrocatalytic activity of the Pt shell across a wide range. The electrochemical results presented in **Figure 7a-b** demonstrated the superior performance of the strained Pt shells, surpassing that of Pt/C catalyst in terms of specific activities and mass activities. Furthermore, **Figure 7c** revealed that the strain-activity relationship for the HER follows a volcanic curve pattern.

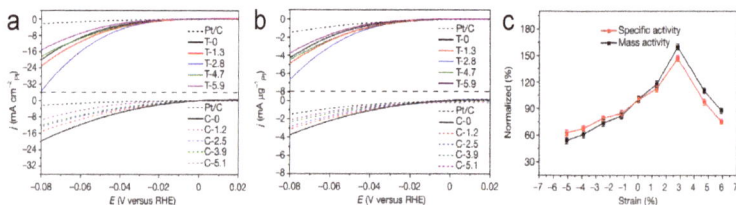

Figure 7. HER polarization curves conducted in 1 M KOH. Normalized against (a) the electrochemical surface areas of the samples and (b) the mass of Pt. (c) Normalized HER catalytic activities of all strained Pt shells versus the unstrained one. Reproduced with permission from ref [19]. Copyright 2021, Springer Nature.

Furthermore, the DFT calculations were performed to assess the impact of lattice strain on the HER. In both acidic and alkaline solutions, an increase in tensile strain promotes hydrolytic dissociation, enhancing the adsorption of OH* and H*, while compressive strain has the opposite effect. These strain-induced changes in binding strength align with the predictions of d-band center theory. Lattice strain also affects the adsorption sites. The adsorption of hydrolytic ionization and OH* is most active at the bridge site. However, as the lattice expands, the adsorption of H* switches from the bridge site to the adjacent hollow site. These findings indicate that on a stretched Pt surface, H* and OH* can be efficiently separated, thus accelerating the HER. This approach holds potential for screening lattice strains to optimize the performance of Pt catalysts and can potentially be extended to other metal catalysts for a wide range of reactions.

2.3.   Transition Metal Alloying

Nano-alloying has been proven to be an effective approach for promoting water dissociation in alkaline HER with reduced Pt usage, thereby enhancing HER properties [9, 20]. Based on the *d*-band center theory, a stronger adsorption capacity is achieved when the *d*-band center is closer to the Fermi level [20, 21]. Therefore, constructing a Pt alloy with 3*d*-transition metals such as Fe, Co, Ni, etc. can effectively reduce the consumption of precious metals, adjust the electronic structure of Pt for enhanced hydrolysis, introduce diverse active sites, and enhance the intrinsic activity for HER. Moreover, due to the significant lattice mismatch between Pt and 3*d*-transition metals, the alloying process induces lattice strain and alters surface properties, as a result, leading to enhanced catalytic efficiency. Therefore, controlling the spatial distribution of alloy nanoparticles on carbon supports with maximized exposure of active sites is desirable.

For instance, Zhang et. al [22] proposed a pyrolysis-replacement-reorganization strategy using Co-based ZIF-67 nanorods (NRs) as the starting material. This strategy led to the synthesis of ultrafine Pt_3Co alloy nanoparticles (sub-10 nm) supported on the inner and outer shells of porous nitrogen-doped carbon nanotubes (designated as $Pt_3Co@NCNT$) with closed ends. The detailed synthesis process is illustrated in **Figure 8**. In the first step, Co nanoparticles are confined within porous N-doped carbon nanotubes (designated as Co@NCNT). Subsequently, a simple galvanic replacement reaction with a K_2PtCl_4 solution converts the Co nanoparticles into pre-Pt_3Co nanoparticles (pre-$Pt_3Co@NCNT$).

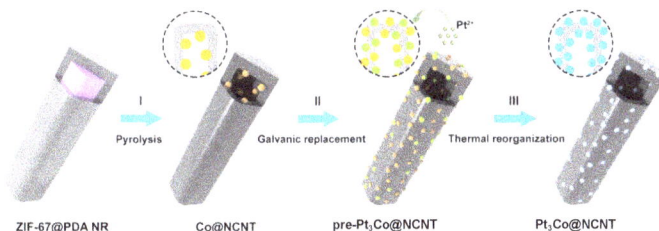

Figure 8. *Illustration of the synthesis of $Pt_3Co@NCNT$. Reproduced with permission from ref [22]. Copyright 2021, Wiley-VCH.*

During the thermal recombination process, the crystallinity of the Pt_3Co nanoalloy is further enhanced. The alloy particles migrate to the inner and outer surfaces of the carbon nanotubes, ensuring maximum exposure of the active site and maintaining strong adhesion to the porous carbon matrix. Due to the optimized electronic structure resulting from alloying and the strong interaction between the metal and the carrier, the obtained $Pt_3Co@NCNT$ electrocatalyst exhibits ultra-high activity and exceptional long-term stability in both acidic and alkaline media.

Figure 9 shows the electrochemical performance of the catalyst in acidic electrolyte, evaluated using a standard three-electrode system. For comparison, the HER properties of commercial Pt/C catalysts with an average Pt particle size of 2-3 nm were measured under the same test conditions. As shown in **Figure 9a**, the polarization curve of the $Pt_3Co@NCNT$, with *iR* compensation at a scanning rate of 5 mV s^{-1}, shows the highest activity in 0.5 M H_2SO_4, requiring an overpotential of only 42 mV to achieve a current density of 10 mA cm^{-2}. In contrast, PtCo@NCNT and Pt/C require overpotentials of 64 and 47 mV, respectively, to reach the same current density. At an overpotential of 70 mV, Pt$_3$Co@NCNT catalyst can generate a current density of up to 100 mA cm^{-2}, significantly surpassing PtCo@NCNT (13.3 mA cm^{-2}) and the benchmark Pt/C catalyst (34.6 mA cm^{-2}). To further understand the enhancement in HER activity, the turnover frequency (TOF) was calculated to investigate the intrinsic activity of the catalyst, as illustrated in **Figure 9b**. Specifically, at an overpotential of 100 mV, the Pt$_3$Co@NCNT catalyst exhibits a TOF of 1.95 s^{-1}, which is 6.5 times and 2.3 times larger than that of PtCo@NCNT and Pt/C under the same conditions.

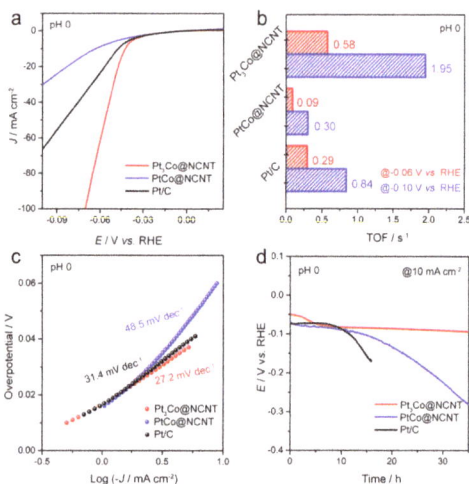

Figure 9. Electrocatalytic HER performance of Pt$_3$Co@NCNT and PtCo@NCNT in 0.5 M H$_2$SO$_4$. (a) LSV curves, (b) a bar graph of comparative TOF values at two different overpotentials, (c) Tafel plots, and (d) chronopotentiometric curves recorded at 10 mA cm^{-2} of Pt$_3$Co@NCNT, PtCo@NCNT, and Pt/C catalyst. Reproduced with permission from ref [22]. Copyright 2021, Wiley-VCH.

As shown in **Figure 9c**, the Pt$_3$Co@NCNT catalyst exhibits faster kinetic and a lower Tafel slope of 27.2 mV dec^{-1} compared to PtCo@NCNT (48.5 mV dec^{-1}) and Pt/C catalyst (31.4 mV dec^{-1}). The HER performance of Pt$_3$Co@NCNT in 0.5 M H$_2$SO$_4$ is also comparable to recently reported Pt-based catalysts. The chronopotentiometric test shows that the Pt$_3$Co@NCNT catalyst exhibits significantly higher stability than PtCo@NCNT during 35 hours of operation, and both catalysts show improved stability compared to Pt/C, as shown in **Figure 9d**. After the stability test, the morphologies of Pt$_3$Co@NCNT and PtCo@NCNT catalysts remain stable without significant changes. This observation highlights the crucial influence of the designed carbon support and catalyst composition on electrocatalytic performance.

It is worth noting that the Pt$_3$Co@NCNT catalyst also exhibits excellent activity in an alkaline medium (1.0 M KOH). As shown in **Figure 10a**, to achieve a current density of 10 mA cm^{-2}, the Pt$_3$Co@NCNT catalyst requires only an overpotential of 36 mV, which is 17 mV and 30 mV lower than PtCo@NCNT and Pt/C catalysts, respectively. At an overpotential of 100 mV, Pt$_3$Co@NCNT catalyst can generate a current density of 100 mA cm^{-2}, which is 2.4 times and 5 times larger than that of PtCo@NCNT and Pt/C, respectively. To further assess the improvement in the intrinsic activity of the as-prepared catalyst, the TOF in alkaline media was also calculated, as shown in **Figure 10b**. Specifically, at an overpotential of 60 mV, the TOF of Pt$_3$Co@NCNT is 0.3 s^{-1}, which is much higher than PtCo@NCNT (0.13 s^{-1}) and Pt/C catalyst (0.11 s^{-1}). In addition, at an overpotential of 100 mV, the TOF of Pt$_3$Co@NCNT reaches 0.94 s^{-1}, which is about 1.2 times and 2.8 times higher than PtCo@NCNT (0.43 s^{-1}) and Pt/C catalyst (0.25 s^{-1}), respectively.

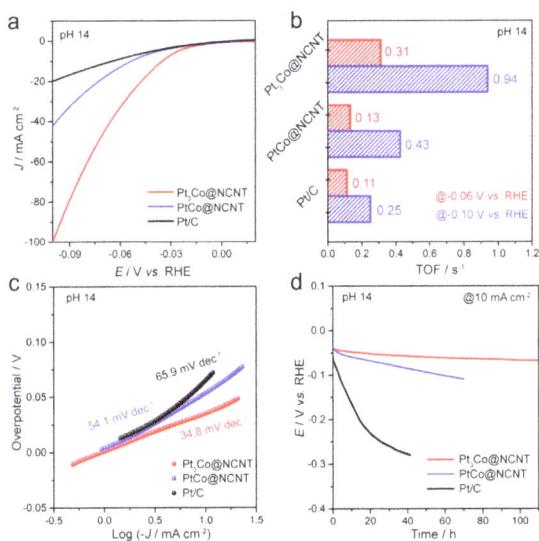

Figure 10. *Electrocatalytic HER performance of Pt$_3$Co@NCNT and PtCo@NCNT in 1.0 M KOH. (a) LSV curves, (b) a bar graph of comparative TOF values at two different overpotentials, (c) Tafel plots, and (d) chronopotentiometric curves recorded at 10 mA cm^{-2} of Pt$_3$Co@NCNT, PtCo@NCNT, and Pt/C. Reproduced with permission from ref [22]. Copyright 2021, Wiley-VCH.*

The Tafel slope of Pt$_3$Co@NCNT catalyst is measured to be 34.8 mV dec^{-1}, which is much smaller than that of PtCo@NCNT (54.1 mV dec^{-1}) and Pt/C catalyst (65.9 mV dec^{-1}), as indicated in **Figure 10c**. With its excellent HER activity in 1.0 M KOH, the Pt$_3$Co@NCNT catalyst outperforms many reported Pt-based catalysts [23-26]. The catalyst exhibits minimal activity loss, as shown in **Figure 10d**, even under continuous operation for 100 hours at a current density of 10 mA cm^{-2}. The outstanding stability of the catalyst can be attributed to the strong metal-support interaction and the composition of the nano-alloy. However, the PtCo@NCNT catalyst experiences a more pronounced decline in activity during a 70 hour chronopotentiometry test, while commercially available Pt/C catalysts demonstrate a sharp decrease in activity within the first 20 hours.

DFT calculations show that the adsorption-free energy of Pt$_3$Co for hydrogen intermediates is close to thermodynamic neutrality. In addition, the alloying process produces various active sites, greatly improving the intrinsic activity of the HER. At the same time, the intricately designed nitrogen-coordinated porous carbon shell structure provides both chemisorption and physical limiting effects. This novel structure is beneficial for preventing the ultra-small Pt$_3$Co NPs from aggregation during the reaction process, thus ensuring their exceptional long-term stability against HER. This work provides a new approach for accurately controlling the composition and spatial distribution of alloy nanoparticles in carbon support, thereby improving the efficiency and stability of alloys in energy-related fields.

2.4.   Surface Modification

Enhancing the metal-support interaction can significantly reduce the aggregation of precious metal-based nanoparticles on the support and promote synergistic effects to enhance catalytic activity. In conventional designs, noble metal particles are typically either attached to or embedded within the surface of porous carbon supports. However, achieving high utilization efficiency and catalytic stability simultaneously can be challenging. Therefore, an alternative strategy to improve catalyst stability involves changing the support material. For example, Zhang et al. [27] utilized non-noble TiN nanowires grown on carbon cloth (TiN/CC) as the support for a precious metal catalyst. Pt nanoparticles were loaded onto the TiN substrate using an electrodeposition method, resulting in the formation of a catalyst comprising Pt nanoparticles and a TiN skeleton (Pt/TiN/CC), as schematically illustrated in **Figure 11**. Specifically, Pt NPs were deposited on the TiN nanowires through electrochemical cyclic deposition in a three-electrode system using 0.5 M H$_2$SO$_4$. The TiN/CC (1 × 1 cm^2), Pt wire, and SCE were used as the working electrode, counter electrode, and reference electrode, respectively. Electrochemical deposition was carried out by cyclic voltammetry (CV) for 500, 750, 1000, and 1500 scanning cycles in the potential range of -1.0 to -0.2 V *vs.* SCE at a scanning rate of 100 mV s^{-1}, resulting in the formation of Pt$_x$/TiN/CC, where x represents the number of CV cycles (500, 750, 1000, and 1500).

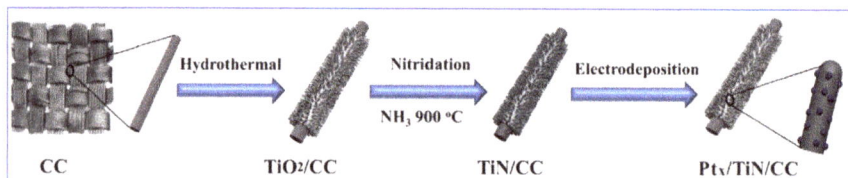

Figure 11. Schematic illustration of electrode preparation. Reproduced with permission from ref [27]. Copyright 2022, American Chemical Society.

Materials Research Forum LLC

https://doi.org/10.21741/9781644903070

The surface of the bare TiN/CC nanowires appears smooth, as shown in the SEM image in **Figure 12a**. However, after electrochemical deposition of Pt nanoparticles for different scanning cycles, the nanowires become slightly rough, with a uniform distribution of numerous nanoparticles on their surface, as illustrated in **Figure 12b-e**. There is a gradual increase in Pt deposition as the number of scanning cycles increases. The uniform loading of Pt nanoparticles on the TiN nanowires facilitates electron transfer from the plasmonic TiN nanowires. However, when the scanning cycle reaches 1,500, the Pt nanoparticles become larger and eventually agglomerate, potentially hindering the generation and transfer of electrons. Generally, smaller Pt nanoparticles loading may reduce the efficiency of hot-electron transfer from TiN nanowires to Pt nanoparticles, thereby decreasing the HER activity. The XRD patterns in **Figure 12f** reveal that the TiO_2/CC nanowires are converted into TiN/CC nanowires after calcination under NH_3 for 3 hous. The peaks observed at 36.2°, 42.2°, 61.7°, 74.3°, and 78.4° are associated with the (111), (200), (220), (311), and (222) crystal planes of cubic TiN (JCPDS card No. 38–1420). Notably, a new peak appears at 39.76° in the XRD patterns of Pt_{1000}/TiN/CC and Pt_{1500}/TiN/CC, indicating the formation of cubic Pt with the (111) crystal plane (JCPDS card No. 04-0802) on the TiN substrate.

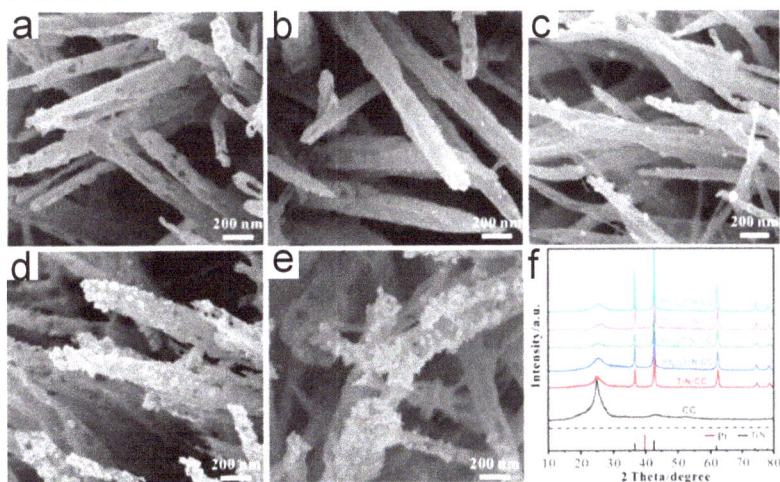

Figure 12. SEM images of (a) TiN/CC, (b) Pt_{500}/TiN/CC, (c) Pt_{750}/TiN/CC, (d) Pt_{1000}/TiN/CC, and (e) Pt_{1500}/TiN/CC NWs. (f) XRD of the samples. Reproduced with permission from ref [27]. Copyright 2022, American Chemical Society.

The plasmonic Pt_x/TiN/CC catalyst is composed of non-precious TiN as a plasma booster and Pt nanoparticles as a co-catalyst. The HER characteristics of the Pt_x/TiN/CC catalysts were investigated under both light illumination and without it. The catalyst exhibited higher catalytic activity during light irradiation due to surface plasmon resonance (SPR) enhancement. The variation in current densities among the samples prepared with different CV cycles emphasizes the significance of Pt anchoring, which captures hot electrons generated by SPR on the TiN nanowires, followed by the reduction of protons to H_2 on the surface of Pt nanoparticles. Under light illumination, the Pt_{1000}/TiN/CC catalyst showed an overpotential of 16 mV to generate a current density of 10 mA cm^{-2}. In comparison, the control experiment without light illumination required an overpotential of 33 mV to achieve the same current density. This decrease in overpotential by 17 mV indicates a significant improvement in the hydrogen evolution process. Furthermore, a constant-time potentiometric test was conducted at -0.04 V (without *iR* correction), and the catalyst demonstrated stability over 10 hours. This approach of loading a small amount of precious platinum on non-precious metal nitrides not only reduces the platinum usage but also maintains superior catalytic stability.

3. Pd-based Catalysts

Alongside Pt-based catalysts, Pd is also a significant member of the precious metal family and finds applications in various fields, particularly in important catalytic reactions [28-30]. Pd, like Pt, exhibits excellent HER capability. Additionally, Pd is more abundant than Pt and is priced at only one-fifth of Pt. Pd has a smaller atomic size, allowing it to adsorb H_2 not only from the gas phase but also from the electrolyte [31]. However, due to the strong palladium-based hydrogen bond, the adsorbed hydrogen is not easily desorbed, which hampers reaction kinetics and stability [32]. Therefore, it is crucial to adjust the hydrogen adsorption energy appropriately to enhance the performance of Pd-based electrocatalysts in HER.

3.1. Zero-Valent Pd Catalysts

Zero-valent Pd (Pd^0) complexes are commonly observed as active intermediates in various organic catalytic transformations, but their application in HER is relatively uncommon [33, 34]. Li et al. [35] conducted a study on the HER activity and stability of Pd^0 supported on ultrathin graphdiyne (GDY) nanosheets, and obtained Pd^0/GDY catalyst. The Pd^0/GDY catalyst was prepared using an electrochemical deposition method, in which the GDY working electrode was immersed into $PdCl_2$ solution (0.2 mM $PdCl_2$ in 0.5 M H_2SO_4 solvent) and immediately subject to galvanostatic conditions at a current density of 2 mA cm^{-2} for 10 s. The high-resolution XPS spectra of the Pd-$3d$ orbitals showed the peaks at 334.9 and 340.2 eV originate from Pd-$3d_{5/2}$ and Pd-$3d_{3/2}$, respectively. This confirmed the presence of metallic Pd in the catalyst, as depicted in **Figure 13a**. The results strongly evidence the successful anchoring of Pd^0 atoms on GDY. The Pd^0/GDY catalyst owns both sp-hybrid and sp^2 hybrid structures. These structures facilitate the chelation of monometallic atoms and promote strong charge transfer between the metal atoms and the GDY support. It would presumably be beneficial for improving the intrinsic catalytic activity of GDY-based catalysts [36, 37].

The as-prepared Pd^0/GDY required an overpotential of 55 mV to generate a current density of 10 mA cm^{-2}, which is smaller than those of 20 wt% Pt/C (62 mV) and pure GDY (481 mV). Meanwhile, Pd^0/GDY presented a Tafel slope of 47 mV dec^{-1}, which was lower than that of the GDY (212 mV dec^{-1}) and those of the reported and state-of-art bulk catalysts [38-41]. This value for the Tafel slope suggests that the HER process of Pd^0/GDY catalyst proceeds through a Volmer-Heyrovsky mechanism, in which hydrogen desorption is the RDS. As a result, The

Pd^0/GDY catalyst exhibited remarkable stability. It was able to operate continuously for 1000 cycles in acidic solutions, ranging from -0.6 to 0.2 V *vs.* RHE. Impressively, chronoamperometric experiments conducted at a constant potential demonstrated a minimal decrease in current density over a period of 72 hours, as shown in the inset of **Figure 13b**. This observation confirms the excellent stability of the Pd^0/GDY catalyst. Similarly, Zhang et. al [30] used an in situ growth strategy to anchor the single-crystal Pd^0 quantum dots on the GDY and achieved GDY-Pd1 catalyst, which can reach the current densities of 500 and 1000 mA cm^{-2} at small overpotentials of only 201 and 261 mV, respectively, with excellent long-term stability. Therefore, the high catalytic activity of the Pd^0 catalyst can effectively promote the decomposition of water to produce hydrogen and show good chemical stability [42, 43].

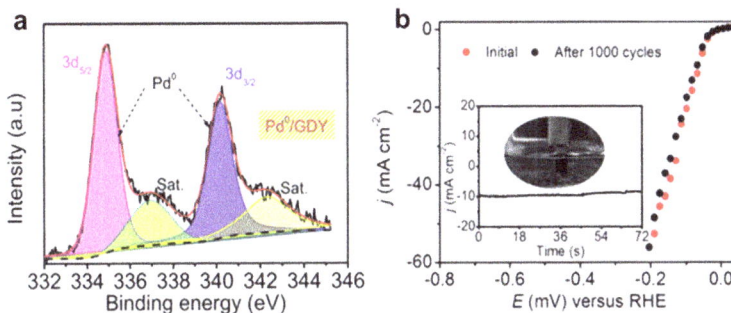

Figure 13. (a) XPS Pd-3d spectrum. (b) Polarization curves of Pd^0/GDY before and after 1,000 cycles tests with inset showing the photo of the electrode and the time-dependent current density curve of Pd^0/GDY obtained at -58 mV vs. RHE. Reproduced with permission from ref [35]. Copyright 2019, Elsevier.

3.2. Pd-based Compounds

The combination of metals with non-metallic elements, such as sulfur (S) and phosphorus (P), can regulate the physical, chemical properties, and electronic structure of the metals [44, 45]. Zhang et al. [46] constructed a heterogeneous structure of $Pd_3P_{0.95}$/Pd_4S through the interface engineering. The $Pd_3P_{0.95}$/Pd_4S heterostructure exhibited several favorable characteristics. Firstly, it possessed strong surface antioxidant capability, allowing for improved stability in harsh environments. Additionally, the heterostructure provided a higher number of active sites and excellent electrical conductivity. The synergistic effect between Pd_4S and $Pd_3P_{0.95}$ resulted in a reduced dissociation energy barrier for water molecules and optimized electron distribution at the heterogeneous interface. These factors were conducive to the adsorption and desorption of intermediate substances involved in the HER. As a result, the $Pd_3P_{0.95}$/Pd_4S heterostructure exhibited very high electrocatalytic activity and required an overpotential of only 28 mV to reach the current density of 10 mA cm^{-2}, which slightly surpassed the benchmark commercial 20% Pt/C (30 mV) and was much lower than those of Pd_4S (60 mV), $Pd_3P_{0.95}$ (80 mV), and commercial Pd powder (200 mV) as shown in **Figure 14a.** The Tafel slope shown in **Figure 14b** is consistent well with the result in polarization curves. $Pd_3P_{0.95}$/Pd_4S heterostructure possessed a Tafel slope of 27.8 mV dec^{-1}, which was slightly smaller than that of commercial 20% Pt/C catalyst (31 5 mV dec^{-1}), but much smaller than that of Pd$_4$S (45 mV dec^{-1}) and Pd$_3$P$_{0.95}$ (56.3 mV dec^{-1}). The Tafel slope value suggests that the HER process of Pd$_3$P$_{0.95}$/Pd$_4$S heterostructure proceeds through a Volmer-Tafel mechanism and the recombination of active hydrogen was a is the RDS, indicating that the HER reaction kinetic has been significantly promoted for the heterostructure. Furthermore, the Pd-based catalyst demonstrated remarkable durability. It could continuously operate for 20 hours in HER at high current densities without attenuation under both acidic and alkaline conditions.   This highlights the potential of constructing Pd-based compounds and their heterogeneous structures with strong surface oxidation resistance as a promising approach for the development of robust HER electrocatalysts.

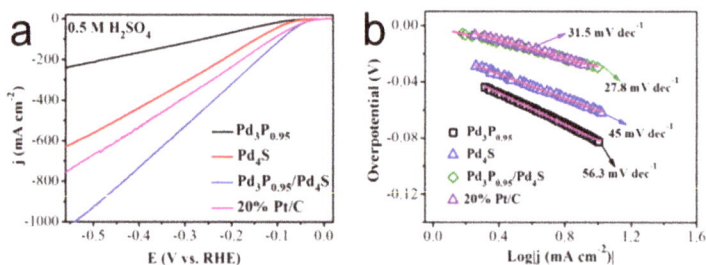

Figure 14. (a) LSV curves and (b) Tafel plots for Pd$_3$P$_{0.95}$, Pd$_4$S, Pd$_3$P$_{0.95}$/Pd$_4$S, and 20% Pt/C catalysts obtained in 0.5 M H$_2$SO$_4$. Reproduced with permission from ref [46]. Copyright 2022, Wiley-VCH.

3.3.   Pd-based Alloys

Pd-based alloys often show superior catalytic performance compared to elemental materials due to the synergistic coupling between multiple elements. While thermodynamically stable face-centered cubic (fcc) crystal phases have been extensively studied for Pd and Pd-based alloys, recent research has shown that amorphous Pd-based nanostructures can display enhanced catalytic performance compared to their crystalline fcc counterparts [47-49]. Therefore, the development of non-conventional Pd-based nano-alloy materials, distinct from the fcc crystalline phase, is crucial for achieving high-performance catalysts.

Ge et al. [50] reported a convenient and universal wet chemical reduction method for synthesizing a series of amorphous polymetallic Pd-based nanomaterials, including bimetallic PdRu, PdRh alloys, and trimetallic PdRuRh alloys. The researchers systematically investigated the effect of phase structure on the electrocatalytic hydrogen evolution properties of these materials. Among the synthesized nanomaterials, the amorphous bimetallic *a*-PdRh catalyst exhibited excellent performance in the electrocatalytic HER, as illustrated in **Figure 15a.** It required a mere overpotential of 20.6 mV to achieve a current density of 10 mA cm^{-2}, surpassing previous catalysts shown in **Figure 15b**. Moreover, the Tafel slopes of series catalysts were analyzed to evaluate their reaction kinetics during the HER progress, as indicated in **Figure 15c**.

The Tafel slope of *a*-PdRh was found to be 41.9 mV dec⁻¹, which is much lower than that of crystalline fcc-PdRh (60.6 mV dec⁻¹) and mono-metallic *a*-Pd (66.7 mV dec⁻¹) catalyst. This suggests that the *a*-PdRh catalyst exhibits faster kinetics in the HER than its crystalline fcc-PdRh and mono-metallic *a*-Pd counterparts.

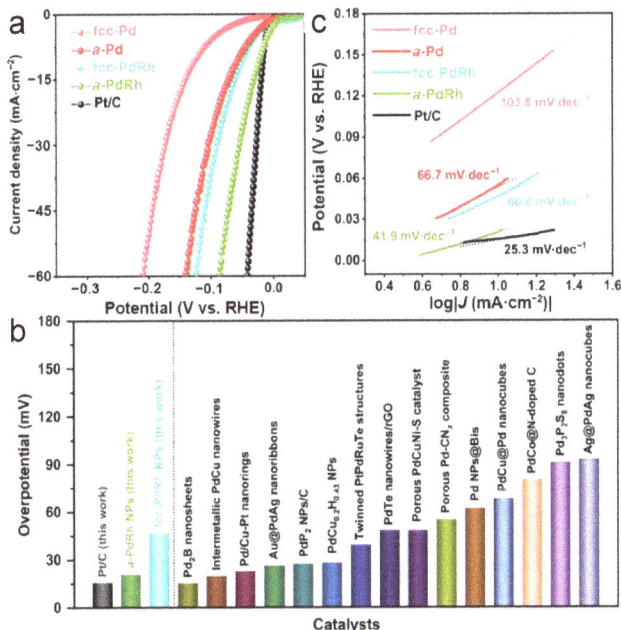

Figure 15. Electrocatalytic performance of as-synthesized Pd-based nanomaterials and commercial Pt/C catalyst towards HER in 0.5 M H₂SO₄ aqueous solution. (a) LSV curves; (b) Comparison of overpotential at the current density of 10 mA cm⁻² for a-PdRh NPs, fcc-PdRh NPs, Pt/C, and some previously reported Pd-based electrocatalysts; (c) Tafel slopes. Reproduced with permission from ref [50]. Copyright 2023, Springer Nature.

Indeed, the work discussed highlights the significance of composition and structure in determining the catalytic reaction path and activity. The ability to engineer the phase of Pd-based nanomaterials offers a new approach to designing highly efficient Pd-based catalysts with desirable catalytic properties.

4. Other Precious Metal-based Catalysts

Ruthenium (Ru) has emerged as a promising candidate for superior HER catalysts, exhibiting properties comparable to platinum-based catalysts in terms of hydrolytic ionization and OH* chemisorption. Similar to Pt-based catalysts, Ru-based catalysts are often supported on carbon-based scaffolds to investigate their HER activity and stability. However, many Ru/C composites have shown poor durability and stability [51-53].

To address this issue, alloying Ru with transition metals has proven to be an effective strategy for improving its stability. For instance, Harzandi et.al [54] synthesized two types of separated Ru-based metals, namely Cu/Ru@GN and Ru@GN catalysts, by dispersing Ru and Cu on an N-doped graphite matrix (GN) through immiscible drive under continuous beam irradiation. In the Cu/Ru@GN catalyst, the Ru nanoparticles (NPs) did not aggregate, as depicted in **Figure 16a-d**. Additionally, the pure Ru NPs in the Ru@GN catalyst were exposed to light, leading to the formation of larger Ru crystals. The Ru NPs and Cu single atoms (SAs) synergistically enhanced the conductivity and charge transfer on the GN matrix, resulting in fast kinetics. Moreover, due to the immiscibility of Cu and Ru, the Cu SAs coordinated with the N atoms bridged on the Ru surface, introducing new active sites and preventing the aggregation of Ru NPs, thus providing long-term stability. Furthermore, the Cu/Ru@GN catalyst exhibited excellent electrochemical activity, which only needs approximately 11 mV to drive 10 mA cm^{-2} in 1.0 M KOH. The Cu/Ru@GN catalyst exhibited excellent stability, operating continuously for over 600 hours in 0.5 M H_2SO_4 and over 2 days in 1.0 M KOH, respectively.

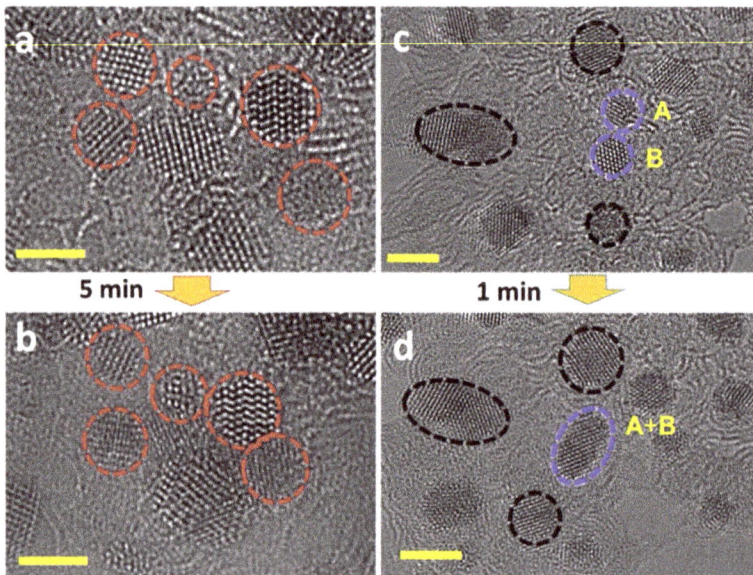

Figure 16. HR-TEM images of (a-b) Cu/Ru@GN and (c-d) Ru@GN. Reproduced with permission from ref [54]. Copyright 2020, Elsevier.

Rhodium (Rh)-based catalysts have demonstrated excellent HER activity in various forms, such as Rh nanoparticles loaded on silicon nanowires, Rh alloyed with transition metals, and Rh-based phosphides with phosphorus defects [55-57]. In addition, precious metal oxides, such as Rh_2O_3 and RuO_2, have shown promise for highly alkaline HER due to the synergistic effect of metal cations and nonmetallic anions, resulting in high surface energy and effective proton adsorption [58, 59]. However, noble metal oxides commonly undergo reduction reactions during the HER process, posing challenges to their long-term stability in alkaline HER applications.

One approach to improving the phase stability of noble metal oxides is by applying compressive strain, which effectively enhances the energy barrier for the phase transition from noble metal oxides (NMOs) to metals [60, 61]. For example, Li et al. [62] utilized rhodium oxide (RhO_2) as the model material and designed a strawberry-like nanoparticle (SLNP) catalyst to investigate the phase stability and HER properties under compressive strain. The lattice mismatch between the RhO_2 cluster and the Rh substrate induces a strong compressive strain, stabilizing rhodium oxide at a reduction potential of -0.3 V *vs.* RHE. Notably, the SLNP catalysts exhibited excellent alkaline HER activity, requiring an overpotential of 14 mV to generate the current density of 10 mA cm^{-2} with a small Tafel slope of only 30 mV dec^{-1}, which is superior to the commercial Pt/C catalyst. Furthermore, the catalyst performs outstanding long-term durability, remaining stable for over 50 hours of operation.

DFT calculations were employed to investigate the role of compressive strain in stabilizing RhO_2 and improving its catalytic properties. As shown in **Figure 17a**, the Rh-O bonds were observed to become more rigid under compressive stress. Specifically, the bond-breaking energy (E_{OV}) increases by ≈ 0.31 and ≈ 0.48 eV around O_1 and O_2 sites, respectively, under 5% compressive strain. The results highlight the ability of compressive strain to stabilize RhO_2 in the HER process. The effect of compressive strain on the HER activity of RhO_2 was further examined. In alkaline electrolytes, the HER process involves water adsorption, hydrolytic dissociation, hydrogen adsorption (H*), hydrogen recombination, and H_2 desorption. From a reaction kinetics perspective, the hydrolytic ionization energy barrier (ΔG_B) at unstrained RhO_2 (0.35 eV) is significantly lower than that of Rh (1.24 eV) and Pt (1.05 eV), as indicated in **Figure 17b**.

In addition, the thermodynamic reaction barrier (ΔG_T) of RhO_2 is found to be 0.27 eV, which is lower than that of Rh (0.75 eV) and Pt (0.57 eV), indicating that RhO_2 has a more favorable thermodynamic landscape for the HER process. In particular, the application of 5% compressive strain further reduces the ΔG_T of RhO_2 to 0.083 eV, as depicted in **Figure 17c**. Meanwhile, compressive strain plays a crucial role in promoting the adsorption of water molecules on the surface of RhO_2, leading to a decrease in the adsorption energy of H* (ΔG_{H*}) to 0.001 eV, as illustrated in **Figure 17d**. The electronic structure analysis reveals that compressive strain facilitates the transfer of electrons from the adsorbed water molecules to RhO_2, resulting in a decrease in the electron density of the water molecules and a weakening of the H-O bond. This phenomenon facilitates water dissociation, which is a crucial step in the HER process. Theoretical calculations thus demonstrate that compressive strain not only stabilizes the chemical state of precious metal-based oxide catalysts, such as RhO_2, but also enhances their catalytic performance. The improved stability and catalytic activity under compressive strain contribute to the excellent alkaline HER capability exhibited by these catalysts.

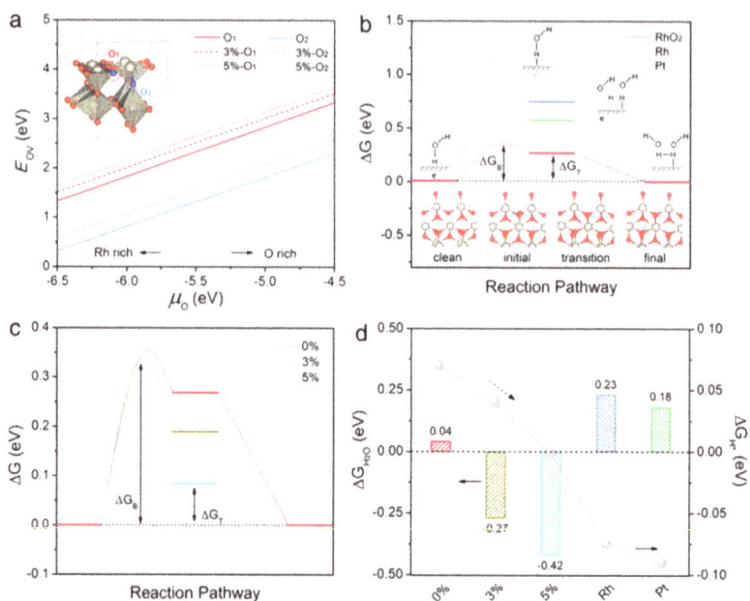

Figure 17. *Theoretical calculations of RhO₂, Rh, and Pt catalyst in HER. (a) The Rh-O bond breaking energy for RhO₂ system in strain-free state and with applied 3% and 5% biaxial in-plane compressive strain, respectively, at O₁ and O₂ sites, where O₁ and O₂ denote two kinds of inequitable oxygen positions. Inset shows the side view of RhO₂ (110) facet. (b) Gibbs free energy diagram of RhO₂ (110), Rh (111), and Pt (111) facets, including initial, transition, and final state of water dissociation in HER. ΔG_B denotes the water dissociation free energy barrier, and ΔG_T denotes the thermodynamic barrier. Inset shows the atomic configurations of water dissociation step on RhO₂ (110) facet. (c) Gibbs free energy diagram of RhO₂ (110) facet in free-strain sate and with applied 3% and 5% biaxial in-plane compressive strain in HER. (d) Calculated adsorption energies of H₂O (bar) and H* (scatter) on the surface of RhO₂ (110) facet in strain-free state, with applied 3% or 5% biaxial in-plane compressive strains, respectively. Reproduced with permission from ref [62]. Copyright 2020, Wiley-VCH.*

5.   Conclusion and Prospects

Precious metal-based catalysts have shown great potential for the HER. Precious metals, such as Pt, Pd, Ru, Rh, etc. and their compounds as well as alloys have been extensively studied and utilized as catalysts for superior HER. These catalysts exhibit low overpotentials, with fast reaction kinetics and high hydrogen production efficiency. However, the high cost and limited availability of precious metals pose challenges for their widespread commercial applications.

Therefore, extensive research efforts have been devoted to reducing the precious metal loading, improving the utilization efficiency, and exploring alternative non-precious metal catalysts with comparable or even better performance.

(i) Catalyst design and optimization. Efforts will continue to optimize the composition, structure, and morphology of precious metal-based catalysts to enhance their catalytic activity and stability. Strategies such as alloying, surface modification, and nanostructuring will be employed to improve the catalytic performance.

(ii) Ultra-low mass loading. It is an important area of research aimed at reducing the cost and increasing the efficiency of catalysts. Achieving high catalytic activity and stability with minimal use of precious metals is crucial for the large-scale commercialization of HER technologies.

(iii) Catalyst support engineering. The choice of suitable catalyst supports can significantly enhance the performance of precious metal-based catalysts. Designing and developing advanced support materials with high surface area, good conductivity, and excellent stability will be crucial for improving the overall catalytic activity and durability.

(iv) Catalyst scalability and cost reduction. Developing scalable synthesis methods and cost-effective catalyst fabrication processes are essential for the practical application of precious metal-based catalysts. Efforts will be focused on exploring scalable production techniques, such as electrodeposition, chemical vapor deposition, and template-assisted synthesis, to achieve large-scale synthesis of catalyst materials at reduced cost.

(v) Precious metal substitution and non-precious metal catalysts. Exploring alternative catalyst materials that can replace or reduce the reliance on precious metals is of great interest. Non-precious metal catalysts, such as transition metal-based compounds, carbon-based materials, and earth-abundant elements, are being extensively investigated for their HER performance. Developing efficient and cost-effective non-precious metal catalysts is essential for achieving large-scale commercialization of HER technologies.

References

[1] X. Peng, Y. Yan, X. Jin, C. Huang, W. Jin, B. Gao, P.K. Chu, Recent advance and prospectives of electrocatalysts based on transition metal selenides for efficient water splitting, Nano Energy, 78 (2020) 105234. https://doi.org/10.1016/j.nanoen.2020.105234

[2] X. Peng, X. Jin, B. Gao, Z. Liu, P.K. Chu, Strategies to improve cobalt-based electrocatalysts for electrochemical water splitting, J. Catal., 398 (2021) 54-66. https://doi.org/10.1016/j.jcat.2021.04.003

[3] S. Chu, Y. Cui, N. Liu, The path towards sustainable energy, Nat. Mater., 16 (2017) 16-22. https://doi.org/10.1038/nmat4834

[4] L. Xiong, Y. Qiu, X. Peng, Z. Liu, P.K. Chu, Electronic structural engineering of transition metal-based electrocatalysts for the hydrogen evolution reaction, Nano Energy, 104 (2022) 107882. https://doi.org/10.1016/j.nanoen.2022.107882

[5]   Y. Luo, Z. Zhang, M. Chhowalla, B. Liu, Recent advances in design of electrocatalysts for high-current-density water splitting, Adv. Mater., 34 (2022) 2108133. https://doi.org/10.1002/adma.202108133

[6]   G.-F. Chen, T.Y. Ma, Z.-Q. Liu, N. Li, Y.-Z. Su, K. Davey, S.-Z. Qiao, Efficient and Stable bifunctional electrocatalysts Ni/NixMy (M = P, S) for overall water splitting, Adv. Funct. Mater., 26 (2016) 3314-3323. https://doi.org/10.1002/adfm.201505626

[7]   J. Greeley, T.F. Jaramillo, J. Bonde, I. Chorkendorff, J.K. Nørskov, Computational high-throughput screening of electrocatalytic materials for hydrogen evolution, Nat. Mater., 5 (2006) 909-913. https://doi.org/10.1038/nmat1752

[8]   X.-K. Wan, H.B. Wu, B.Y. Guan, D. Luan, X.W. Lou, Confining sub-nanometer pt clusters in hollow mesoporous carbon spheres for boosting hydrogen evolution activity, Adv. Mater., 32 (2020) 1901349. https://doi.org/10.1002/adma.201901349

[9]   W. Xu, J. Chang, Y. Cheng, H. Liu, J. Li, Y. Ai, Z. Hu, X. Zhang, Y. Wang, Q. Liang, Y. Yang, H. Sun, A multi-step induced strategy to fabricate core-shell Pt-Ni alloy as symmetric electrocatalysts for overall water splitting, Nano Res., 15 (2022) 965-971. https://doi.org/10.1007/s12274-021-3582-x

[10] N.M. Marković, B.N. Grgur, P.N. Ross, Temperature-dependent hydrogen electrochemistry on platinum low-index single-crystal surfaces in acid solutions, J. Phys. Chem. B, 101 (1997) 5405-5413. https://doi.org/10.1021/jp970930d

[11] T.J. Schmidt, P.N. Ross, N.M. Markovic, Temperature dependent surface electrochemistry on Pt single crystals in alkaline electrolytes: Part 2. The hydrogen evolution/oxidation reaction, J. Electroanal. Chem., 524-525 (2002) 252-260. https://doi.org/10.1016/S0022-0728(02)00683-6

[12] J. Durst, A. Siebel, C. Simon, F. Hasché, J. Herranz, H. Gasteiger, New insights into the electrochemical hydrogen oxidation and evolution reaction mechanism, Energy Environ. Sci., 7 (2014) 2255-2260. https://doi.org/10.1039/C4EE00440J

[13] N.M. Marković, P.N. Ross, Surface science studies of model fuel cell electrocatalysts, Surf. Sci. Rep., 45 (2002) 117-229. https://doi.org/10.1016/S0167-5729(01)00022-X

[14] D. Liu, X. Li, S. Chen, H. Yan, C. Wang, C. Wu, Y.A. Haleem, S. Duan, J. Lu, B. Ge, P.M. Ajayan, Y. Luo, J. Jiang, L. Song, Atomically dispersed platinum supported on curved carbon supports for efficient electrocatalytic hydrogen evolution, Nat. Energy, 4 (2019) 512-518. https://doi.org/10.1038/s41560-019-0402-6

[15] L. Zhang, J.M.T.A. Fischer, Y. Jia, X. Yan, W. Xu, X. Wang, J. Chen, D. Yang, H. Liu, L. Zhuang, M. Hankel, D.J. Searles, K. Huang, S. Feng, C.L. Brown, X. Yao, Coordination of atomic Co-Pt coupling species at carbon defects as active sites for oxygen reduction reaction, J. Am. Chem. Soc., 140 (2018) 10757-10763. https://doi.org/10.1021/jacs.8b04647

[16] R. Lin, X. Cai, H. Zeng, Z. Yu, Stability of high-performance Pt-based catalysts for

oxygen reduction reactions, Adv. Mater., 30 (2018) 1705332.
https://doi.org/10.1002/adma.201705332

[17] M. Mavrikakis, B. Hammer, J.K. Nørskov, Effect of strain on the reactivity of metal surfaces, Phys. Rev. Lett., 81 (1998) 2819-2822.
https://doi.org/10.1103/PhysRevLett.81.2819

[18] B. Hammer, J.K. Nørskov, Theoretical surface science and catalysis-calculations and concepts, Advances in Catalysis [M], Academic Press2000, pp. 71-129.
https://doi.org/10.1016/S0360-0564(02)45013-4

[19] T. He, W. Wang, F. Shi, X. Yang, X. Li, J. Wu, Y. Yin, M. Jin, Mastering the surface strain of platinum catalysts for efficient electrocatalysis, Nature, 598 (2021) 76-81.
https://doi.org/10.1038/s41586-021-03870-z

[20] S. Sun, X. Zhou, B. Cong, W. Hong, G. Chen, Tailoring the d-band centers endows $(Ni_xFe_{1-x})_2P$ nanosheets with efficient oxygen evolution catalysis, ACS Catal., 10 (2020) 9086-9097. https://doi.org/10.1021/acscatal.0c01273

[21] Q. Hu, K. Gao, X. Wang, H. Zheng, J. Cao, L. Mi, Q. Huo, H. Yang, J. Liu, C. He, Subnanometric Ru clusters with upshifted D band center improve performance for alkaline hydrogen evolution reaction, Nat. Commun., 13 (2022) 3958.
https://doi.org/10.1038/s41467-022-31660-2

[22] S.L. Zhang, X.F. Lu, Z.-P. Wu, D. Luan, X.W. Lou, Engineering platinum-cobalt nano-alloys in porous nitrogen-doped carbon nanotubes for highly efficient electrocatalytic hydrogen evolution, Angew. Chem. Int. Ed., 60 (2021) 19068-19073.
https://doi.org/10.1002/anie.202106547

[23] K. Jiang, B. Liu, M. Luo, S. Ning, M. Peng, Y. Zhao, Y.-R. Lu, T.-S. Chan, F.M.F. de Groot, Y. Tan, Single platinum atoms embedded in nanoporous cobalt selenide as electrocatalyst for accelerating hydrogen evolution reaction, Nat. Commun., 10 (2019) 1743. https://doi.org/10.1038/s41467-019-09765-y

[24] X. Li, Y. Fang, J. Wang, H. Fang, S. Xi, X. Zhao, D. Xu, H. Xu, W. Yu, X. Hai, C. Chen, C. Yao, H.B. Tao, A.G.R. Howe, S.J. Pennycook, B. Liu, J. Lu, C. Su, Ordered clustering of single atomic Te vacancies in atomically thin $PtTe_2$ promotes hydrogen evolution catalysis, Nat. Commun., 12 (2021) 2351. https://doi.org/10.1038/s41467-021-22681-4

[25] P. Wang, X. Zhang, J. Zhang, S. Wan, S. Guo, G. Lu, J. Yao, X. Huang, Precise tuning in platinum-nickel/nickel sulfide interface nanowires for synergistic hydrogen evolution catalysis, Nat. Commun., 8 (2017) 14580. https://doi.org/10.1038/ncomms14580

[26] H. Zhang, P. An, W. Zhou, B.Y. Guan, P. Zhang, J. Dong, X.W. Lou, Dynamic traction of lattice-confined platinum atoms into mesoporous carbon matrix for hydrogen evolution reaction, Sci. Adv., 4 (2018) eaao6657. https://doi.org/10.1126/sciadv.aao6657

[27] J. Zhang, A. Reda Woldu, X. Peng, Y. Song, H. Xia, F. Lu, P.K. Chu, L. Hu, Plasmon-enhanced hydrogen evolution on Pt-anchored titanium nitride nanowire arrays, Appl.

Surf. Sci., 598 (2022) 153745. https://doi.org/10.1016/j.apsusc.2022.153745

[28] W. Zhang, X. Jiang, Z. Dong, J. Wang, N. Zhang, J. Liu, G.-R. Xu, L. Wang, Porous Pd/NiFeO$_X$ nanosheets enhance the pH-universal overall water splitting, Adv. Funct. Mater., 31 (2021) 2107181. https://doi.org/10.1002/adfm.202107181

[29] L. Li, Y. Ji, X. Luo, S. Geng, M. Fang, Y. Pi, Y. Li, X. Huang, Q. Shao, Compressive strain in N-doped palladium/amorphous-cobalt (II) interface facilitates alkaline hydrogen evolution, Small, 17 (2021) 2103798. https://doi.org/10.1002/smll.202103798

[30] D. Zhang, X. Zheng, L. Qi, Y. Xue, F. He, Y. Li, Controlled growth of single-crystal Pd quantum dots on 2D carbon for large current density hydrogen evolution, Adv. Funct. Mater., 32 (2022) 2111501. https://doi.org/10.1002/adfm.202111501

[31] Y. Zhan, X. Zhou, H. Nie, X. Xu, X. Zheng, J. Hou, H. Duan, S. Huang, Z. Yang, Designing Pd/O co-doped MoS$_x$ for boosting the hydrogen evolution reaction, J. Mater. Chem. A, 7 (2019) 15599-15606. https://doi.org/10.1039/C9TA02997D

[32] J. Zhu, L. Hu, P. Zhao, L.Y.S. Lee, K.-Y. Wong, Recent advances in electrocatalytic hydrogen evolution using nanoparticles, Chem. Rev., 120 (2020) 851-918. https://doi.org/10.1021/acs.chemrev.9b00248

[33] S. Kozuch, S. Shaik, A. Jutand, C. Amatore, Active anionic zero-valent palladium catalysts: characterization by density functional calculations, Chem. Eur. J., 10 (2004) 3072-3080. https://doi.org/10.1002/chem.200306056

[34] A.M. Kluwer, C.J. Elsevier, M. Bühl, M. Lutz, A.L. Spek, Zero-valent palladium complexes with monodentate nitrogen σ-donor ligands, Angew. Chem. Int. Ed., 42 (2003) 3501-3504. https://doi.org/10.1002/anie.200351189

[35] H. Yu, Y. Xue, B. Huang, L. Hui, C. Zhang, Y. Fang, Y. Liu, Y. Zhao, Y. Li, H. Liu, Y. Li, Ultrathin nanosheet of graphdiyne-supported palladium atom catalyst for efficient hydrogen production, iScience, 11 (2019) 31-41. https://doi.org/10.1016/j.isci.2018.12.006

[36] Y. Xue, Y. Liu, H. Liao, X. Zhan, X. Fang, H. Deng, F. Wang, W. Huang, Y. Liang, W. Wei, Y. Huang, Z. Liao, M. Shehata, X. Wang, S. Wu, Evaluation of electrophysiological mechanisms of post-surgical atrial tachycardias using an automated ultra-high-density mapping system, JACC-Clin. Electrophy, 4 (2018) 1460-1470. https://doi.org/10.1016/j.jacep.2018.07.002

[37] Y. Xue, Y. Li, J. Zhang, Z. Liu, Y. Zhao, 2D graphdiyne materials: challenges and opportunities in energy field, Sci. China Chem., 61 (2018) 765-786. https://doi.org/10.1007/s11426-018-9270-y

[38] W. Chen, J. Pei, C.-T. He, J. Wan, H. Ren, Y. Zhu, Y. Wang, J. Dong, S. Tian, W.-C. Cheong, S. Lu, L. Zheng, X. Zheng, W. Yan, Z. Zhuang, C. Chen, Q. Peng, D. Wang, Y. Li, Rational design of single molybdenum atoms anchored on N-doped carbon for effective hydrogen evolution reaction, Angew. Chem. Int. Ed., 56 (2017) 16086-16090.

https://doi.org/10.1002/anie.201710599

[39]   N. Cheng, S. Stambula, D. Wang, M.N. Banis, J. Liu, A. Riese, B. Xiao, R. Li, T.-K.
        Sham, L.-M. Liu, G.A. Botton, X. Sun, Platinum single-atom and cluster catalysis of the
        hydrogen evolution reaction, Nat. Commun., 7 (2016) 13638.
        https://doi.org/10.1038/ncomms13638

[40]   L. Hui, Y. Xue, B. Huang, H. Yu, C. Zhang, D. Zhang, D. Jia, Y. Zhao, Y. Li, H. Liu, Y.
        Li, Overall water splitting by graphdiyne-exfoliated and -sandwiched layered double-
        hydroxide nanosheet arrays, Nat. Commun., 9 (2018) 5309.
        https://doi.org/10.1038/s41467-018-07790-x

[41]   H. Zhu, G. Gao, M. Du, J. Zhou, K. Wang, W. Wu, X. Chen, Y. Li, P. Ma, W. Dong, F.
        Duan, M. Chen, G. Wu, J. Wu, H. Yang, S. Guo, Atomic-scale core/shell structure
        engineering induces precise tensile strain to boost hydrogen evolution catalysis, Adv.
        Mater., 30 (2018) 1707301. https://doi.org/10.1002/adma.201707301

[42]   L. Xiao, Z. Wang, J. Guan, Optimization strategies of covalent organic frameworks and
        their derivatives for electrocatalytic applications, Adv. Funct. Mater., (2023) 2310195.
        https://doi.org/10.1002/adfm.202310195

[43]   L. Jiang, K. Liu, S.-F. Hung, L. Zhou, R. Qin, Q. Zhang, P. Liu, L. Gu, H.M. Chen, G.
        Fu, N. Zheng, Facet engineering accelerates spillover hydrogenation on highly diluted
        metal nanocatalysts, Nat. Nanotechnol., 15 (2020) 848-853.
        https://doi.org/10.1038/s41565-020-0746-x

[44]   L. Yue, H. Guo, X. Wang, T. Sun, H. Liu, Q. Li, M. Xu, Y. Yang, W. Yang, Non-
        metallic element modified metal-organic frameworks as high-performance electrodes for
        all-solid-state asymmetric supercapacitors, J. Colloid Interf. Sci., 539 (2019) 370-378.
        https://doi.org/10.1016/j.jcis.2018.12.079

[45]   E. Pessard, F. Morel, A. Morel, D. Bellett, Modelling the role of non-metallic inclusions
        on the anisotropic fatigue behaviour of forged steel, Int. J. Fatigue 33 (2011) 568-577.
        https://doi.org/10.1016/j.ijfatigue.2010.10.012

[46]   G. Zhang, A. Wang, L. Niu, W. Gao, W. Hu, Z. Liu, R. Wang, J. Chen, Interfacial
        engineering to construct antioxidative $Pd_4S/Pd_3P_{0.95}$ heterostructure for robust hydrogen
        production at high current density, Adv. Energy Mater., 12 (2022) 2103511.
        https://doi.org/10.1002/aenm.202103511

[47]   W. Yang, L. Zhu, M. Yang, W. Xu, Synthesis of amorphous/crystalline hetero-phase
        nanozymes with peroxidase-like activity by coordination-driven self-assembly for
        biosensors, Small, 19 (2023) 2204782. https://doi.org/10.1002/smll.202204782

[48]   M. Duval, V. Deboos, A. Hallonet, G. Sagorin, A. Denicourt-Nowicki, A. Roucoux,
        Selective palladium nanoparticles-catalyzed hydrogenolysis of industrially targeted
        epoxides in water, J. Catal., 396 (2021) 261-268.
        https://doi.org/10.1016/j.jcat.2021.02.027

[49]   X. Zhou, Y. Ma, Y. Ge, S. Zhu, Y. Cui, B. Chen, L. Liao, Q. Yun, Z. He, H. Long, L. Li,
        B. Huang, Q. Luo, L. Zhai, X. Wang, L. Bai, G. Wang, Z. Guan, Y. Chen, C.-S. Lee, J.
        Wang, C. Ling, M. Shao, Z. Fan, H. Zhang, Preparation of Au@Pd core–shell nanorods
        with fcc-2H-fcc heterophase for highly efficient electrocatalytic alcohol oxidation, J. Am.
        Chem. Soc., 144 (2022) 547-555. https://doi.org/10.1021/jacs.1c11313

[50]   Y. Ge, J. Ge, B. Huang, X. Wang, G. Liu, X.-H. Shan, L. Ma, B. Chen, G. Liu, S. Du, A.
        Zhang, H. Cheng, Q. Wa, S. Lu, L. Li, Q. Yun, K. Yuan, Q. Luo, Z.J. Xu, Y. Du, H.
        Zhang, Synthesis of amorphous Pd-based nanocatalysts for efficient alcoholysis of
        styrene oxide and electrochemical hydrogen evolution, Nano Res., 16 (2023) 4650-4655.
        https://doi.org/10.1007/s12274-022-5101-0

[51]   D.H. Kweon, M.S. Okyay, S.-J. Kim, J.-P. Jeon, H.-J. Noh, N. Park, J. Mahmood, J.-B.
        Baek, Ruthenium anchored on carbon nanotube electrocatalyst for hydrogen production
        with enhanced Faradaic efficiency, Nat. Commun., 11 (2020) 1278.
        https://doi.org/10.1038/s41467-020-15069-3

[52]   F. Bao, Z. Yang, Y. Yuan, P. Yu, G. Zeng, Y. Cheng, Y. Lu, J. Zhang, H. Huang,
        Synergistic cascade hydrogen evolution boosting via integrating surface oxophilicity
        modification with carbon layer confinement, Adv. Funct. Mater., 32 (2022) 2108991.
        https://doi.org/10.1002/adfm.202108991

[53]   J.-S. Li, M.-J. Huang, Y.-W. Zhou, X.-N. Chen, S. Yang, J.-Y. Zhu, G.-D. Liu, L.-J. Ma,
        S.-H. Cai, J.-Y. Han, RuP$_2$-based hybrids derived from MOFs: highly efficient pH-
        universal electrocatalysts for the hydrogen evolution reaction, J. Mater. Chem. A, 9
        (2021) 12276-12282. https://doi.org/10.1039/D1TA01868J

[54]   A.M. Harzandi, S. Shadman, M. Ha, C.W. Myung, D.Y. Kim, H.J. Park, S. Sultan, W.-S.
        Noh, W. Lee, P. Thangavel, W.J. Byun, S.-h. Lee, J.N. Tiwari, T.J. Shin, J.-H. Park, Z.
        Lee, J.S. Lee, K.S. Kim, Immiscible bi-metal single-atoms driven synthesis of
        electrocatalysts having superb mass-activity and durability, Appl. Catal. B-Environ., 270
        (2020) 118896. https://doi.org/10.1016/j.apcatb.2020.118896

[55]   L. Zhu, H. Lin, Y. Li, F. Liao, Y. Lifshitz, M. Sheng, S.-T. Lee, M. Shao, A
        rhodium/silicon co-electrocatalyst design concept to surpass platinum hydrogen evolution
        activity at high overpotentials, Nat. Commun., 7 (2016) 12272.
        https://doi.org/10.1038/ncomms12272

[56]   Y. Zhao, J. Bai, X.-R. Wu, P. Chen, P.-J. Jin, H.-C. Yao, Y. Chen, Atomically ultrathin
        RhCo alloy nanosheet aggregates for efficient water electrolysis in broad pH range, J.
        Mater. Chem. A, 7 (2019) 16437-16446. https://doi.org/10.1039/C9TA05334D

[57]   H. Xin, Z. Dai, Y. Zhao, S. Guo, J. Sun, Q. Luo, P. Zhang, L. Sun, N. Ogiwara, H.
        Kitagawa, B. Huang, F. Ma, Recording the Pt-beyond hydrogen production
        electrocatalysis by dirhodium phosphide with an overpotential of only 4.3 mV in alkaline
        electrolyte, Appl. Catal. B-Environ., 297 (2021) 120457.
        https://doi.org/10.1016/j.apcatb.2021.120457

[58]   M.K. Kundu, R. Mishra, T. Bhowmik, S. Barman, Rhodium metal–rhodium oxide (Rh-Rh$_2$O$_3$) nanostructures with Pt-like or better activity towards hydrogen evolution and oxidation reactions (HER, HOR) in acid and base: Correlating its HOR/HER activity with hydrogen binding energy and oxophilicity of the catalyst, J. Mater. Chem. A, 6 (2018) 23531-23541. https://doi.org/10.1039/C8TA07028H

[59]   R. Jiang, D.T. Tran, J. Li, D. Chu, Ru@RuO$_2$ core-shell nanorods: a highly active and stable bifunctional catalyst for oxygen evolution and hydrogen evolution reactions, Energy Environ. Mater., 2 (2019) 201-208. https://doi.org/10.1002/eem2.12031

[60]   W.J. Huang, R. Sun, J. Tao, L.D. Menard, R.G. Nuzzo, J.M. Zuo, Coordination-dependent surface atomic contraction in nanocrystals revealed by coherent diffraction, Nat. Mater., 7 (2008) 308-313. https://doi.org/10.1038/nmat2132

[61]   T. Ling, D.-Y. Yan, H. Wang, Y. Jiao, Z. Hu, Y. Zheng, L. Zheng, J. Mao, H. Liu, X.-W. Du, M. Jaroniec, S.-Z. Qiao, Activating cobalt(II) oxide nanorods for efficient electrocatalysis by strain engineering, Nat. Commun., 8 (2017) 1509. https://doi.org/10.1038/s41467-017-01872-y

[62]   Z. Li, Y. Feng, Y.-L. Liang, C.-Q. Cheng, C.-K. Dong, H. Liu, X.-W. Du, Stable rhodium (IV) oxide for alkaline hydrogen evolution reaction, Adv. Mater., 32 (2020) 1908521. https://doi.org/10.1002/adma.201908521

CHAPTER 4

Single Atom Catalysts for Electrocatalytic Hydrogen Production

Xiang Peng*

Hubei Key Laboratory of Plasma Chemistry and Advanced Materials, Engineering Research Center of Phosphorus Resources Development and Utilization of Ministry of Education, School of Materials Science and Engineering, Wuhan Institute of Technology, Wuhan 430205, China

xpeng@wit.edu.cn

Abstract

The size and location of metal catalysts have a significant impact on catalytic reactions. Reducing the size of metal catalyst particles to small dimensions and dispersing them on suitable supports is an effective strategy to expose more catalytic active sites, thereby maximizing the utilization of the limited surface area available for reactions. In recent years, there has been growing interest in the design of single metal atom catalysts (SACs) supported on suitable carriers. SACs offer great potential for enhancing the efficiency and selectivity of hydrogen evolution reaction.

Keywords

Single Atom Catalyst, Carrier, Active Center, Coordination Regulation, Metal-Organic Framework

1.   Introduction

The size and location of metal catalysts have a significant impact on catalytic reactions [1, 2]. Reducing the size of metal catalyst particles to small dimensions and dispersing them on suitable supports is an effective strategy to expose more catalytic active sites, thereby maximizing the utilization of the limited surface area available for reactions [3-5]. This approach enhances the catalytic efficiency and performance of the catalyst. In recent years, there has been growing interest in the design of single metal atom catalysts (SACs) supported on suitable carriers [6-8]. SACs consist of isolated single metal atoms as the catalytic active sites, supported by support materials. These catalysts possess unique properties, such as low coordination numbers, high chemical activity, and maximized utilization of each metal atom [9-11]. These characteristics make SACs highly attractive for various catalytic applications.

The application of SACs in hydrogen production and hydrogen energy storage has particularly garnered attention and research [12-14]. SACs offer great potential for enhancing the efficiency and selectivity of hydrogen evolution reaction (HER). The design of efficient SACs for hydrogen evolution involves careful considerations such as the selection of suitable carriers, the choice of active metal centers, and the regulation of coordination structures. The design of efficient single atom catalysts for HER requires careful considerations regarding the choice of support material, selection of active metal centers, and regulation of coordination structures. By optimizing these factors, it is possible to develop highly efficient and selective s SACs for hydrogen production and energy storage applications.

2.   Carrier selection

The choice of carrier material is a pivotal decision in the design of single-atom hydrogen evolution catalysts, leading researchers into a vast realm of exploration [15, 16]. In the field of catalysis, the selection of a carrier plays an indispensable role in determining the catalyst's performance, stability, and effectiveness in hydrogen evolution [17-19]. It provides a stable framework for anchoring and dispersing the metal atoms, while also interacting with reactants and intermediates during the catalytic process, thereby influencing the overall catalytic performance.

2.1.   Carbon-based Carriers

Carbon-based carriers, such as grapheme [20-22], carbon black [23-25], carbon nanotubes [26-28] and porous carbon materials [29-31], have received significant attention in the design of single-atom hydrogen evolution catalysts. These materials possess exceptional conductivity, high chemical inertness, and extensive surface area, making them ideal platforms for anchoring SACs. Researchers are actively exploring the structural advantages of carbon-based carriers to enhance the efficiency and stability of hydrogen evolution.

For example, Li et al. [32] employed a hard template and heat treatment approach to anchor single Co atoms onto nitrogen-rich carbon nanoboxes (Co@CNB-N$_4$). This innovative strategy resulted in an inexpensive and promising bifunctional catalyst for overall water splitting. **Figure 1** illustrates the preparation process of the Co@CNB-N$_4$ catalyst. First, Fe$_2$O$_3$ nanoboxes with a cubic template were synthesized using the classic hydrothermal method.

After the synthesis of Fe$_2$O$_3$ nanoboxes, a coating of polyvinyl pyrrolidone (PVP) was applied to create Fe$_2$O$_3$/PVP nanoboxes. Subsequently, cobalt atoms were introduced using a hydrothermal method. Under a N$_2$ atmosphere, the nanoboxes were calcined, resulting in the formation of Co@Fe$_2$O$_3$/PVP carbon-based nanoboxes. Finally, the Fe$_2$O$_3$ core was removed by etching with an oxalic acid solution, yielding the Co@CNB-N$_4$ product. For comparison, a sample of N-rich nanoboxes (CNB-N$_4$) was prepared using a similar method but without the introduction of cobalt atoms. The microstructure of Co@CNB-N$_4$ was examined using field emission scanning electron microscopy (FE-SEM). As depicted in **Figure 2a-c**, the sample exhibited a uniformly sized and shaped regular nanobox structure, which can be attributed to the Fe$_2$O$_3$ nanobox serving as a template. Nitrogen adsorption-desorption analysis was employed to characterize the microscopic surface area and porosity of the prepared samples. The Brunauer-Emmett-Teller (BET) characterization results in **Figure 2d** revealed that Co@CNB-N$_4$ exhibited the highest specific surface area, measuring 78 m^2 g^{-1}. The pore size distribution of the sample, as shown in **Figure 2e**, indicated an average pore diameter of 2.9 nm, indicating a mesoporous structure for these catalysts. These findings demonstrate that Co@CNB-N$_4$ possesses sufficient microscale surface area and a mesoporous nanostructure, providing abundant reaction sites and mass transfer channels. Consequently, these characteristics actively facilitate the diffusion and permeation of various transition species during the catalytic process.

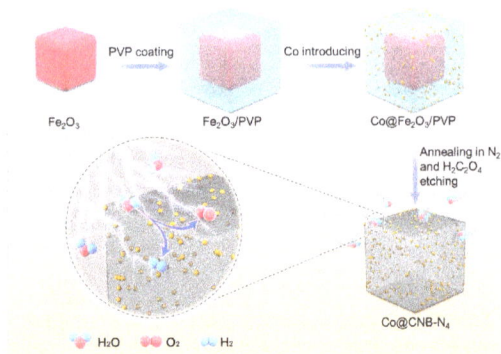

Figure 1. *Schematic illustration of the synthesis process of Co single atoms anchored on carbon nanoboxes (Co@CNB-N₄). Reproduced with permission from ref [32]. Copyright 2023, Elsevier.*

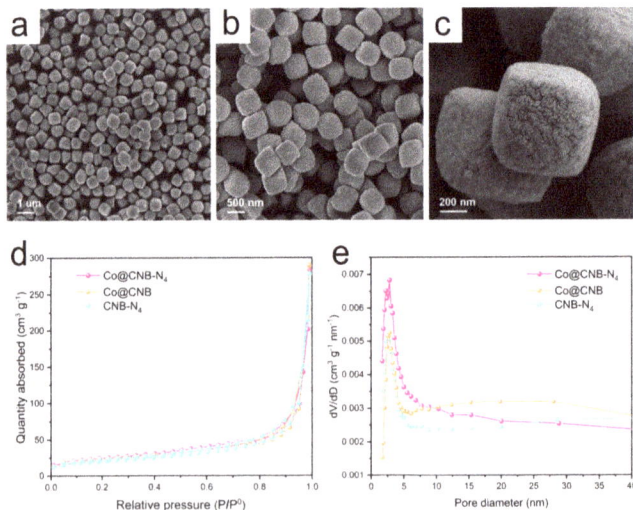

Figure 2. *(a-c) FE-SEM images of Co@CNB-N₄. (d) N₂ adsorption-desorption isotherms of the Co@CNB-N₄, Co@CNB, and CNB-N₄. (e) The pore size distribution calculated from the adsorption branch. Reproduced with permission from ref [32]. Copyright 2023, Elsevier.*

Based on the analysis of the material's chemical composition, microstructure, and theoretical calculations, the Co@CNB-N$_4$ catalyst demonstrated superior catalytic performance in HER within an alkaline medium. To evaluate its performance, a series of catalysts, including Co@CNB-N$_4$, Co@CNB, and CNB-N$_4$, were tested alongside noble metal catalysts (Pt/C, 20wt%) for HER catalytic activity in a 0.1 M KOH alkaline electrolyte. As illustrated in **Figure 3a-b**, Co@CNB-N$_4$ exhibited significantly better HER catalytic activity compared to other samples. It required the overpotentials of 45 mV, 85 mV, and 163 mV to achieve the current densities of 10, 20, and 50 mA cm^{-2}, respectively. Notably, even when compared to the commercial Pt/C catalyst, Co@CNB-N$_4$ showed remarkably close HER catalytic performance. The lower HER catalytic activity observed in CNB-N$_4$ without the introduction of Co atoms can be attributed to the limited exposure of active sites in aggregated materials, further highlighting the crucial role of dispersed single Co atoms as active centers in the HER catalysis.

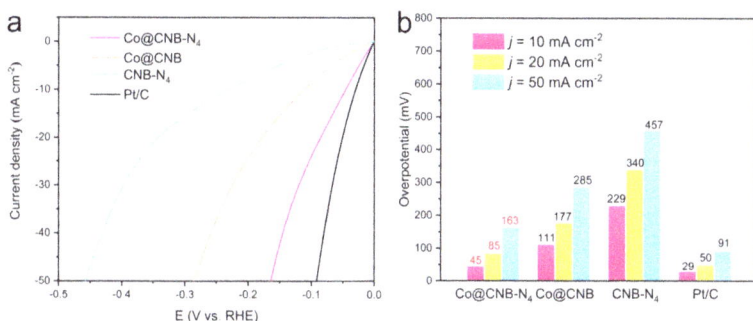

Figure 3. Electrocatalytic alkaline HER performances of the catalysts in 0.1 M KOH electrolyte: (a) HER polarization curves of Co@CNB-N$_4$, Co@CNB, CNB-N$_4$ and Pt/C. (b) The comparison of overpotentials required to achieve the current densities of 10, 20 and 50 mA cm^{-2} for various catalysts. Reproduced with permission from ref [32]. Copyright 2023, Elsevier.

The Tafel slopes derived from the linear sweep voltammetry (LSV) curves were used to validate the reaction kinetics of the prepared catalysts. Co@CNB-N$_4$ exhibited a Tafel slope of approximately 117.5 mV dec^{-1}, which is significantly lower than that of Co@CNB (approximately 205 mV dec^{-1}) and CNB-N$_4$ (approximately 302.5 mV dec^{-1}), as illustrated in **Figure 4a**. This indicates that in an alkaline environment, the rate-determining step (RDS) in the HER reaction of the Co@CNB-N$_4$ catalyst is the Volmer step rather than the Heyrovsky step. Moreover, the charge transfer resistance (R_{ct}) for Co@CNB-N$_4$ is approximately 6.04 Ω, which is smaller than that of Co@CNB (approximately 9.66 Ω) and CNB-N$_4$ (approximately 19.62 Ω), as shown in **Figure 4b**. This is mainly attributed to the enhanced electronic structure resulting from introduced Co atoms and N elements, which enhance the charge transfer capability and conductivity during the HER process.

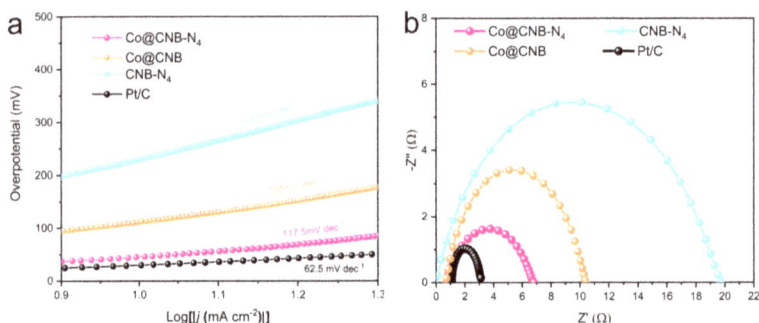

Figure 4. *Electrocatalytic alkaline HER performances of the catalysts in 0.1 M KOH electrolyte: (a) Tafel slopes. (b) Nyquist plots of the catalysts. Reproduced with permission from ref [32]. Copyright 2023, Elsevier.*

Electrochemical stability is crucial for widespread applications [33-35]. Through multiple cyclic voltammetry (CV) and chronopotentiometry tests in an alkaline electrolyte, the excellent stability of Co@CNB-N$_4$ was confirmed. As depicted in **Figure 5a**, after 5000 cycles, negligible loss in HER catalytic performance and current density was observed, with the morphology of the material remaining unchanged. Additionally, the *i-t* curve (inset **Figure 5a**) demonstrated that even after 50 hours of operation at -0.085 V *vs.* RHE, the current density remained close to 10 mA cm^{-2}, with only around a 0.7% variation in Co atomic content. These results indicate outstanding electrochemical stability of the sample, attributed to the special bonding between Co and the doped carbon nano substrates.

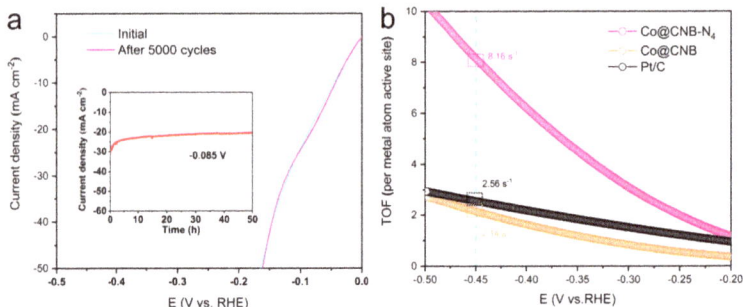

Figure 5. *(a) Stability test of Co@CNB-N$_4$ through cyclic potential scanning and chronoamperometry method (inset). (b) TOF plots of the metal-based catalysts. Reproduced with permission from ref [32]. Copyright 2023, Elsevier.*

To explore the electrochemically active surface area (ECSA) and active site count of Co@CNB-N$_4$ in detail, various scans at different speeds were conducted, and a series of CV curves were analyzed [36-38]. By fitting the current difference (Δj) at E = 0.07 V *vs.* RHE into a linear model, the double-layer capacitance (C_{dl}) was obtained. This C_{dl} is directly linked to ECSA, which was calculated straightforwardly for the samples. Co@CNB-N$_4$ exhibited the highest capacitance among the prepared samples, aligning with its specific surface area and suggesting a connection between ECSA and specific surface area. Further analysis revealed that Co@CNB-N$_4$ maintained the largest ECSA value, while Co@CNB and CNB-N$_4$ had significantly lower ECSA values, respectively, supporting the conclusions above.

At an overpotential of 450 mV, turnover frequencies (TOFs) were calculated for the catalysts, as shown in **Figure 5b**. Co@CNB-N$_4$ exhibited TOFs (~8.16 s^{-1}) 3.81 and 3.18 times higher than Co@CNB (~2.14 s^{-1}) and Pt/C (~2.56 s^{-1}), respectively. Additionally, the exchange current density (j_0), crucial for evaluating electrocatalytic performance, was determined from Tafel plots. Remarkably, Co@CNB-N$_4$ showed a satisfying j_0 value (~0.0021 mA cm^{-2}), slightly lower than Pt/C (j_0 = 0.0064 mA cm^{-2}) but significantly higher than the other samples. These findings suggest that uniformly dispersed SACs with non-metallic elements as their coordination environment possess a greater number of intrinsic active sites compared to the initial substrate.

2.2. Metal-Organic Frameworks Carriers

Metal-Organic Frameworks (MOFs) have recently emerged as a topic of significant interest and discussion in the field of catalysis [39-41]. Their distinctive porous structure and adjustable chemical properties present exciting opportunities for incorporating and utilizing SACs. Scientists and researchers are actively exploring ways to harness the unique structure and functionality of MOFs to enhance catalytic performance, particularly in the context of HER [42-44].

Cheng et al. [45] employed a low-temperature solvothermal immersion method to successfully synthesize stable Mo and W binary SACs. Initially, they utilized a solvothermal approach to grow FeNi bimetallic amino-functionalized MOF (FN-MOF) in situ on the surface of nickel foam (NF). Subsequently, the FN-MOF-NF composite was immersed in a mixed solution of Mo and W precursors, resulting in the formation of the desired product, MOF-Mo$_{SA}$W$_{SA}$. For comparison, unary single-atom systems with FN-MOFs containing only Mo single atoms (sample MOF-Mo$_{SA}$) and only W single atoms (sample MOF-W$_{SA}$), as well as a plain FN-MOF (sample MOF) were fabricated. The morphology of the MOF-Mo$_{SA}$W$_{SA}$ was observed using scanning electron microscopy (SEM), as depicted in **Figure 6a**. The SEM image revealed that MOF-Mo$_{SA}$W$_{SA}$ exhibited an interwoven nanosheet structure with an average thickness of approximately 230 nm, completely covering the NF framework. Upon closer examination, it was observed that each nanosheet consisted of stacked nanoslices, with each slice measuring approximately 22 nm in thickness (see inset at the top-right of **Figure 6a**).

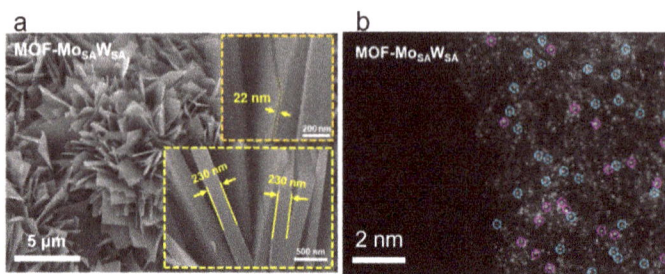

Figure 6. (a) SEM images of MOF-Mo$_{SA}$W$_{SA}$. Inset shows enlarged views of nano-slabs and nano-sheets. (b) HAADF-STEM image of MOF-Mo$_{SA}$W$_{SA}$. Green and pink circles mark the positions of Mo and W SAs, respectively. Reproduced with permission from ref [45]. Copyright 2023, Elsevier.

Furthermore, **Figure 6b** depicted the high-angle annular dark-field scanning transmission electron microscope (HAADF-STEM) image of MOF-Mo$_{SA}$W$_{SA}$. In this image, the bright spots, indicated by green and pink circles, corresponded to Mo and W atoms, respectively, while the unmarked spots represented Fe and Ni atoms. Two significant conclusions can be drawn from **Figure 6b**. Firstly, the Mo and W atoms exhibited a well-dispersed pattern without aggregation, confirming their isolated atom nature within the MOF-Mo$_{SA}$W$_{SA}$ structure. Secondly, the Mo and W single atoms appeared separately rather than closely paired, indicating the coexistence of distinct Mo and W SACs within MOF-Mo$_{SA}$W$_{SA}$.

Figure 7a-b showed the LSV curves and Tafel plots of the samples in HER, allowing for a comparison of their HER performances. The four catalysts, namely MOF-W$_{SA}$, MOF-Mo$_{SA}$, MOF-Mo$_{SA}$W$_{SA}$, and Pt/C, were evaluated in terms of the overpotentials required to generate a current density of 10 mA cm^{-2}. The overpotentials for these catalysts were determined as 216 mV, 165 mV, 111 mV, and 57 mV, respectively. Similarly, to achieve a higher current density of 500 mA cm^{-2}, the overpotentials ranged from 434 mV for FN-MOF to 297 mV for MOF-Mo$_{SA}$W$_{SA}$. It is evident that the three MOFs loaded with single atoms exhibit significantly improved efficiency in hydrogen evolution compared to the pristine MOF, indicating a positive synergy resulting from the introduction of single atoms into the MOF structure.

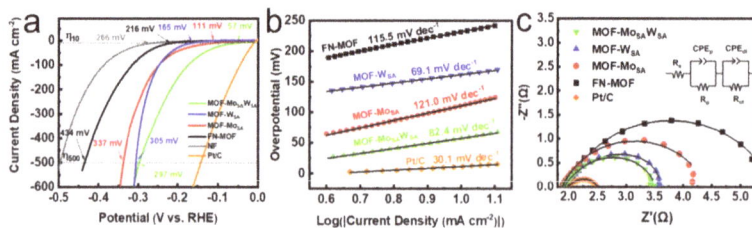

Figure 7. (a) LSV curves. (b) Tafel plots. (c) Nyquist plots for HER recorded at -0.280 V vs. RHE. Reproduced with permission from ref [45]. Copyright 2023, Elsevier.

Furthermore, the Tafel slope of MOF-Mo$_{SA}$W$_{SA}$, as shown in **Figure 7b**, was determined to be 82.4 mV dec^{-1}, indicating that the RDS is Heyrovsky-Volmer step. The electrochemical impedance spectroscopy (EIS) data was shown in **Figure 7c**. The diameters of the semi-arcs displayed a decreasing order of FN-MOF, MOF-Mo$_{SA}$, MOF-W$_{SA}$, MOF-Mo$_{SA}$W$_{SA}$, and Pt/C, corresponding to the R_{ct} values of 3.41, 2.30, 1.44, 1.39, and 0.40 Ω, respectively. Among the four MOF-based samples, MOF-Mo$_{SA}$W$_{SA}$ shows the smallest diameter of the semi-arc and thus the smallest R_{ct} value, indicating its superior interfacial charge transfer kinetics. This observation aligns with the trends observed from the LSV curves in HER.

2.3. Transition Metal Compounds Carriers

Transition metal compounds, particularly metal oxides like titanium dioxide and iron oxide, exhibit notable advantages as carrier materials for single-atom hydrogen evolution catalysts. These compounds boast high surface areas and abundant active sites, rendering them highly promising for supporting catalysts in the HER. Consequently, the selection of metal compounds as carrier materials will play a pivotal role in the application and advancement of single-atom hydrogen evolution catalysts.

Metal chalcogenides, such as MoS$_2$ and CoSe$_2$, belong to a class of highly significant materials in the realm of energy storage and conversion. These compounds possess distinctive structural features and electronic properties [46]. In recent years, metal chalcogenides have garnered significant attention and utilization in enhancing electrochemical water splitting due to their remarkable intrinsic activity and excellent electrical conductivity [47, 48]. Moreover, metal chalcogenides have emerged as promising hosts for anchoring and stabilizing isolated metal sites, leading to the synthesis of advanced SACs based on these materials. These SACs have demonstrated great potential in facilitating water splitting reactions [49, 50].

For example, Lou et al. [51] have decorated the isolated Ni atoms onto hierarchical MoS$_2$ nanosheets supported on multichannel carbon matrix (MCM) nanofibers (denoted as MCM@MoS$_2$-Ni). The synthesized progress is illustrated in **Figure 8**.

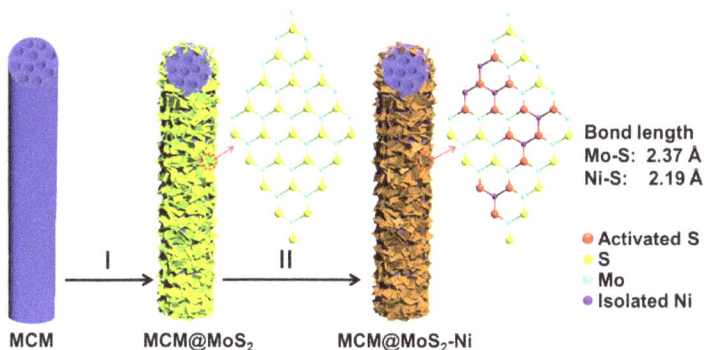

Figure 8. *Schematic illustration of the synthetic process for MCM@MoS$_2$-Ni. Reproduced with permission from ref [51]. Copyright 2018, Wiley-VCH.*

The FE-SEM cross-sectional image shown in **Figure 9a** provides a direct view of the multi-channel structure of MCM@MoS$_2$-Ni. Additionally, high-resolution TEM (HR-TEM) images reveal visible lattice fringes of interconnected MoS$_2$ nanosheets, as depicted in **Figure 9b**. The measured interlayer distance of 0.62 nm corresponds to the (002) plane of hexagonal MoS$_2$. Further examination through atomic-resolution images and Fast Fourier Transform (FFT) filtered images clearly reveals the presence of defects in the hexagonal structure of the Ni-modified MoS$_2$ basal plane, as illustrated in **Figure 9c-d**. These observations confirm the successful decoration of isolated Ni species onto the MoS$_2$ nanosheets.

Both experimental investigations and theoretical calculations demonstrate that the single-atom Ni modification leads to a substantial increase in activated S atoms within the MoS$_2$ basal plane, accompanied by a significantly modified electronic structure. This modification plays a crucial role in enhancing the water dissociation process. Additionally, the distinctive tubular structure of MCM@MoS$_2$-Ni contributes to the improved performance in HER. Notably, the MCM@MoS$_2$-Ni catalyst exhibits a significantly reduced overpotential of 161 mV to achieve a current density of 10 mA cm^{-2}.

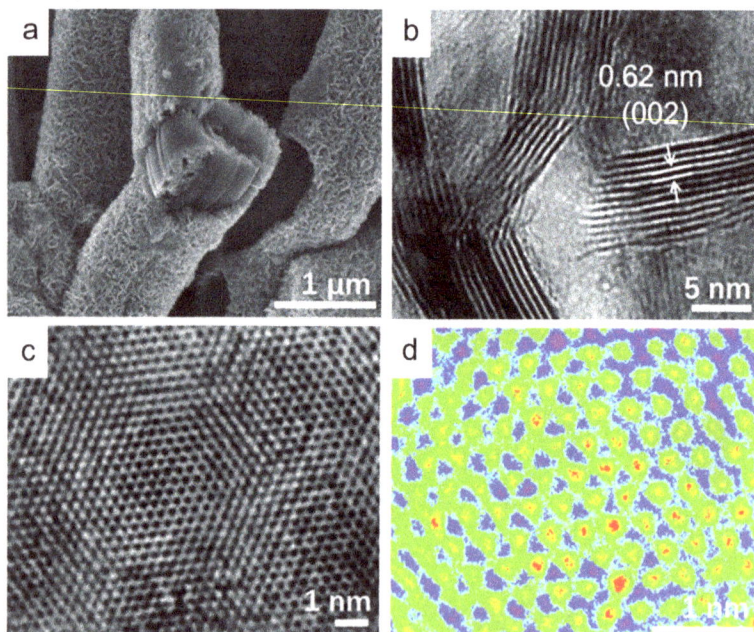

Figure 9. (a) FE-SEM image, (b) HR-TEM image, (c) Atomic resolution picture, and (d) FFT-filtered atomic resolution image of the Ni-decorated MoS$_2$ nanosheets. Reproduced with permission from ref [51]. Copyright 2018, Wiley-VCH.

In the study conducted by Liu et al. [52], they successfully fabricated a structure called SA Co-D $1T$ MoS$_2$, where an atomic cobalt array was covalently bound to distorted $1T$ MoS$_2$ nanosheets. The study revealed that the phase transformation from $2H$ to D-$1T$ MoS$_2$, induced by lattice mismatch strain, is a critical factor in generating highly active SACs for the HER. Through active-site blocking experiments and density functional theory (DFT) calculations, it was inferred that the remarkable electrocatalytic performance of SA Co-D $1T$ MoS$_2$ is strongly associated with an ensemble effect. This effect arises from the synergistic interaction between the Co adatom and the S atoms of the D-$1T$ MoS$_2$ support, achieved by tuning the hydrogen binding mode at the interface. These findings further confirmed the positive influence of MoS$_2$ on the electrocatalytic performance of HER.

In their study, Zhang et al. [53] devised a straightforward optothermal reaction method for synthesizing single-atom platinum (Pt) catalysts anchored on VS$_2$ nanosheets. The entire process for preparing Pt-decorated VS$_2$/CP (Pt/VS$_2$/CP) with potential for scalable production is depicted **Figure 10**. First, VS$_2$ nanosheets were directly synthesized on carbon cloth (CC) through a hydrothermal treatment. Subsequently, different structured Pt catalysts were anchored onto the VS$_2$/CC using a simple and cost-effective optothermal process. The structures of the Pt catalysts were precisely controlled by adjusting the concentration of chloroplatinic acid (CA) with concentrations of 0.1, 0.5, 1, 5, and 10 mM utilized. Accordingly, the Pt/VS$_2$/CC catalysts were labeled as 0.1Pt/VS$_2$/CP, 0.5Pt/VS$_2$/CP, 1Pt/VS$_2$/CP, 5Pt/VS$_2$/CP, and 10Pt/VS$_2$/CP, respectively.

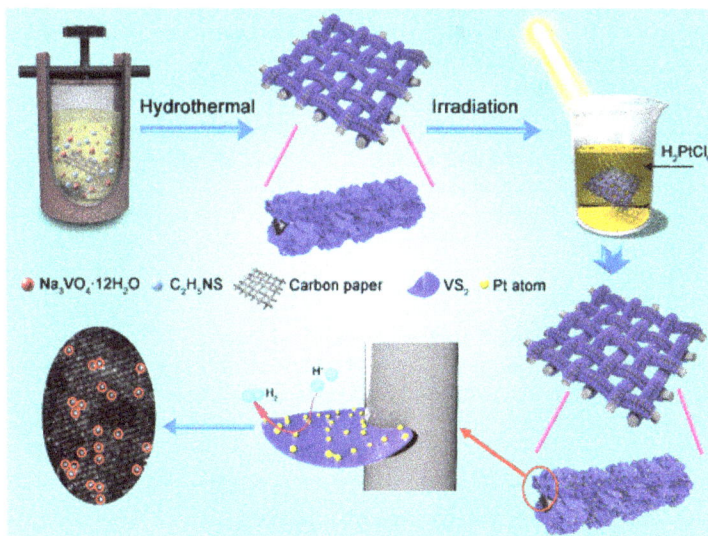

Figure 10*. Experimental flowchart for the synthesis processes of Pt/VS$_2$/CP. Reproduced with permission from ref [53]. Copyright 2020, American Chemical Society.*

The single atomic Pt-decorated VS_2 catalyst demonstrates remarkable performance, exhibiting a mass activity of 22.88 A mg_{Pt}^{-1} at 200 mV, which is 12 times higher than that of the commercial Pt/C catalyst. Moreover, electrochemical measurements revealed that the $Pt/VS_2/CP$ catalyst exhibited an ultralow overpotential of 77 mV to reach the current density of 10 mA cm^{-2}, indicating excellent catalytic activity, fast kinetics for the HER as well as superb electrochemical stability. Theoretical investigations further elucidated that the presence of the decorated Pt atoms significantly modified the absorption behavior of H atoms and the charge-transfer kinetics, contributing to the enhanced performance of the Pt/VS_2 catalyst. These research findings underscore the tremendous potential of metal chalcogenides as hosts for anchoring and stabilizing isolated metal sites, emphasizing their role in the development of efficient and stable single-atom hydrogen evolution catalysts.

In summary, different types of carriers exhibit distinct characteristics in the design of single-atom hydrogen evolution catalysts, each presenting unique advantages and challenges. A comprehensive understanding of these carrier traits and careful selection are crucial in constructing efficient and stable single-atom hydrogen evolution catalysts. The choice of carrier material undoubtedly plays an indispensable and crucial role in catalyst design.

3.   Active Center Metal Selection

The selection of active sites in single-atom hydrogen evolution catalysts is of utmost importance as it greatly influences their performance and catalytic activity [54, 55]. The choice of active sites primarily involves considering the type of metal and its interaction with the carrier material [56, 57]. In recent years, significant efforts have been made by researchers to identify and optimize the active sites that can enhance the efficiency and stability of catalysts.

The selection of the metal type plays a crucial role in determining the active sites of hydrogen evolution catalysts. Precious metals, including Pt and others, have traditionally been favored as active sites due to their excellent catalytic activity. However, their high cost and limited availability have prompted researchers to explore alternative options. Non-precious metals such as iron and cobalt have emerged as promising alternatives due to their lower cost and abundance in the earth's crust. By selecting the appropriate metal type, researchers can directly influence the catalyst's activity and selectivity in the HER.

In the study by Cao et al. [55], a single-atom cobalt site catalyst (Co_1/PCN) was synthesized through pyrolysis. This catalyst consists of dispersed cobalt atoms immobilized as structurally uniform Co_1-N_4 segments within a phosphorus-doped carbon nitride (PCN) framework. The electrocatalytic activity of the Co_1/PCN catalyst was evaluated using chronoamperometric measurements in N_2-saturated 1.0 M KOH electrolyte. For comparison, the HER performances of pristine PCN and commercial Pt/C catalysts were also measured under the same conditions.

Figure 11a displays the HER polarization curves obtained from LSV measurements. It is evident that both pristine PCN and Co_1/CN exhibit inferior HER activity, with overpotentials of 268 mV and 138 mV, respectively, at a current density of 10 mA cm^{-2}. However, upon anchoring the Co single atoms, the Co_1/PCN catalyst demonstrates superior HER catalytic activity, delivering a current density of 10 mA cm^{-2} at a significantly reduced overpotential of 89 mV. Furthermore, the Tafel slope of the Co_1/PCN catalyst is as low as 52 mV dec^{-1}, as illustrated in **Figure 11b**, comparable to that for Pt/C (38 mV dec^{-1}). This suggests the Volmer-Heyrovsky mechanism is the dominant HER pathway on the Co_1/PCN catalyst.

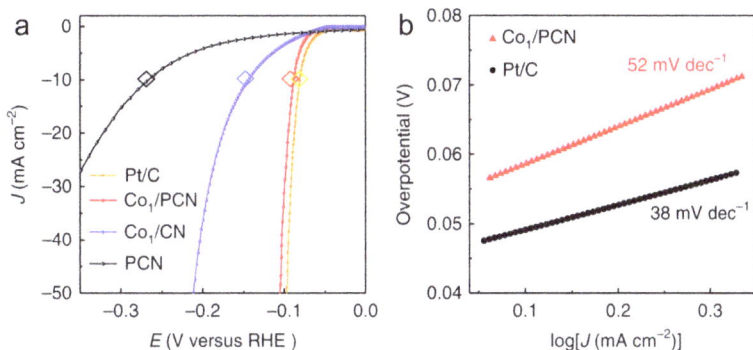

Figure 11. (a) LSV curves of the Co_1/PCN, Co_1/CN, PCN, and Pt/C catalysts in 1.0 M KOH electrolyte. (b) Tafel plots of Co_1/PCN and Pt/C catalysts. Reproduced with permission from ref [55]. Copyright 2019, Springer Nature.

To gain insights into the nature of the active Co sites under HER conditions, they conducted operando X-ray absorption fine structure (XAFS) measurements while simultaneously monitoring the electrochemical reaction using a standard electrochemical workstation, as depicted in **Figure 12a**. In this setup, a porous CC was utilized as the working electrode, providing full contact between the catalyst and the electrolyte. This ensured that all Co sites detected by the XAFS measurements actively participated in the electrocatalytic reaction. During the XAFS measurements, the potential of the working electrode gradually decreased from 0 V to -0.1 V *vs.* the reversible hydrogen electrode (RHE). Under open-circuit conditions, operando XAFS data were collected at two representative potentials near the starting potential and the overpotential for current densities of 0.5 and 10 mA cm^{-2}, respectively (-0.04 and -0.1 V, respectively).

In **Figure 12b**, the operando X-ray absorption near-edge structure (XANES) spectra at the Co K-edge of the Co_1/PCN catalyst are shown at different applied potentials, along with reference data for CoO, Co_3O_4, and CoOOH. The spectra reveal important information about the oxidation state of Co in the catalyst. Comparing the ex-situ sample to the open-circuit condition, the absorption edge shifts to the higher energy side by 0.5 eV, accompanied by a broadening of the white line peak. These observations indicate an increase in the Co oxidation state during the electrochemical reaction.

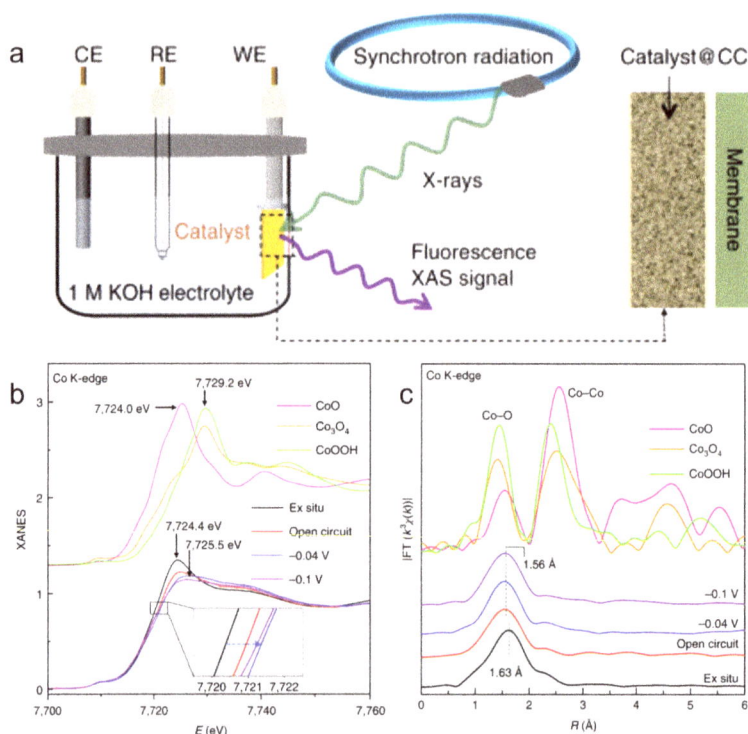

Figure 12. *(a) Schematic of the operando electrochemical cell set-up for identifying the active site of the Co_1/PCN catalyst. CE, counter electrode; WE, working electrode; RE, reference electrode. (b) Operando XANES spectra recorded at the Co K-edge of Co_1/PCN, at different applied voltages from the open circuit condition to -0.1 V during electrocatalytic HER. Inset shows the magnified preedge XANES region. (c) K^3-weighted Fourier transform (FT) spectra. The shaded region highlights the variations in the peak position and intensity. Reproduced with permission from ref [55]. Copyright 2019, Springer Nature.*

Figure 12c demonstrates the Fourier-transformed peak intensity of the sample, showcasing a decrease followed by an increase under open-circuit conditions and applied potentials, respectively. This trend indicates a change in the coordination environment of the Co sites as the electrochemical conditions vary. Overall, the operando studies provide compelling evidence that the structures of single-atom site catalysts, such as Co_1/PCN, can undergo significant changes under realistic catalytic conditions. The atomically dispersed Co site is highly susceptible to a high valence induced by chemisorbed OH^- under electrochemical conditions. These high-valence transition-metal oxide species exhibit enhanced reactivity towards water adsorption and oxygen-related reactions. Therefore, the formation of a high-valence HO-Co1-N2 site under working conditions is likely responsible for the high alkaline HER activity. Furthermore, operando techniques enable us to monitor the preferential adsorption of H_2O on Co sites during the HER, providing valuable insights into the reaction mechanism and the interaction between the catalyst and reactants.

To gain a deeper understanding of the reaction mechanism, theoretical studies based on DFT calculations were conducted. The obtained Tafel slope results revealed that the catalyst can facilitate the release of hydrogen molecules through the Heyrovsky reaction. Based on these findings, a proposed catalytic cycle for alkaline HER on HO-Co$_1$/PCN is presented in **Figure 13a**. The catalytic cycle begins with the adsorption of H_2O on Co, as shown in step 1. This H_2O molecule then undergoes dissociation into OH* adsorbed on Co and H* adsorbed on nearby N sites, as depicted in step 2. The asterisks denote the positions of adsorption. Subsequently, another proton from an adjacent H_2O molecule reacts with the first H* to produce H_2, as illustrated in steps III-V in **Figure 13a**.

It is well-known that the first slow step in alkaline HER involves the catalytic dissociation of a water molecule, generating H* ($H_2O + e^- = H^* + OH^-$). In light of this, the adsorption energies of H_2O and H on the HO-Co$_1$/PCN surface were considered. Comparative studies were conducted with a model of a finished Pt plate representing commercial Pt/C, with a (111) facet. The adsorption energy of H_2O molecules on HO-Co$_1$/PCN was found to be 0.94 eV, higher than the adsorption energy on Pt(111) (0.46 eV). This indicates a more favorable adsorption of H_2O on HO-Co$_1$/PCN, as shown in **Figure 13b**. Furthermore, the kinetics of water dissociation into OH* and H*, corresponding to steps I and II in **Figure 13a**, were computed to better understand the reaction mechanism and rate-determining steps involved in alkaline HER on HO-Co$_1$/PCN.

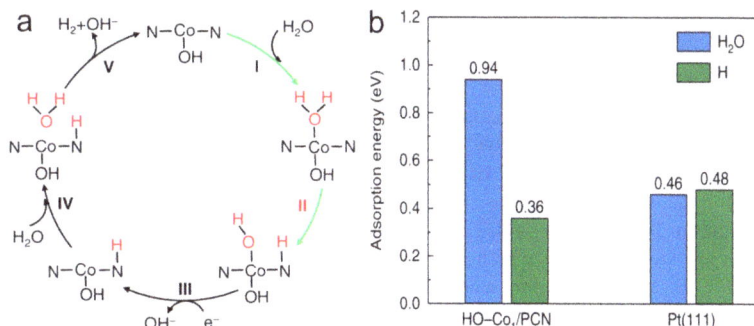

Figure 13. (a) Alkaline HER mechanism on HO-Co$_1$/PCN. (b) Calculated adsorption energies of H_2O and H on the surface of HO-Co$_1$ PCN and Pt(111). Reproduced with permission from ref [55]. Copyright 2019, Springer Nature.

Electrocatalytic Hydrogen Production: Catalysts and Applications   Materials Research Forum LLC
Materials Research Foundations **165** (2024)   https://doi.org/10.21741/9781644903070

4.   Coordination Structure Regulation

The coordination structure plays a crucial role in the selection of active sites within a catalyst [58, 59]. By exerting control over the atomic coordination environment, it becomes possible to influence the catalyst's activity, selectivity, and stability. One noteworthy example pertains to single-atom metals, which exhibit diverse coordination modes. They can either bind individually to the catalyst support or form coordinated structures with multiple atoms. These structural disparities exert a direct influence on both the catalytic activity and electronic structure of the metal sites, ultimately shaping their performance in HER.

Lyu et al. [60] developed a novel carbon-free heterogeneous nanostructure with single iron atoms dispersed based on a molybdenum mineral hydrogel precursor. The synthesis process involved the formation of a two-dimensional wrinkled iron-phosphomolybdic acid mineral hydrogel nanosheet (FePMoG) as the precursor in a single step. Subsequently, through a low-temperature phosphorization process, this precursor was directly converted into carbon-free nanosheets with dispersed single iron atoms within a molybdenum-based matrix (Fe/SAs@Mo-based-HNSs).

Figure 14a shows the resulting structure of Fe/SAs@Mo-based-HNSs, which exhibited a two-dimensional porous nanosheet morphology. The HR-TEM image in **Figure 14b** exhibited distinct lattice fringes, corresponding to different crystal planes, namely MoO_2 (−102), MoO_2 (020), MoP (100), MoP (101), MoP_2 (111), and MoP_2 (131). These lattice fringes confirmed the heterostructure of Fe/SAs@Mo-based-HNSs. Moreover, the high-angle annular dark-field scanning transmission electron microscopy (HAADF-STEM) image of Fe/SAs@Mo-based-HNSs, along with the corresponding energy-dispersive X-ray spectroscopy (EDS) elemental mapping shown in **Figure 14c**, revealed a uniform distribution of Fe, Mo, P, and O atoms. Importantly, the Fe atoms were observed to be independently dispersed within the lattice of the nanosheets. This finding confirmed the occurrence of an in-situ transformation during the phosphorization process and the successful dispersion of iron atoms at the single-atom level.

To assess the electrocatalytic performance of Fe/SAs@Mo-based-HNSs, the HER was investigated in a 1 M KOH solution using a typical three-electrode system. **Figure 15a** The displays the LSV curves, demonstrating the excellent HER performance of Fe/SAs@Mo-based-HNSs. To produce the current density of 10 mA cm^{-2}, Fe/SAs@Mo-based-HNSs exhibited an overpotential of 38.5 mV, surpassing other FeMoP samples and even the commercial 20 wt% Pt/C overpotential. As the current density reached 200 mA cm^{-2}, the overpotential of Fe/SAs@Mo-based-HNSs slightly increased to 109.9 mV, which remained significantly lower than the overpotential of 299.3 mV for 20 wt% Pt/C catalyst.

Tafel plots in **Figure 15b** were examined to gain insights into the electrocatalytic HER mechanism of. The Tafel slope of Fe/SAs@Mo-based-HNSs (35.6 mV dec^{-1}) was smaller than that of the overall FeMoP-500 (83.3 mv dec^{-1}) and other FeMoP samples, approaching the value of commercial 20% Pt/C catalyst (36.1 mV dec^{-1}). This suggests that the HER peocess on Fe/SAs@Mo-based-HNSs is predominantly governed by a rapid Heyrovsky step, indicating strong electronic coupling between the single dispersed atoms and the heterogeneous structures.

The stability of Fe/SAs@Mo-based-HNSs was found to be remarkable. During a durability test lasting 600 hours, it maintained a current density of 200 mA cm^{-2} in the HER, demonstrating exceptional reliability. **Figure 15c** illustrates that the overpotential of Fe/SAs@Mo-based-HNSs decreased by only around 30 mV throughout the stability test. Additionally, the material retained its porous, edge-disordered heterogeneous structure even after the durability test, confirming its outstanding stability during the HER process.

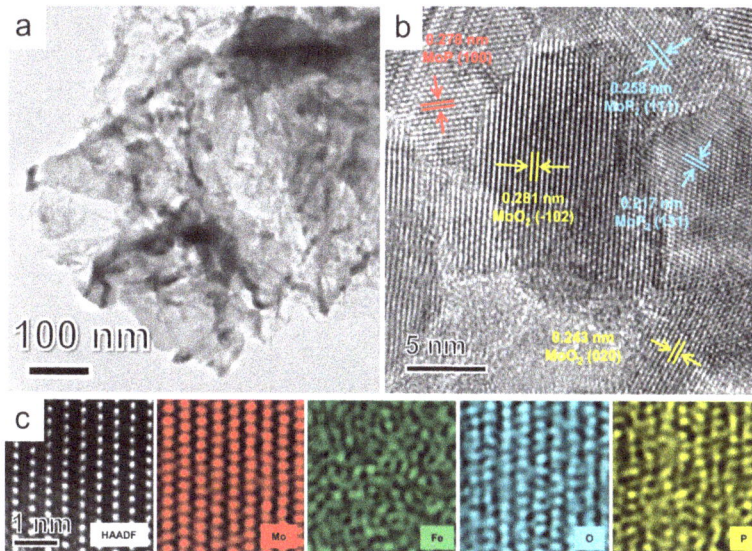

Figure 14. *(a) TEM and (b) HR-TEM images of the catalyst. (c) HAADF-STEM image and EDS elemental mapping of Fe (green), Mo (violet), P (orange), and O (cyan). Reproduced with permission from ref [60]. Copyright 2022, Springer Nature.*

When a metal single atom is anchored to the scaffold, the interaction between the metal and scaffold, along with the heterogeneity of the scaffold, gives rise to tunable electronic properties of the single atom. These properties include oxidation state, highest occupied molecular orbital (HOMO), lowest unoccupied molecular orbital (LUMO), spin state, and more, which collectively contribute to controllable catalytic properties. For example, the hydrolytic activity of Rh-doped rutile VO$_2$ is higher than that of Rh-doped monoclinic VO$_2$ due to the higher occupancy of individual Rh atoms on the rutile VO$_2$ scaffold. Similarly, Mn-C$_3$N$_4$ exhibits high solar-driven water decomposition activity, primarily reliant on its spin state, with an optimized e$_g$ occupancy of approximately 0.95. Among these electronic properties, oxidation state is frequently considered a significant parameter associated with catalytic activity. However, the specific relationship between oxidation state and catalytic activity remains unclear.

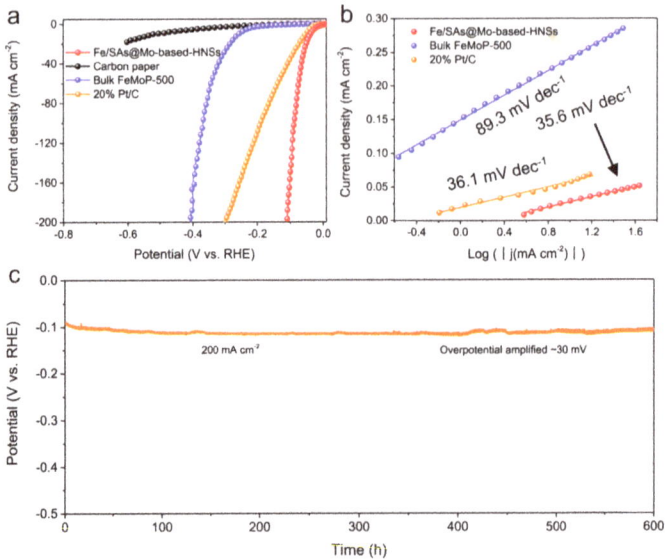

Figure 15. *Electrocatalytic performance of Fe/SAs@Mo-based-HNSs in HER in 1.0 M KOH solution: (a) Polarisation curves with iR correction. (b) Tafel plots. (c) Stability test at a constant current density of 200 mA cm^{-2}. Reproduced with permission from ref [60]. Copyright 2022, Springer Nature.*

To delve into the relationship between oxidation state and catalytic activity, Cao et al. [61] focused on osmium (Os) due to its rich valence states ranging from -2 to +8. They developed a synthetic strategy for fabricating atomically dispersed Os sites supported on N and S co-doped carbon materials (Os/CNS), as depicted inn **Figure 16**. The fabrication process began by introducing N and S atoms into carbon black using urea and 2,2'-bithienyl as the nitrogen and sulfur sources, respectively. Subsequently, Pluronic F127, a triblock copolymer, was thoroughly mixed with the CNS support in 30 mL of deionized water. Next, the osmium trichloride (OsCl$_3$) solution was slowly added to the suspension mentioned above. Following a 6-hour stirring period, the mixture was subjected to centrifugation to separate the solid products from the liquid phase. Finally, the dried products were subsequently calcined in a tubular furnace at 400 °C and an atomically dispersed Os/CNS sample was successfully fabricated after washing with acid solution for 12 hours. Moreover, the researchers prepared other atomically dispersed Os single atoms supported on different carriers. By altering the doping heteroatom in the carbon support while maintaining other conditions unchanged, they successfully synthesized samples supported on pure carbon black (C), N-doped graphite carbon (CN), and S-doped graphite carbon (CS).

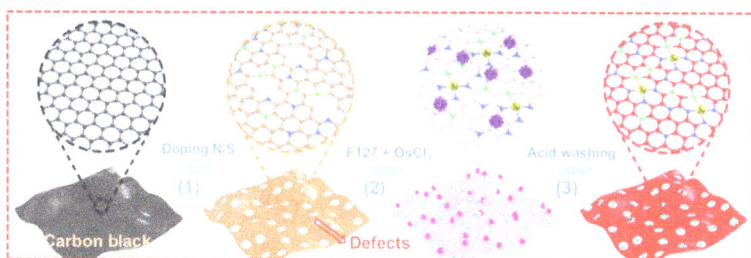

Figure 16. *Synthesis route of Os atomically dispersed catalysts. Reproduced with permission from ref [61]. Copyright 2022, Springer Nature.*

The atomic information of these catalysts was observed using an aberration-corrected high-angle toroidal dark-field scanning transmission electron microscope (AC-HAADF-STEM). The AC-HAADF-STEM images in **Figure 17a-b** showcase Os/CNS on different scales, revealing the uniform anchoring of atomically dispersed Os sites to the CNS support. This clear visualization confirms the successful fabrication of the catalyst. Subsequently, **Figure 17c** presents a TEM image of Os/CNS after aberration correction. Notably, no significant presence of large nanoparticles is observed, underscoring the atomically dispersed nature of the Os sites. Furthermore, the CNS carrier exhibits prominent lattice fringes, indicating its excellent crystallinity and suggesting the presence of exceptional electronic conductivity.

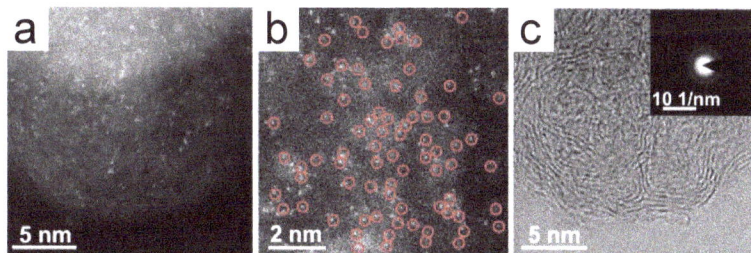

Figure 17. *(a-b) Aberration corrected HAADF-STEM images of Os/CNS in different scale bars. (c) Aberration corrected TEM image of Os/CNS. The inset shows the SAED pattern of the sample. Reproduced with permission from ref [61]. Copyright 2022, Springer Nature.*

To reveal the impact of coordination structure on catalytic performance, the electrochemical hydrogen evolution activity of the Os catalyst was evaluated in a 0.5 M H_2SO_4 solution. The overpotentials required to achieve a current density of 10 mA cm^{-2} were measured using the LSV curves for Os/CNS, Os/C, Os/CN, Os/CS, and 20% Pt/C catalyst, yielding values of 22, 132, 45, 63, and 13 mV, respectively, as indicated in **Figure 18a**. When the current density is 1 mA cm^{-2}, the initial potentials for Os/CNS, Os/C, Os/CN, Os/CS, and 20% Pt/C catalysts were 5, 47, 11, 18, and 3 mV, respectively. Therefore, HER activities of different Os SACs in this study followed the order: Os/CNS > Os/CN > Os/CS > Os/C.

In addition, the Tafel slopes in an acidic medium were determined and illustrated in **Figure 18b**. The Tafel slopes for Os/CNS, Os/C, Os/CN, Os/CS, and 20% Pt/C catalysts were measured as 41, 113, 56, 67, and 30 mV dec^{-1}, respectively. These slopes indicate that Os-N$_3$S$_1$ in Os/CNS exhibited the fastest HER kinetics among the various single atom samples with different coordination structures. Importantly, Os/CNS, Os/C, Os/CN, Os/CS operated through the Heyrovsky-Volmer mechanism. Moreover, stability is stability is a crucial parameter for assessing the HER performance of the catalysts. Long-term durability measurements were performed on the Os/CNS sample, including chronoamperometric curves and accelerated linear potential sweeps, in a 0.5 M H$_2$SO$_4$ solution. The results demonstrated that the current density of the Os/CNS sample exhibited negligible decrease after more than 50 hours of testing, as shown in **Figure 18c**. Additionally, the LSV curves of Os/CNS showed minimal decline in acidic conditions even after 3000 cycles, as indicated in the inset of **Figure 18c**.

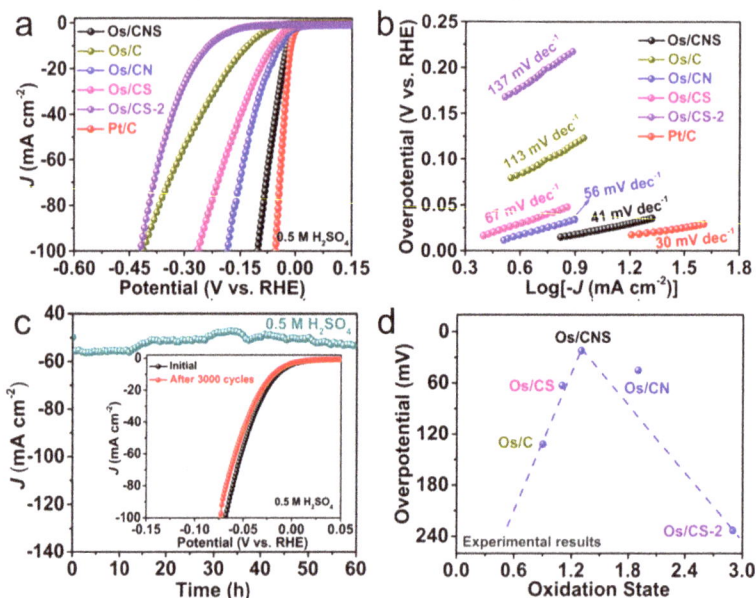

Figure 18. *(a) LSV curves. (b) Tafel plots. (c) The chronoamperometric curves of Os/CNS with inset showing the polarization curves measured before and after 3000 CV cycles. (d) The relationship between oxidation state and overpotential obtained from experiments for different Os SACs. Reproduced with permission from ref [61]. Copyright 2022, Springer Nature.*

The correlation between the analyzed oxidation states of Os monatoms and HER overpotential in the acidic medium was depicted in the volcanic curve shown in **Figure 18d**. It was observed that different coordination environments induced distinct oxidation states of Os, which directly influenced the catalytic performance. Notably, when the oxidation state of Os/CNS was +1.3, the catalyst exhibited high catalytic activity. These experimental findings suggest a volcanic curve relationship between the oxidation state of the isolated metal atom and HER activity. Understanding the role of oxidation states in coordination-dependent activity provides valuable insights for the rational design of SACs through coordination environment engineering.

5.   Conclusion and Prospects

SACs have emerged as a focal point in sustainable energy research, attracting widespread attention due to their remarkable catalytic performance. These catalysts play a pivotal role in crucial domains such as water electrolysis and fuel cells. In the context of hydrogen evolution through water electrolysis, there is a growing body of research dedicated to single-atom hydrogen evolution catalysts. This chapter emphasizes the careful selection of suitable carriers, the choice of metals for active sites, and the regulation of coordination environments. Through a combination of experimental investigations and theoretical explorations, researchers are striving to develop more efficient and stable single-atom catalysts for hydrogen evolution.

Despite significant achievements, there are still challenges and unknowns that need to be addressed.

(i) Enhanced Activity: Future investigations will aim to further enhance the catalytic activity of SACs in HER. This will involve exploring new metal choices, optimizing the configurations of active sites, and designing novel coordination environments to maximize catalytic performance. The integration of experimental studies with computational modeling and theoretical calculations will guide these efforts.

(ii) Improved Stability: Stability is a crucial factor for the practical application of SACs. Researchers will focus on enhancing the long-term stability of these catalysts by addressing issues such as metal atom migration, aggregation, and deactivation during catalysis. Strategies may involve the development of protective coatings, the exploration of different support materials, and the optimization of synthesis and preparation methods to enhance catalyst durability.

(iii) Scalability and Cost-Effectiveness: As SACs progress towards practical applications, scalability and cost-effectiveness become crucial considerations. Future research will focus on the development of scalable synthesis methods capable of producing these catalysts in large quantities without compromising their unique atomic dispersion. Additionally, efforts will be made to explore economically viable alternatives to expensive and scarce metals, aiming to reduce the overall cost of catalyst fabrication.

(iv) Advancements in In Situ Characterization Techniques: The progress of in situ and operando characterization techniques will be crucial for understanding the dynamic behavior of SACs under realistic reaction conditions. Real-time monitoring of catalyst structures, active site transformations, and surface species evolution will provide valuable insights into the reaction mechanisms and performance of these catalysts. Researchers will explore and develop advanced characterization tools to gain a comprehensive understanding of the catalysts' behavior during hydrogen production.

(v) Elucidating Reaction Mechanisms: A thorough understanding of the intricate reaction mechanisms involved in SACs is essential for further optimization and rational design. Future research will involve in-depth investigations into intermediate species, reaction kinetics, and surface interactions during the hydrogen evolution process. This knowledge will contribute to the development of more efficient and selective SACs.

References

[1]   Z.W. Seh, J. Kibsgaard, C.F. Dickens, I. Chorkendorff, J.K. Nørskov, T.F. Jaramillo, Combining theory and experiment in electrocatalysis: Insights into materials design, Science, 355 (2017) eaad4998. https://doi.org/10.1126/science.aad4998

[2]   D. Deng, K.S. Novoselov, Q. Fu, N. Zheng, Z. Tian, X. Bao, Catalysis with two-dimensional materials and their heterostructures, Nat. Nanotechnol., 11 (2016) 218-230. https://doi.org/10.1038/nnano.2015.340

[3]   S. Qiao, Q. He, Q. Zhou, Y. Zhou, W. Xu, H. Shou, Y. Cao, S. Chen, X. Wu, L. Song, Interfacial electronic interaction enabling exposed Pt(110) facets with high specific activity in hydrogen evolution reaction, Nano Res., 16 (2023) 174-180. https://doi.org/10.1007/s12274-022-4654-2

[4]   Y. Jiao, Y. Zheng, K. Davey, S.-Z. Qiao, Activity origin and catalyst design principles for electrocatalytic hydrogen evolution on heteroatom-doped graphene, Nat. Energy, 1 (2016) 16130. https://doi.org/10.1038/nenergy.2016.130

[5]   D. Lancet, I. Pecht, Spectroscopic and immunochemical studies with nitrobenzoxadiazolealanine, a fluorescent dinitrophenyl analog, Biochemistry, 16 (1977) 5150-5157. https://doi.org/10.1021/bi00642a031

[6]   G. Liu, A.W. Robertson, M.M.-J. Li, W.C.H. Kuo, M.T. Darby, M.H. Muhieddine, Y.-C. Lin, K. Suenaga, M. Stamatakis, J.H. Warner, S.C.E. Tsang, MoS$_2$ monolayer catalyst doped with isolated Co atoms for the hydrodeoxygenation reaction, Nat. Chem., 9 (2017) 810-816. https://doi.org/10.1038/nchem.2740

[7]   J. Durst, A. Siebel, C. Simon, F. Hasché, J. Herranz, H.A. Gasteiger, New insights into the electrochemical hydrogen oxidation and evolution reaction mechanism, Energy Environ. Sci., 7 (2014) 2255-2260. https://doi.org/10.1039/C4EE00440J

[8]   L. Zhang, N. Jin, Y. Yang, X.-Y. Miao, H. Wang, J. Luo, L. Han, Advances on axial coordination design of single-atom catalysts for energy electrocatalysis: A review, Nano-Micro Lett., 15 (2023) 228. https://doi.org/10.1007/s40820-023-01196-1

[9]   R. Lang, X. Du, Y. Huang, X. Jiang, Q. Zhang, Y. Guo, K. Liu, B. Qiao, A. Wang, T. Zhang, Single-atom catalysts based on the metal-oxide interaction, Chem. Rev., 120 (2020) 11986-12043. https://doi.org/10.1021/acs.chemrev.0c00797

[10]   F. Mo, Q. Zhou, W. Xue, W. Liu, S. Xu, Z. Hou, J. Wang, Q. Wang, The optimized catalytic performance of single-atom catalysts by incorporating atomic clusters or nanoparticles: In-depth understanding on their synergisms, Adv. Energy Mater., 13

(2023) 2301711. https://doi.org/10.1002/aenm.202301711

[11]   Z. Pu, I.S. Amiinu, R. Cheng, P. Wang, C. Zhang, S. Mu, W. Zhao, F. Su, G. Zhang, S. Liao, S. Sun, Single-atom catalysts for electrochemical hydrogen evolution reaction: recent advances and future perspectives, Nano-Micro Lett., 12 (2020) 21. https://doi.org/10.1007/s40820-019-0349-y

[12]   N. Cheng, S. Stambula, D. Wang, M.N. Banis, J. Liu, A. Riese, B. Xiao, R. Li, T.-K. Sham, L.-M. Liu, G.A. Botton, X. Sun, Platinum single-atom and cluster catalysis of the hydrogen evolution reaction, Nat. Commun., 7 (2016) 13638. https://doi.org/10.1038/ncomms13638

[13]   L. Zhang, L. Han, H. Liu, X. Liu, J. Luo, Potential-cycling synthesis of single platinum atoms for efficient hydrogen evolution in neutral media, Angew. Chem. Int. Ed., 56 (2017) 13694-13698. https://doi.org/10.1002/anie.201706921

[14]   B. Xu, D. Li, Q. Zhao, S. Feng, X. Peng, P.K. Chu, Electrochemical reduction of nitrate to ammonia using non-precious metal-based catalysts, Coord. Chem. Rev., 502 (2024) 215609. https://doi.org/10.1016/j.ccr.2023.215609

[15]   Y. Hu, G. Luo, L. Wang, X. Liu, Y. Qu, Y. Zhou, F. Zhou, Z. Li, Y. Li, T. Yao, C. Xiong, B. Yang, Z. Yu, Y. Wu, Single Ru atoms stabilized by hybrid amorphous/crystalline FeCoNi layered double hydroxide for ultraefficient oxygen evolution, Adv. Energy Mater., 11 (2021) 2002816. https://doi.org/10.1002/aenm.202002816

[16]   K. Liu, X. Zhao, G. Ren, T. Yang, Y. Ren, A.F. Lee, Y. Su, X. Pan, J. Zhang, Z. Chen, J. Yang, X. Liu, T. Zhou, W. Xi, J. Luo, C. Zeng, H. Matsumoto, W. Liu, Q. Jiang, K. Wilson, A. Wang, B. Qiao, W. Li, T. Zhang, Strong metal-support interaction promoted scalable production of thermally stable single-atom catalysts, Nat. Commun., 11 (2020) 1263. https://doi.org/10.1038/s41467-020-14984-9

[17]   L. Li, N. Zhang, Atomic dispersion of bulk/nano metals to atomic-sites catalysts and their application in thermal catalysis, Nano Res., 16 (2023) 6380-6401. https://doi.org/10.1007/s12274-022-5335-x

[18]   B. Wang, Y. Fu, F. Xu, C. Lai, M. Zhang, L. Li, S. Liu, H. Yan, X. Zhou, X. Huo, D. Ma, N. Wang, X. Hu, X. Fan, H. Sun, Copper single-atom catalysts-A rising star for energy conversion and environmental purification: Synthesis, modification, and advanced applications, small, 20 (2024) 2306621. https://doi.org/10.1002/smll.202306621

[19]   P. Aggarwal, D. Sarkar, K. Awasthi, P.W. Menezes, Functional role of single-atom catalysts in electrocatalytic hydrogen evolution: Current developments and future challenges, Coord. Chem. Rev., 452 (2022) 214289. https://doi.org/10.1016/j.ccr.2021.214289

[20]   A. Ali, P.K. Shen, Nonprecious metal's graphene-supported electrocatalysts for hydrogen evolution reaction: Fundamentals to applications, Carbon Energy, 2 (2020) 99-121. https://doi.org/10.1002/cey2.26

[21]   H. Fei, J. Dong, D. Chen, T. Hu, X. Duan, I. Shakir, Y. Huang, X. Duan, Single atom electrocatalysts supported on graphene or graphene-like carbons, Chem. Soc. Rev., 48 (2019) 5207-5241. https://doi.org/10.1039/C9CS00422J

[22]   S. Tang, L. Xu, K. Dong, Q. Wang, J. Zeng, X. Huang, H. Li, L. Xia, L. Wang, Curvature effect on graphene-based Co/Ni single-atom catalysts, Appl. Surf. Sci., 615 (2023) 156357. https://doi.org/10.1016/j.apsusc.2023.156357

[23]   Z. Ji, M. Perez-Page, J. Chen, R.G. Rodriguez, R. Cai, S.J. Haigh, S.M. Holmes, A structured catalyst support combining electrochemically exfoliated graphene oxide and carbon black for enhanced performance and durability in low-temperature hydrogen fuel cells, Energy, 226 (2021) 120318. https://doi.org/10.1016/j.energy.2021.120318

[24]   S. Shen, P. Gao, H. Chen, Z. Tang, J. Li, H. Xiu, J. Yang, Graphited carbon black curled nanoribbons simultaneously boosted stability and electrocatalytic activity of 1T-MoS_2/MoO_3 toward hydrogen evolution, J. Alloys Compd., 949 (2023) 169831. https://doi.org/10.1016/j.jallcom.2023.169831

[25]   L. Wang, J. Zhang, L. Zheng, J. Yang, Y. Li, X. Wan, X. Liu, X. Zhang, R. Yu, J. Shui, Carbon black-supported FM-N-C (FM = Fe, Co, and Ni) single-atom catalysts synthesized by the self-catalysis of oxygen-coordinated ferrous metal atoms, J. Mater. Chem. A, 8 (2020) 13166-13172. https://doi.org/10.1039/D0TA01208D

[26]   S. Jessl, J. Rongé, D. Copic, M.A. Jones, J. Martens, M. De Volder, Honeycomb-shaped carbon nanotube supports for $BiVO_4$ based solar water splitting, Nanoscale, 11 (2019) 22964-22970. https://doi.org/10.1039/C9NR06737J

[27]   L. Najafi, S. Bellani, R. Oropesa-Nuñez, M. Prato, B. Martín-García, R. Brescia, F. Bonaccorso, Carbon nanotube-supported $MoSe_2$ holey flake:Mo_2C ball hybrids for bifunctional pH-universal water splitting, ACS Nano, 13 (2019) 3162-3176. https://doi.org/10.1021/acsnano.8b08670

[28]   K. Qu, Y. Zheng, Y. Jiao, X. Zhang, S. Dai, S.-Z. Qiao, Polydopamine-inspired, dual heteroatom-doped carbon nanotubes for highly efficient overall water splitting, Adv. Energy Mater., 7 (2017) 1602068. https://doi.org/10.1002/aenm.201602068

[29]   T. Jiang, Y. Zhang, S. Olayiwola, C. Lau, M. Fan, K. Ng, G. Tan, Biomass-derived porous carbons support in phase change materials for building energy efficiency: a review, Mater. Today Energy, 23 (2022) 100905. https://doi.org/10.1016/j.mtener.2021.100905

[30]   C. Yang, Y. Wang, M. Liang, Z. Su, X. Liu, H. Fan, T.J. Bandosz, Towards improving H_2S catalytic oxidation on porous carbon materials at room temperature: A review of governing and influencing factors, recent advances, mechanisms and perspectives, Appl. Catal. B-Environ., 323 (2023) 122133. https://doi.org/10.1016/j.apcatb.2022.122133

[31]   R. Shi, C. Tian, X. Zhu, C.-Y. Peng, B. Mei, L. He, X.-L. Du, Z. Jiang, Y. Chen, S. Dai, Achieving an exceptionally high loading of isolated cobalt single atoms on a porous carbon matrix for efficient visible-light-driven photocatalytic hydrogen production,

Chem. Sci., 10 (2019) 2585-2591. https://doi.org/10.1039/C8SC05540H

[32] T. Li, S. Ren, C. Zhang, L. Qiao, J. Wu, P. He, J. Lin, Y. Liu, Z. Fu, Q. Zhu, W. Pan, B. Wang, Z. Chen, Cobalt single atom anchored on N-doped carbon nanoboxes as typical single-atom catalysts (SACs) for boosting the overall water splitting, Chem. Eng. J., 458 (2023) 141435. https://doi.org/10.1016/j.cej.2023.141435

[33] H. Xiong, A.K. Datye, Y. Wang, Thermally stable single-atom heterogeneous catalysts, Adv. Mater., 33 (2021) 2004319. https://doi.org/10.1002/adma.202004319

[34] G. Bae, S. Han, H.-S. Oh, C.H. Choi, Operando stability of single-atom electrocatalysts, Angew. Chem. Int. Ed., 62 (2023) e202219227. https://doi.org/10.1002/anie.202219227

[35] C.B. Hiragond, N.S. Powar, J. Lee, S.-I. In, Single-atom catalysts (SACs) for photocatalytic CO$_2$ reduction with H$_2$O: Activity, product selectivity, stability, and surface chemistry, Small, 18 (2022) 2270157. https://doi.org/10.1002/smll.202201428

[36] C. Wei, S. Sun, D. Mandler, X. Wang, S.Z. Qiao, Z.J. Xu, Approaches for measuring the surface areas of metal oxide electrocatalysts for determining their intrinsic electrocatalytic activity, Chem. Soc. Rev., 48 (2019) 2518-2534. https://doi.org/10.1039/C8CS00848E

[37] X. Fu, Z. Zhao, C. Wan, Y. Wang, Z. Fan, F. Song, B. Cao, M. Li, W. Xue, Y. Huang, X. Duan, Ultrathin wavy Rh nanowires as highly effective electrocatalysts for methanol oxidation reaction with ultrahigh ECSA, Nano Res., 12 (2019) 211-215. https://doi.org/10.1007/s12274-018-2204-8

[38] T. Uenishi, M. Ibe, Effect of electrochemical surface area on carbon dioxide electrolysis using anionic electrolyte membrane electrode assembly, Fuel, 346 (2023) 128309. https://doi.org/10.1016/j.fuel.2023.128309

[39] Y. Shen, T. Pan, L. Wang, Z. Ren, W. Zhang, F. Huo, Programmable logic in metal-organic frameworks for catalysis, Adv. Mater., 33 (2021) 2007442. https://doi.org/10.1002/adma.202007442

[40] D. Li, H.-Q. Xu, L. Jiao, H.-L. Jiang, Metal-organic frameworks for catalysis: State of the art, challenges, and opportunities, EnergyChem, 1 (2019) 100005. https://doi.org/10.1016/j.enchem.2019.100005

[41] H. Konnerth, B.M. Matsagar, S.S. Chen, M.H.G. Prechtl, F.-K. Shieh, K.C.W. Wu, Metal-organic framework (MOF)-derived catalysts for fine chemical production, Coord. Chem. Rev., 416 (2020) 213319. https://doi.org/10.1016/j.ccr.2020.213319

[42] J. Guo, Y. Qin, Y. Zhu, X. Zhang, C. Long, M. Zhao, Z. Tang, Metal-organic frameworks as catalytic selectivity regulators for organic transformations, Chem. Soc. Rev., 50 (2021) 5366-5396. https://doi.org/10.1039/D0CS01538E

[43] V. Pascanu, G. González Miera, A.K. Inge, B. Martín-Matute, Metal-organic frameworks as catalysts for organic synthesis: A critical perspective, J. Am. Chem. Soc., 141 (2019) 7223-7234. https://doi.org/10.1021/jacs.9b00733

[44]   D. Yang, B.C. Gates, Catalysis by metal organic frameworks: Perspective and suggestions for future research, ACS Catal., 9 (2019) 1779-1798. https://doi.org/10.1021/acscatal.8b04515

[45]   C.-C. Cheng, T.-Y. Lin, Y.-C. Ting, S.-H. Lin, Y. Choi, S.-Y. Lu, Metal-organic frameworks stabilized Mo and W binary single-atom catalysts as high performance bifunctional electrocatalysts for water electrolysis, Nano Energy, 112 (2023) 108450. https://doi.org/10.1016/j.nanoen.2023.108450

[46]   H. Xu, B. Huang, Y. Zhao, G. He, H. Chen, Engineering heterostructured Pd-Bi_2Te_3 doughnut/Pd hollow nanospheres for ethylene glycol electrooxidation, Inorg. Chem., 61 (2022) 4533-4540. https://doi.org/10.1021/acs.inorgchem.2c00296

[47]   Z. Zhang, Y. Wang, X. Leng, V.H. Crespi, F. Kang, R. Lv, Controllable edge exposure of MoS_2 for efficient hydrogen evolution with high current density, ACS Appl. Energy Mater., 1 (2018) 1268-1275. https://doi.org/10.1021/acsaem.8b00010

[48]   X. Ren, Q. Ma, H. Fan, L. Pang, Y. Zhang, Y. Yao, X. Ren, S. Liu, A Se-doped MoS_2 nanosheet for improved hydrogen evolution reaction, Chem. Commun., 51 (2015) 15997-16000. https://doi.org/10.1039/C5CC06847A

[49]   Y. Luo, S. Zhang, H. Pan, S. Xiao, Z. Guo, L. Tang, U. Khan, B.-F. Ding, M. Li, Z. Cai, Y. Zhao, W. Lv, Q. Feng, X. Zou, J. Lin, H.-M. Cheng, B. Liu, Unsaturated single atoms on monolayer transition metal dichalcogenides for ultrafast hydrogen evolution, ACS Nano, 14 (2020) 767-776. https://doi.org/10.1021/acsnano.9b07763

[50]   T. Yang, T.T. Song, J. Zhou, S. Wang, D. Chi, L. Shen, M. Yang, Y.P. Feng, High-throughput screening of transition metal single atom catalysts anchored on molybdenum disulfide for nitrogen fixation, Nano Energy, 68 (2020) 104304. https://doi.org/10.1016/j.nanoen.2019.104304

[51]   H. Zhang, L. Yu, T. Chen, W. Zhou, X.W. Lou, Surface modulation of hierarchical MoS_2 nanosheets by ni single atoms for enhanced electrocatalytic hydrogen evolution, Adv. Funct. Mater., 28 (2018) 1807086. https://doi.org/10.1002/adfm.201807086

[52]   K. Qi, X. Cui, L. Gu, S. Yu, X. Fan, M. Luo, S. Xu, N. Li, L. Zheng, Q. Zhang, J. Ma, Y. Gong, F. Lv, K. Wang, H. Huang, W. Zhang, S. Guo, W. Zheng, P. Liu, Single-atom cobalt array bound to distorted 1T MoS_2 with ensemble effect for hydrogen evolution catalysis, Nat. Commun., 10 (2019) 5231. https://doi.org/10.1038/s41467-019-12997-7

[53]   J. Zhu, L. Cai, X. Yin, Z. Wang, L. Zhang, H. Ma, Y. Ke, Y. Du, S. Xi, A.T.S. Wee, Y. Chai, W. Zhang, Enhanced electrocatalytic hydrogen evolution activity in single-atom Pt-decorated VS_2 nanosheets, ACS Nano, 14 (2020) 5600-5608. https://doi.org/10.1021/acsnano.9b10048

[54]   F. Li, G.-F. Han, Y. Bu, S. Chen, I. Ahmad, H.Y. Jeong, Z. Fu, Y. Lu, J.-B. Baek, Unveiling the critical role of active site interaction in single atom catalyst towards hydrogen evolution catalysis, Nano Energy, 93 (2022) 106819. https://doi.org/10.1016/j.nanoen.2021.106819

[55]   L. Cao, Q. Luo, W. Liu, Y. Lin, X. Liu, Y. Cao, W. Zhang, Y. Wu, J. Yang, T. Yao, S. Wei, Identification of single-atom active sites in carbon-based cobalt catalysts during electrocatalytic hydrogen evolution, Nat. Catal., 2 (2019) 134-141. https://doi.org/10.1038/s41929-018-0203-5

[56]   J. Zhang, Y. Zhao, X. Guo, C. Chen, C.-L. Dong, R.-S. Liu, C.-P. Han, Y. Li, Y. Gogotsi, G. Wang, Single platinum atoms immobilized on an MXene as an efficient catalyst for the hydrogen evolution reaction, Nat. Catal., 1 (2018) 985-992. https://doi.org/10.1038/s41929-018-0195-1

[57]   D. Liu, X. Li, S. Chen, H. Yan, C. Wang, C. Wu, Y.A. Haleem, S. Duan, J. Lu, B. Ge, P.M. Ajayan, Y. Luo, J. Jiang, L. Song, Atomically dispersed platinum supported on curved carbon supports for efficient electrocatalytic hydrogen evolution, Nat. Energy, 4 (2019) 512-518. https://doi.org/10.1038/s41560-019-0402-6

[58]   T. Sun, W. Zang, H. Yan, J. Li, Z. Zhang, Y. Bu, W. Chen, J. Wang, J. Lu, C. Su, Engineering the coordination environment of single cobalt atoms for efficient oxygen reduction and hydrogen evolution reactions, ACS Catal., 11 (2021) 4498-4509. https://doi.org/10.1021/acscatal.0c05577

[59]   X.-P. Yin, H.-J. Wang, S.-F. Tang, X.-L. Lu, M. Shu, R. Si, T.-B. Lu, Engineering the coordination environment of single-atom platinum anchored on graphdiyne for optimizing electrocatalytic hydrogen evolution, Angew. Chem. Int. Ed., 57 (2018) 9382-9386. https://doi.org/10.1002/anie.201804817

[60]   F. Lyu, S. Zeng, Z. Jia, F.-X. Ma, L. Sun, L. Cheng, J. Pan, Y. Bao, Z. Mao, Y. Bu, Y.Y. Li, J. Lu, Two-dimensional mineral hydrogel-derived single atoms-anchored heterostructures for ultrastable hydrogen evolution, Nat. Commun., 13 (2022) 6249. https://doi.org/10.1038/s41467-022-33725-8

[61]   D. Cao, H. Xu, H. Li, C. Feng, J. Zeng, D. Cheng, Volcano-type relationship between oxidation states and catalytic activity of single-atom catalysts towards hydrogen evolution, Nat. Commun., 13 (2022) 5843. https://doi.org/10.1038/s41467-022-33589-y

Electrocatalytic Hydrogen Production: Catalysts and Applications   Materials Research Forum LLC
Materials Research Foundations **165** (2024)   https://doi.org/10.21741/9781644903070

CHAPTER 5

Transition Metal Compound Catalysts for HER

Xiang Peng*

Hubei Key Laboratory of Plasma Chemistry and Advanced Materials, Engineering Research Center of Phosphorus Resources Development and Utilization of Ministry of Education, School of Materials Science and Engineering, Wuhan Institute of Technology, Wuhan 430205, China

xpeng@wit.edu.cn

Abstract

The research on hydrogen evolution catalysts has primarily focused on inorganic catalysts, yielding significant results. Subsequently, the exploration of non-precious metal catalysts for hydrogen evolution reaction has experienced explosive growth. This category includes transition metal chalcogenides, phosphites, carbides, nitrides, oxides, and others. Although there still exists a performance gap between non-precious metal catalysts and the precious metal counterparts, the optimization of synthesis-structure-performance correlations represents a crucial aspect of non-precious metal hydrogen evolution catalyst research.

Keywords

Transition Metal Compounds, Hydrogen Evolution Reaction, Water Splitting, Catalyst, Hydrogen Production

1. Introduction

Hydrogen energy possesses high calorific value and exhibits strong environmental friendliness, positioning it as one of the most promising energy sources in the future energy and industry system. The electrocatalytic hydrogen evolution reaction (HER), serving as a green and sustainable approach to hydrogen production, has garnered extensive research attention in recent years [1]. Currently, the primary challenge in this field lies in the development of hydrogen evolution catalysts that are highly efficient, cost-effective, and demonstrate exceptional performance. While Pt-based materials stand as the leading HER catalysts, Pt/C is commonly employed as the benchmark. Nevertheless, the limited natural availability and high cost of Pt hinder its widespread application [2]. On the other hand, non-precious metal-based electrocatalysts, abundant in the Earth's crust, inexpensive, and exhibiting relatively low overpotential, have been extensively investigated as electrocatalytic materials for HER.

In recent years, research on hydrogen evolution catalysts has primarily focused on inorganic catalysts, yielding significant results. Among these materials, MoS_2 has emerged as one of the earliest and most prominent inorganic nano-hydrogen evolution catalysts. Subsequently, the exploration of non-precious metal catalysts for HER has experienced explosive growth. This category includes transition metal chalcogenides [3, 4], phosphites [5], carbides [2, 6, 7], nitrides

[8-11], oxides [12-15], and others. Although there still exists a performance gap between non-precious metal catalysts and the precious metal counterparts, the optimization of synthesis-structure-performance correlations represents a crucial aspect of non-precious metal hydrogen evolution catalyst research. In this context, this chapter focuses on the structure-activity relationships of transition metal carbides, nitrides, phosphites, oxides, sulfides, and selenides, thus paving the way for achieving a clean and sustainable energy supply.

2. Transition Metal Carbides

2.1. Structure and Properties of Transition Metal Carbides

In recent years, there has been a significant advancement in the field of non-precious metal compounds used as hydrogen evolution catalysts. Among these compounds, transition metal carbides (TMCs) have garnered substantial attention from researchers due to their remarkable characteristics such as high electrical conductivity, pH tolerance, activity, and stability [16]. Generally, a wide range of transition metals can form various carbides. **Figure 1** illustrates the most stable carbides formed by transition metals along with their corresponding stoichiometric ratios [17]. These carbides exhibit three primary crystal structures: face-centered cubic (fcc), hexagonal close-packed (hcp), and simple hexagonal (hex) [18].

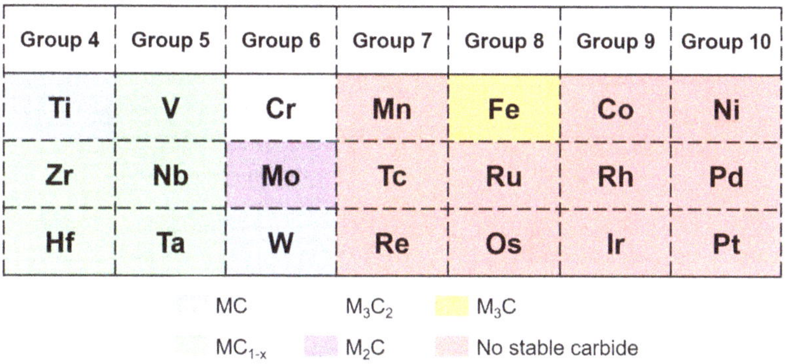

Group 4	Group 5	Group 6	Group 7	Group 8	Group 9	Group 10
Ti	V	Cr	Mn	Fe	Co	Ni
Zr	Nb	Mo	Tc	Ru	Rh	Pd
Hf	Ta	W	Re	Os	Ir	Pt

MC M_3C_2 M_3C

MC_{1-x} M_2C No stable carbide

Figure 1. Typical transition metals for the TMCs. Reproduced from ref [17]. Copyright 2013, American Chemical Society.

The presence of smaller carbon atoms occupying the interstitial positions within the metal lattice leads to a structural alteration in the parent metal, resulting in distinctive properties. According to the Engel-Brewer metal theory [17], transition metal carbides are interstitial compounds that form through the interaction between the *s-p* orbitals of carbon atoms and the *s-p-d* orbitals of metal atoms. The introduction of carbon atoms into the lattice increases the bond length between metal atoms, causing electron transfer from the metal atoms to the carbon atoms. This modification in the *d*-band electron density of the metal at the Fermi energy level imparts unique adsorption/desorption properties to TMCs as compared to pure metal.

2.2.   Transition Metal Carbides for HER

Molybdenum carbide (MoC_x) is a highly investigated material in the field of carbide electrocatalytic hydrogen evolution catalysts. Qiang et al. [2] presented a novel approach for producing high-quality nanostructured Mo_2C utilizing a carbon template through a salt-assisted chemical vapor deposition (CVD) process. One-dimensional whisker carbon nanotube (WCNT) powders are used as the carbon source and template, while a combination of MoO_3 and NaCl forms the metal source. X-ray diffraction (XRD) patterns of the original pure carbon nanotubes and the Mo_2C-900 sample show a broad diffraction peak at approximately 26°, associated with graphitic carbon, as indicated in **Figure 2a**. However, upon subjecting the WCNTs to a thermal reaction at 900 °C in the presence of MoO_2Cl_2, the carbon peak is replaced by a series of peaks at 34.4°, 37.9°, 39.4°, 52.1°, 61.5°, 69.6°, 74.6°, and 75.5°, corresponding to the (100), (002), (101), (102), (110), (103), (112), and (201) facets of hexagonally close-packed Mo_2C (JCPDS card No. 35–0787), indicating the formation of crystalline Mo_2C.

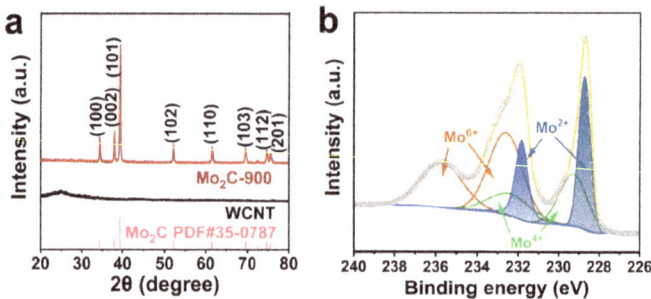

Figure 2. *(a) XRD patterns of the Mo_2C-900 and WCNT. (b) High-resolution XPS of the Mo-3d spectrum of Mo_2C-900. Reproduced with permission from ref [2]. Copyright 2022, Elsevier.*

Figure 2b displays the high-resolution X-ray photoelectron spectroscopy (XPS) of the Mo-3d spectrum, which can be deconvoluted into three double peaks. The peaks observed at 228.7/231.8 eV can be attributed to the Mo^{2+} species, indicating the presence of Mo_2C species, consistent with previous studies [19]. The presence of double peaks at 229.3/232.4 eV and 232.7/235.8 eV can be attributed to the surface oxidation of Mo_2C upon exposure to air [20, 21]. These peaks specifically correspond to Mo^{4+} (MoO_2) and Mo^{6+} (MoO_3) species, respectively.

The linear sweep voltammetry (LSV) curves demonstrate that Mo_2C-900 exhibits a significantly smaller overpotential of 115 mV to generate a current density of 10 mA cm^{-2}, outperforming Mo_2C-800 (164 mV), as shown in **Figure 3a-b**. The Tafel slope of Mo_2C-900 is 58 mV dec^{-1}, lower than that of Mo_2C-800 (63 mV dec^{-1}) and bare WCNT (206 mV dec^{-1}), suggesting the Volmer-Heyrovsky HER mechanism of the as-prepared Mo_2C catalyst [22].

The low-temperature magnesium thermic reaction (MTR) is commonly employed in the synthesis of carbides, such as silicon carbide (SiC), as it eliminates the need for high temperatures and expensive carbonaceous gases [23, 24]. In the work by Peng et al. [7], a hierarchical nanosheet structure called vanadium carbide nanosheet (VC-NS) was successfully produced. This structure consists of isolated vanadium carbide (VC) nanoparticles enclosed within a highly conductive mesoporous graphitic carbon network.

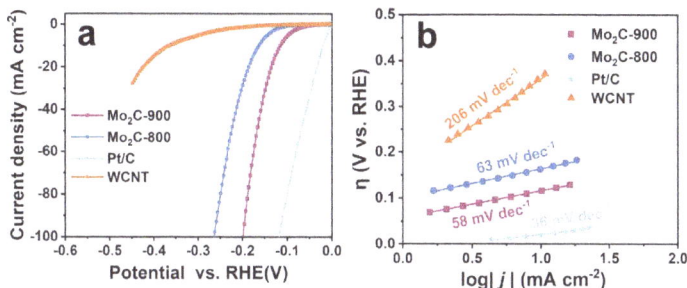

Figure 3. (a) LSV curves and (b) Tafel plots of the prepared Mo_2C catalysts. Reproduced with permission from ref [2]. Copyright 2022, Elsevier.

The synthesis process involved a hydrothermal reaction followed by a low-temperature MTR, as schematically illustrated in **Figure 4**. The VC nanosheets were synthesized using the as-prepared $V_2O_5 \cdot nH_2O$ nanosheets (V_2O_5-NS) through a low-temperature MTR process. A mixture containing 182 mg of V_2O_5-NS, 336 mg of sodium bicarbonate ($NaHCO_3$), and 300 mg of magnesium powders was combined and enclosed within a stainless steel sealed container. The container was then heated to subjected 700 °C for 5 hours. During the MTR process, Mg reacts with V_2O_5, reducing it to metallic vanadium with enhanced reactivity. Simultaneously, carbon dioxide (CO_2), produced from the thermal degradation of $NaHCO_3$, undergoes carbon conversion. The deposited carbon forms a conductive network on the nanosheet surface, utilizing the V_2O_5 nanosheets as a template. The highly active metallic vanadium then reacts with carbon to form VC.

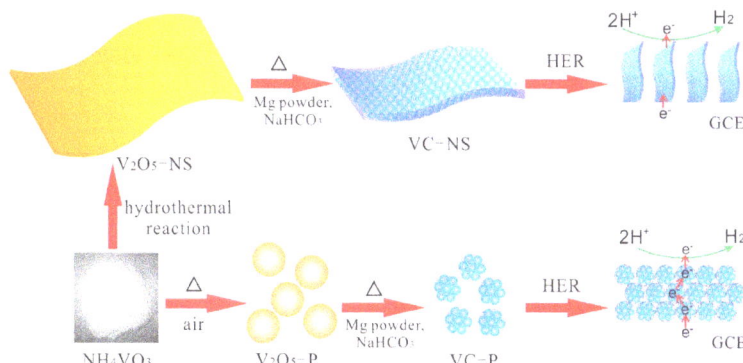

Figure 4. Schematically illustration of the preparation of the VC/C catalyst. Reproduced with permission from ref [7]. Copyright 2016, Elsevier.

The XRD patterns provide information about the crystal structure and composition of the samples, as shown in **Figure 5a**. The observed diffraction peaks at 37.4°, 43.4°, 63.1°, 75.7°, and 79.7° correspond to the lattice planes (111), (200), (220), (311), and (222) of the cubic VC crystal (JCPDS card No. 73-0476). These results suggest the successful conversion of V_2O_5 to VC. VC-NS exhibits a relatively low-intensity peak at $2\theta=26°$, indicating the presence of carbon species in the final products. Transmission electron microscopy (TEM) is employed to investigate the distribution of carbon and VC within the composite material. TEM images reveal a significant quantity of nanoparticles ranging in diameter from 10 to 30 nm, enclosed within the carbon network, forming a hierarchical nanosheet architecture, as indicated in **Figure 5b**. Lattice fringes observed in the high-resolution TEM (HR-TEM) image in **Figure 5c** exhibit spacings of 0.24 nm and 0.34 nm, corresponding to the (111) planes of VC and graphitic carbon, respectively.

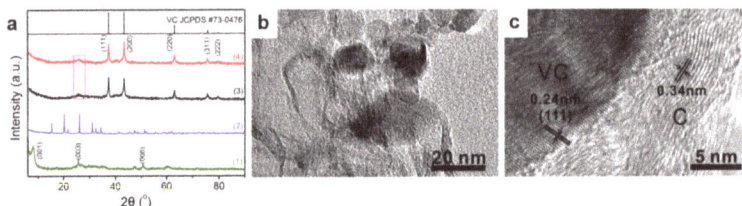

Figure 5. *(a) XRD, (b) TEM, and (c) HR-TEM of the VC/C catalyst. Reproduced with permission from ref [7]. Copyright 2016, Elsevier.*

The utilization of graphitic carbon as a conductive substrate for VC nanoparticles results in the formation of hierarchical nanosheets. These nanosheets possess a large exposed specific surface area, facilitating efficient electron transport and enhanced electrical interaction with the active sites. To produce a current density of 10 mA cm^{-2}, the hierarchical VC-NS catalyst exhibits an overpotential of just 98 mV and a small Tafel slope of 56 mV dec^{-1}, as shown in **Figure 6a-b**. The exceptional stability of the catalyst is demonstrated by a variation of only 10 mV in overpotential after 10,000 cyclic voltammetry (CV) cycles at a current density of 80 mA cm^{-2}, as indicated in **Figure 6c**.

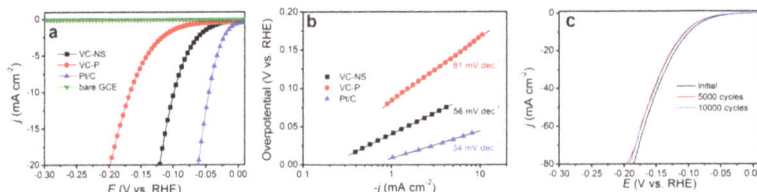

Figure 6. *(a) LSV curves, (b) Tafel plots, and (c) Stability test of the catalysts. Reproduced with permission from ref [7]. Copyright 2016, Elsevier.*

3. Transition Metal Nitrides

3.1. Structure and Physical Properties of Transition Metal Nitrides

Elements belonging to the transition metal nitrides (TMNs) in groups IVB to VIB exhibit distinctive physical and chemical characteristics, often referred to as "interstitial alloys." These alloys are formed through the incorporation of nitrogen atoms into the interstitial positions of the parent metals [25, 26]. This results in the densely packed or nearly densely packed arrangement of metal atoms in TMNs, which contributes to their desirable electrical conductivity [27, 28]. **Figure** 7 illustrates the typical crystal structures observed in TMNs, which include face-centered cubic (fcc), hexagonally closed packed (hcp), and simple hexagonal (hex) structures [18]. These structures are formed when the small nitrogen atoms occupy interstitial positions within the crystal lattice.

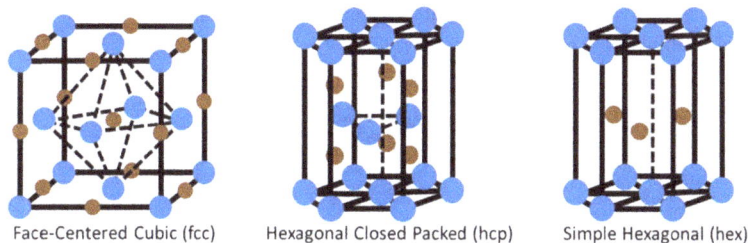

Face-Centered Cubic (fcc)   Hexagonal Closed Packed (hcp)   Simple Hexagonal (hex)

Figure 7. Common crystal structures of TMNs. The blue balls represent transition metal atoms and the brown balls represent nitrogen atoms. Reproduced with permission from ref [18]. Copyright 2016, Wiley-VCH.

TMNs exhibit three distinct types of metal-nitrogen (M-N) bonding topologies: covalent, ionic, and metallic bonds [29]. Covalent bonding contributes to increased hardness, brittleness, and improved stress tolerance in TMNs. On the other hand, the ionic bonding between nitrogen atoms and metals in TMNs leads to a contraction of the metals' d-band, resulting in a transition metal electronic structure similar to noble metals such as Pd and Pt. This similarity imparts exceptional electrocatalytic performance to TMNs [30]. The presence of metallic bonding in transition metal nanoparticles facilitates efficient electron transport and provides excellent resistance to corrosion.

TMNs typically exhibit low electrical resistivity. For example, TiN (27 $\mu\Omega$ cm), VN (65 $\mu\Omega$ cm), NbN (60 $\mu\Omega$ cm), and ZrN (24 $\mu\Omega$ cm) [29]. The electrical resistivity of TMNs is affected by the ratio of non-metal to metal. Specifically, the resistivity decreases as the non-metal content increases at ambient temperature [29]. Superconductivity is particularly prevalent in all face-centered cubic transition metal nitrides of groups IVB and VB [29]. Furthermore, TMNs exhibit exceptional chemical stability and resistance to corrosion when exposed to diluted acidic or alkaline solutions. These features enhance their reliability in electrochemical processes compared to the respective metals or metal alloys.

Materials Research Forum LLC
https://doi.org/10.21741/9781644903070

3.2. Transition Metal Nitrides for HER

TMNs, such as Mo_2N and CoN, have emerged as promising catalysts for the HER [31, 32]. Previous research has shown that the HER activity can be further enhanced by using bimetallic TMNs. Incorporating a second metal atom in bimetallic TMNs not only increases the number of active sites for reactions but also improves electronic conductivity, surpassing that of monometallic TMNs [24, 33]. One example of bimetallic TMNs is nickel molybdenum nitrides (Ni–Mo–N), which exhibit significant HER activity. Park et al. [34] synthesized 7 nm Ni_2Mo_3N nanoparticles on nickel foam (Ni_2Mo_3N/NF) using a cost-effective and straightforward process. The synthesis process as shown **Figure 8** involved annealing Ni foam, $MoCl_5$, and urea in a single step, eliminating the need for a separate Ni precursor. The Ni foam and Mo-urea complex were co-located in an alumina boat and annealed at 600 °C for 3 hours under a flow of nitrogen gas. This process resulted in the formation of Ni_2Mo_3N nanoparticles supported on the nickel foam. The synthesis technique is straightforward, cost-effective, and eco-friendly, as it avoids the use of harmful NH_3 gas.

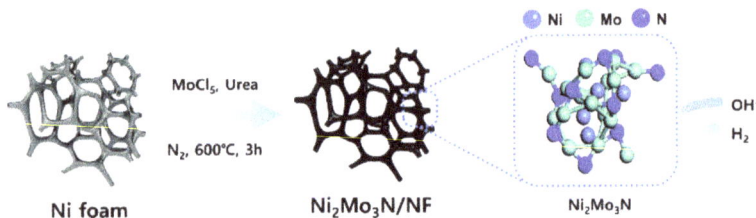

Figure 8. *Schematic illustration of the synthetic method for Ni_2Mo_3N/NF. Reproduced with permission from ref [34]. Copyright 2021, the Royal Society of Chemistry.*

The XRD patterns of Ni_2Mo_3N/NF revealed prominent peaks at 45°, 52°, and 76°, corresponding to cubic Ni (JCPDS card No. 00-004-0850), while the remaining peaks were associated with the cubic Ni_2Mo_3N (JCPDS card No. 01-089-4564) phase. The absence of impurity peaks, such as metal oxides or monometallic nitrides, suggested the successful growth of phase-pure Ni_2Mo_3N on the Ni foam. The XPS analysis showed three oxidation states of Mo: Mo^0 (228.3 and 231.5 eV), Mo^{3+} (228.9 and 232.6 eV), and Mo^{6+} (233.1 and 235.3 eV), with distinctive binding energies, as displayed in **Figure 9a-b**. The Mo^{3+} species indicated the production of metal nitride, while the presence of Mo^{6+} species suggested the development of surface oxide. The N-1s spectrum exhibited peaks at 397.3 eV (attributed to nitrogen species within the metal nitride lattice) and 398.9 eV (corresponding to the N-H group). The Mo-3p and N-1s peaks coincided due to partial overlap.

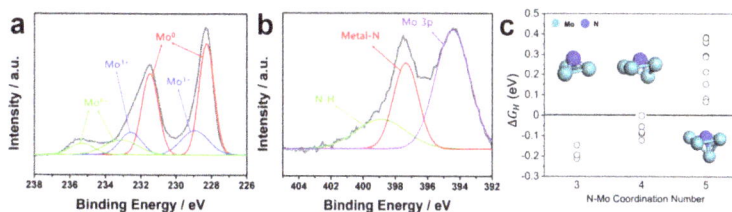

Figure 9. *High-resolution XPS of (a) Mo-3d and (b) N-1s spectra of Ni₂Mo₃N/NF. (c) Overall H-adsorption strength with respect to coordination number of N-Mo bonding. Reproduced with permission from ref [34]. Copyright 2021, the Royal Society of Chemistry.*

Density functional theory (DFT) calculations were performed to investigate the atomic-level source of the HER activity in the Ni₂Mo₃N/NF catalyst. The active site was anticipated to be located at the N atom rather than Ni or Mo based on the significant HER activity observed. ΔG_H values were computed for six distinct N sites over three surface models, resulting in a total of 18 N sites. The H-adsorption energies for the N sites ranged from −0.21 to 0.38 eV, covering the thermal neutral condition required for high HER performance.   The examination of the local configuration of the N active sites revealed a correlation between the N-Mo coordination number and ΔG_H, as indicated in **Figure 9c**. As the coordination number increased, the nitrogen atom became more immersed in the molybdenum network, leading to stabilization and a decrease in the strength of the hydrogen adsorption bond.   The Ni₂Mo₃N/NF catalyst demonstrated exceptional performance in the HER under alkaline media, displaying strong activity with low overpotential.   It achieved the current densities of 10 mA cm⁻² and 100 mA cm⁻² at the overpotentials of 21.3 and 123.8 mV, respectively.   Furthermore, it displayed outstanding stability, making it one of the top-performing catalysts among the latest transition metal nitride-based catalysts.

Although TMNs have been extensively studied, achieving both sufficient activity and corrosion resistance remains a challenge.   Therefore, advanced modification techniques are needed.   One such technique involves adjusting the stoichiometry of TMNs. Jin et al. [35] developed a nickel surface nitride enclosed in a carbon shell (Ni-SN@C) using an unsaturated nitriding method. The Ni-SN@C and carbon-coated nickel nanoparticles (Ni@C) were prepared by heating a complex of Ni and ethylenediaminetetraacetic acid (EDTA) in either NH₃ or argon atmosphere at 500 °C for 2 hours.   Unlike normal TMNs or metal/metal nitride heterostructures, the unsaturated Ni-SN@C did not exhibit a bulk nickel nitride phase. Instead, it primarily consisted of metallic nickel with distinct unsaturated surface nickel-nitrogen bonding within the nitrogen-doped carbon shell (**Figure 10a**). Advanced synchrotron-based spectroscopy confirmed the presence of unsaturated Ni-N bonding and charge redistribution on the catalyst's surface.

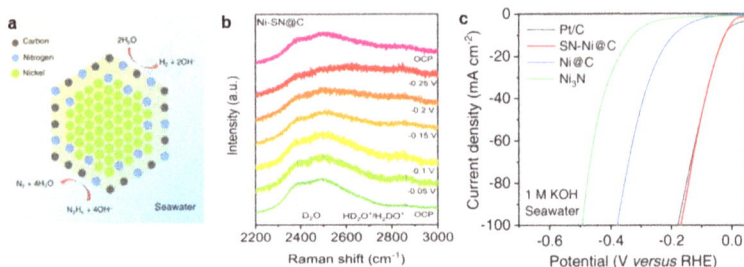

***Figure 10.** (a) Proposed structure of the Ni-SN@C catalyst. (b) In situ Raman spectra in D_2O electrolyte of Ni-SN@C under different operating potentials. (c) Electrochemical activity evaluation in alkaline seawater. Reproduced with permission from ref [35]. Copyright 2021, Wiley-VCH.*

The activity source of Ni-SN@C was investigated using in situ Raman spectroscopy with isotopically tagged electrolytes (**Figure 10b**). The unsaturated Ni-SN@C catalyst demonstrates platinum-like characteristics in HER by facilitating the generation of hydronium ions under alkaline conditions, effectively reducing the energy barrier for the overall process. This innovative approach not only combines the desirable properties of metallic Ni and nickel nitrides in a single catalyst but also induces a redistribution of electrical charge on the catalyst's surface. This charge redistribution enhances the catalyst's corrosion resistance and allows it to exhibit HER activity comparable to that of platinum, particularly in seawater electrolysis. The catalyst achieves a current density of 10 mA cm^{-2} at an overpotential of a mere 23 mV and demonstrates exceptional durability for the HER in alkaline seawater (**Figure 10c**).

4.   Transition Metal Phosphides

4.1.   Structure and Physical Properties of Transition Metal Phosphides

Transition metal phosphides (TMPs) exhibit a unique crystal structure characterized by trigonal prisms comprising transition metal atoms [36], with phosphorus (P) atoms positioned within the interior, as depicted in **Figure 11** [37].   These prisms arrange themselves in diverse configurations, resulting in a variety of crystal structures. The incorporation of P atoms optimizes the properties of TMPs in several ways. Firstly, the metallic phosphide structure, based on trigonal prisms, effectively accommodates P atoms due to their larger atomic radii [38]. Secondly, the introduction of phosphorus into the metal lattice preserves the electronic structure of the metal, minimizing the free energy of adsorbed hydrogen, and thereby favoring hydrogen desorption. Thirdly, the doping of P atoms induces modifications in both the d-band and Fermi energy levels of transition metals, enhancing the adsorption of intermediate active species and improving the electrocatalytic performance. Furthermore, the electronegativity of P atoms serves as an adsorption center for protons, while the transition metals act as active hydride centers, synergistically enhancing reactivity [39, 40]. However, the electronegativity of P atoms also attracts electrons from the surrounding transition metals, affecting the overall electrical conductivity of TMPs. Therefore, precise control over the doping level of P is necessary.

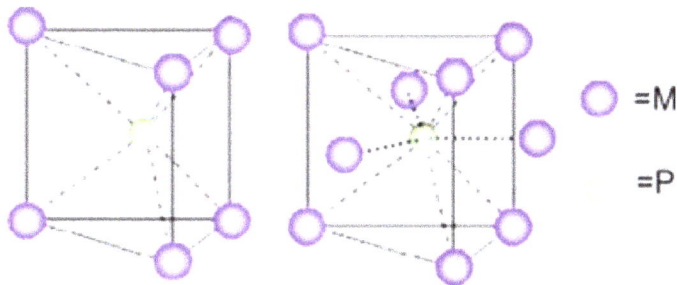

Figure 11. *Triangular prism and tetrakaidecahedral structures in phosphides. Reproduced with permission from ref [37]. Copyright 2021, Wiley-VCH.*

4.2. Transition Metal Phosphides for HER

TMPs have emerged as highly promising materials in the field of energy conversion and storage, owing to their exceptional electrocatalytic performance, cost-effectiveness, and distinctive physicochemical properties. These compounds are denoted by the formula M_xP_y, where the introduction of highly electronegative P atoms has a profound impact on the electron delocalization of metal atoms. This process extracts electrons from neighboring transition metals, leading to alterations in the electronic structure of the metal and significant adjustments in the intrinsic activity of its active sites [41, 42]. In addition, TMPs offer controllable compositional elements and stoichiometric ratios, enabling infinite composition adjustments to modify their electronic structure, as well as physical and chemical properties [43]. During reactions, the P and metal sites in TMPs serve as proton and hydride acceptor sites, respectively. The negatively charged P atoms not only trap protons as bases but also enhance moderate bonding between reaction intermediates and the catalyst surface [44].

Currently, TMPs are predominantly based on Fe, Ni, Co, Cu, Mo, and W [45]. To further enhance their catalytic performance, mainstream modification methods include adjusting the stoichiometric ratios of metal and phosphorus atoms, alloying or doping, interfacial engineering, vacancy and defect engineering, and morphology control [46]. For instance, Dong et al. [46, 47] employed a controlled strategy to prepare Ni_3P and $Ni_{12}P_5$ for modifying $CdS@Ni_3S_2$ electrodes, with experiment and density functional theory calculation results indicating the superior effectiveness of Ni_3P modification with a high P/M ratio. As shown in **Figure 12a**, $CdS@Ni_3S_2/Ni_3P$ requires an overpotential of 13 mV to generate the current density of 10 mA cm^{-2}, which is smaller than that of the $CdS@Ni_3S_2/Ni_{12}P_5$ catalyst. Moreover, compared with Ni_3S_2 and $Ni_3S_2/Ni_{12}P_5$, the significantly increased $|\Delta G_{H*}|$ value of 0.772 eV for the Ni_3S_2/Ni_3P system proves the excellent HER property of $CdS@Ni_3S_2/Ni_3P$ electrode (**Figure 12b**). Recent research has also highlighted the potential of polymetallic phosphides as core catalysts due to strong synergistic interactions between metal atoms, providing new impetus for the future advancement of TMPs catalysts.

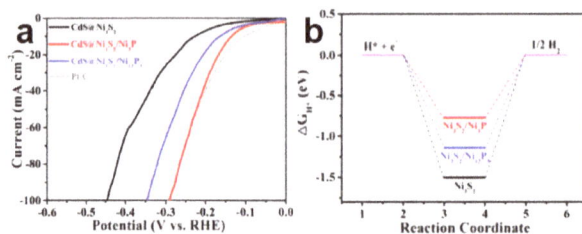

Figure 12. *(a) HER polarization curves, (b) Calculated free-energy diagram of HER.*
Reproduced with permission from ref [47]. Copyright 2021, Wiley-VCH.

Recent studies have demonstrated that morphology-controlled catalysts can enhance the catalytic activity of various metal nanoparticles, including metal phosphides [48-56]. For example, Pan et al. [49] observed that highly porous CoP polyhedra with large surface areas exhibit enhanced HER activities due to the increased number of available catalytic active sites. The hollow polyhedron composed of carbon nanotubes remained and showed high porosity (**Figure 13a**). Moreover, in acidic and alkaline environments, the overpotential of the porous CoP polyhedron to produce the current density of 10 mA cm^{-2} is 140 and 115 mV, respectively. Similarly, under alkaline conditions, Yu et al. [53] found that hollow CoP catalysts loaded on N-doped graphene (CoP$_h$/NG) outperformed solid CoPs/NG in terms of HER electrolytic activities. TEM images (**Figure 13b**) show that the resulting CoP$_h$/NG mixture appears in a flaky form, similar to the original graphene. The nearly monodisperse hollow CoP nanoparticles are fixed on the graphene sheet (**Figure 13c**). The hollow CoP nanoparticles are uniformly dispersed without significant aggregation, which facilitates maximum exposure of the active site to the electrolyte. As a result, the CoP$_h$/NG hybrid exhibits excellent HER activity with an initial potential of 50 mV and an overpotential of only 83 mV at 10 mA cm^{-2}. In addition, CoP$_h$/NG hybrids only require overpotentials of 100 and 140 mV to achieve current densities of 20 and 100 mA cm^{-2}, respectively.

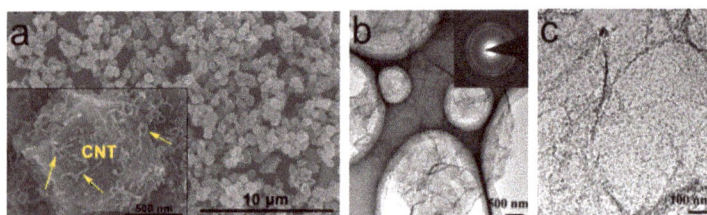

Figure 13. *(a) SEM of the CoP/NCNHP catalyst. Inset in (a): magnified SEM image.*
Reproduced with permission from ref [49]. Copyright 2018, American Chemical Society. (b-c)
TEM of the CoP$_h$/NG hybrids composite. Reproduced with permission from ref [53]. Copyright
2016, the Royal Society of Chemistry.

Materials Research Forum LLC
https://doi.org/10.21741/9781644903070

Recent work by Feng et al. [56] demonstrated that Ni_3S_2 catalysts wrapped by a high-index (210) surface exhibit exceptional activity and stability, attributed to the promotion of reactive intermediate adsorption on these facets. For comparison, the catalytic activities of NF, Ni_3S_2-NPs and Pt/C (20 wt%) were also measured. Ni_3S_2/NF showed significant catalytic activity against HER (**Figure 14a**), much higher than Ni_3S_2 NPs. In fact, Ni_3S_2/NF can produce a current density of 10 mA cm^{-2} at overpotential (η) to 170 mV, and Ni_3S_2 NPs can produce the same current density at higher η (310 mV). Additionally, nanostructures of noble metal phosphides with excellent properties have been investigated [57-62]. For example, Duan et al. [62] discovered that Rh_2P nanocubes display remarkably high HER activity due to P-rich surfaces and defects on the catalyst surface. The polarization curves in **Figure 14b** show increasing HER activities in the order Rh/C < Pt/C < Rh_2P/C, which matches the one established for well-defined thin-film surfaces. Wang et al. [61] reported that w-Rh_2P nanosheets enhance HER performance across a wide pH range. The HER performance of w-Rh_2P NS/C, Pt/C, and Rh NS/C was investigated in 0.1 M $HClO_4$ as well. As shown in **Figure 14c**, the w-Rh_2P NS/C displays the best HER catalytic activity among all three samples in 0.1 M $HClO_4$. Thus, the morphology of nanostructures also plays a crucial role in controlling the catalytic activity of noble metal sulfides and phosphides during electrolysis.

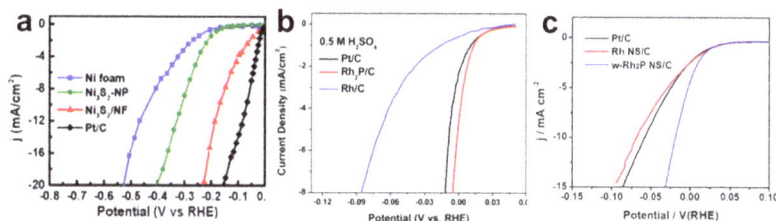

Figure 14. (a) Steady-state current density as a function of applied voltage during HER at pH=7 over NF, Ni_3S_2 nanoparticles, Ni_3S_2/NF, and Pt/C (20 wt%). Reproduced with permission from ref [56]. Copyright 2015, American Chemical Society. (b)Polarization curves for Pt/C, Rh_2P/C, and Rh/C recorded at 5 mV s^{-1} in 0.5 M H_2SO_4. Reproduced with permission from ref [62]. Copyright 2017, American Chemical Society. (c) HER polarization curves of Pt/C, Rh NS/C, and w-Rh_2P NS/C in 0.1 M $HClO_4$. Reproduced with permission from ref [61]. Copyright 2018, Wiley-VCH.

5. Transition Metal Oxides

Due to their unique electronic structure, transition metal oxides (TMOs) possess unfilled 3d orbitals and unstable unpaired electrons, resulting in a range of metastable valence states, which is beneficial for electrochemical processes. Huang et al. [63] employed a solvothermal technique to produce zirconium-controlled $CoFe_2O_4$ (CoFeZr oxides) nanosheets on a three-dimensional nanostructure (3D NF). These nanosheets exhibit remarkable catalytic activity for both OER and HER. Nanometer-thick layers of CoFeZr oxides were uniformly formed on a nanofiber substrate.

The XRD patterns of CoFe and CoFeZr oxides, as depicted in **Figure 15a**, revealed peaks at 18.3°, 30.1°, 35.4°, 43.1°, 57.0°, and 62.6° in the CoFe oxide pattern, corresponding precisely to the (111), (220), (311), (400), (511), and (440) planes, respectively, of the $CoFe_2O_4$ spinel structure (JCPDS card No. 22-1086). The XRD results of CoFeZr oxides exhibited consistency with CoFe oxides. Even with varying levels of Zr control, the crystal structure of CoFe oxides remained intact (**Figure 15b**), indicating that the incorporation of Zr did not affect the overall structure. However, due to the difference in ionic radii, with Zr^{4+} having a larger radius (72 pm) compared to Fe^{3+} (55 pm), the replacement of Fe with Zr could potentially lead to lattice distortion. This lattice deformation impedes the crystalline growth of nanoparticles and results in the formation of nanosheet structures.

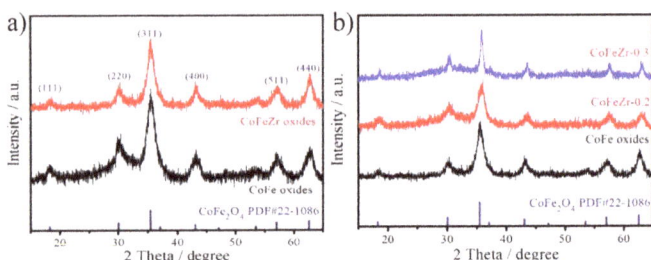

Figure 15. *(a) XRD pattern of CoFe oxides and CoFeZr oxides; (b) XRD patterns of CoFeZr-0.2 and CoFeZr-0.3. Reproduced with permission from ref [63]. Copyright 2019, Wiley-VCH.*

The impact of Zr incorporation on the electronic structure was examined by calculating the density of states. The computed density of states revealed that pure CoFe oxides have a semiconducting band gap of approximately 0.6 eV, as shown in **Figure 16a**. However, the substitution of Zr reduced the band gap due to the presence of an impurity state near the fermi level (**Figure 16b**). Consequently, CoFe oxides regulated by Zr exhibited enhanced electrical conductivity and accelerated electron transport compared to pure CoFe oxides, which can significantly influence their electrochemical performance. The addition of Zr to $CoFe_2O_4$ modified the nanosheet structure, increasing the number of active sites, altering the chemical composition, and modifying the electronic structure, thereby improving the intrinsic activity of these active sites. The 3D CoFeZr oxide nanosheets demonstrated exceptional HER activity with a small overpotential of 104 mV at 10 mA cm^{-2} in alkaline media, as indicated in **Figure 16c**.

However, the observed reactivity of TMOs for HER is considerably lower than that of the most advanced Pt/C catalysts currently available. This is primarily due to the intrinsic limited electrical conductivity of transition metal oxides [64]. To enhance the electrocatalytic activity, heterogeneous atom doping has proven to be a highly successful approach. Ling et al. [65] demonstrated the precise manipulation of the surface and internal electrical properties of CoO nanorods (NRs) through dual-doping with Ni and Zn for HER. A simple and scalable cation exchange technique using ZnO nanorods as sacrificial templates was employed to directly grow Ni, Zn dual-doped Ni, and Zn CoO NRs on conductive carbon fiber paper (CFP). The Ni, Zn dual-doped CoO NRs were formed by partially substituting Zn^{2+} ions with Co^{2+} and Ni^{2+} ions.

By controlling the mass ratio of Ni^{2+} and Co^{2+} precursors and modifying the temperature during the cation exchange process, the stoichiometry could be easily regulated without altering the nanorod shape and microstructure.

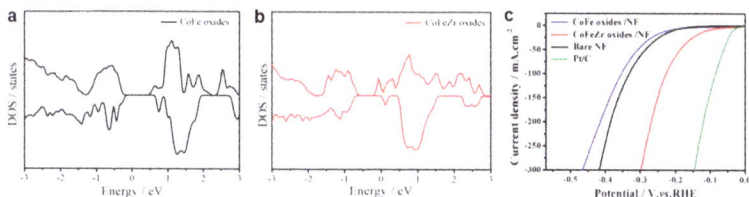

Figure 16. *Computed density of states for pure CoFe oxides (a) and CoFeZr oxides (b). (c) The polarization curves. Reproduced with permission from ref [63]. Copyright 2019, Wiley-VCH.*

The dual doping of Ni and Zn substantially enhanced the HER activity of the host oxide. Ni dopants aggregated around surface oxygen vacancies, creating an optimal electronic surface configuration for binding hydrogen intermediates. On the other hand, Zn dopants dispersed throughout the host oxide, influencing the bulk electronic structure and enhancing electrical conduction. The dual-doped Ni, Zn CoO NRs exhibited a significantly high turnover frequency (TOF) of 9.34 s^{-1} at an overpotential of 200 mV, as shown in **Figure 17a**. This value exceeded the TOF of 5.39 s^{-1} achieved by state-of-the-art Pt/C catalysts. Furthermore, the newly developed Ni, Zn dual-doped CoO NRs showed a 10 mV reduction in overpotential required to achieve a current density of 100 mA cm^{-2} compared to the Pt/C catalyst, as shown in **Figure 17b**. The superior performance of the Ni, Zn dual-doped CoO nanorods can be attributed to their ability to promote water dissociation and maintain an ideal ΔG_{H*} on their surface, along with improved electrical conduction in their bulk structure. The Ni, Zn dual-doped CoO NRs demonstrated current densities of 10 and 20 mA cm^{-2} at overpotentials of 53 and 79 mV, respectively.

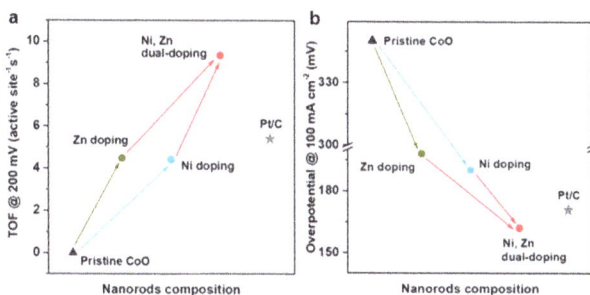

Figure 17. *Experimental TOFs (a) at an overpotential of 200 mV and (b) overpotentials at the current density of 100 mA cm^{-2} for pristine CoO, Ni-doped CoO, Zn-doped CoO, Ni, Zn dual-doped CoO NRs, and Pt/C catalyst. Reproduced with permission from ref [65]. Copyright 2019, Wiley-VCH.*

6.   Transition Metal Sulfides

6.1.   Advantages of Transition Metal Sulfides

Transition metal sulfides (TMSs), such as nickel, iron, and cobalt sulfides, offer several advantages that make them highly desirable for electrocatalysis [66]. Firstly, metal sulfides exhibit excellent catalytic activity due to their abundant reserves, low cost, and favorable catalytic performance [67, 68]. The presence of numerous active sites on their high-specific surface area, well-crystallized, and porous surfaces enhances their catalytic performance [69]. Moreover, the unique electronic structure of metal sulfides, with easily participatory valence and conduction bands, enables efficient participation in water decomposition reactions, further enhancing their catalytic activity [70].

Secondly, TMSs demonstrate exceptional stability. They can form robust nanofiber bundles that withstand higher current densities and resist electrolyte corrosion better than other materials [71]. This enhanced toughness extends their service life and reduces maintenance costs for catalysts [72]. Additionally, metal sulfides exhibit remarkable chemical inertness, minimizing the occurrence of redox reactions during electrochemical processes [73]. This inert behavior protects the catalyst surface, improves overall stability, and helps maintain long-term catalytic performance [74].

Finally, TMSs offer the advantage of low preparation costs. Compared to platinum-based catalysts, metal sulfides are more cost-effective to produce, requiring fewer production processes and offering a greener and more sustainable alternative [75]. These characteristics make metal sulfides a promising choice for sustainable energy technologies. In the realm of green energy, metal sulfides hold great potential as durable, efficient, and affordable electrocatalysts, paving the way for the development of more sustainable energy solutions.

6.2.   Transition Metal Sulfides for HER

The exploration of true active sites is crucial for the development of novel electrocatalysts. Despite the predominance of metal atoms in these systems, understanding the role of sulfur (S) atoms remains a pressing matter. TMSs exhibit a distinctive two-dimensional layered structure on their surfaces, such as the MS_2 (M represents the transition metals) and nickel-sulfur alloy configurations, including Ni_3S_2, NiS_2, NiS_2, and Ni_3S_4 [76]. The influence of S atoms in HER can be categorized into direct and indirect effects [77].

In the HER, certain S atoms and S vacancies exposed on the edges or basal surfaces of TMSs significantly impact the electrocatalytic performance, representing the direct effect of S atoms [78]. Notably, Jaramillo et al. [79] determined through experimental and theoretical findings that S atoms on the edges of MoS_2 serve as active sites in the HER. Additionally, S vacancies on the basal surface of layered $2H$-MoS_2 contribute to HER enhancement, presenting a powerful approach to improve its performance. On the other hand, S vacancies induce the formation of additional active sites, significantly increasing their number and enhancing the overall electrocatalytic performance. Consequently, S vacancies generate new active sites on the inert substrate of $2H$-MoS_2, where the presence of a gap state near the Fermi energy level facilitates the direct binding of hydrogen to exposed Mo sites [80], as illustrated in **Figure 18**. In addition,

Materials Research Forum LLC
https://doi.org/10.21741/9781644903070

the manipulation of the number of S vacancies allows fine control over ΔG_{H^*}. In cases where metal atoms serve as active sites, the indirect effect of S atoms in the HER involves providing a platform for H adsorption and dissociation, thereby enhancing the catalytic activity of metal sites in alkaline solutions.

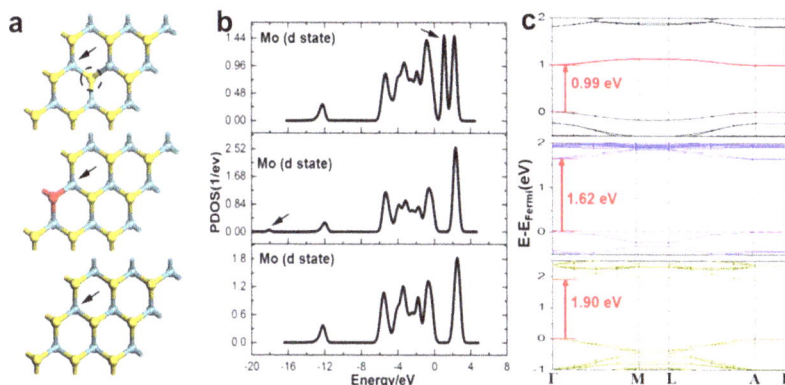

Figure 18. (a) Optimized structures (Cyan, Mo; yellow, S; red, O. The dotted circle represents S-vacancy. The Mo atom with an arrow is investigated for PDOS.), (b) partial density of states (PDOS) of Mo-d, (c) calculated band structures of (3×3×1) 2H-MoS₂ containing 3.7% S vacancies or 3.7% O atoms, and pristine one (from top to down). The Fermi level is set to zero in PDOS (red dotted line). Reproduced with permission from ref [80]. Copyright 2018, American Chemical Society.

MoS₂, a representative metal sulfide, has been extensively investigated for its remarkable electrocatalytic potential in water electrolysis [81]. The preferred stabilized phase of MoS₂ is the 2H phase, while the stand-alone 1T-MoS₂ is unstable [82]. Combining MoS₂ with other active materials not only significantly enhances the electrocatalytic activity but also stabilizes the unstable 1T-MoS₂ by serving as a substrate [83]. Shang et al. [82] successfully embedded 1T-MoS₂ into amorphous CoOOH using a simple one-pot method, resulting in bifunctional catalysts for water electrolysis. The synthesized 1T-MoS₂-CoOOH heterostructures exhibited high activity, with an overpotential of 158 mV for the HER in 0.5 M H₂SO₄ and 345 mV for OER in 1 M KOH at the current density of 10 mA cm⁻², as shown in **Figure 19a-b**. In this heterostructure, 1T-MoS₂ mainly contributed to excellent HER activity, while CoOOH nanosheets provided outstanding OER activity. Importantly, the CoOOH nanosheets also acted as stabilizers for the 1T-MoS₂ monolayer, promoting phase transition and stabilization of the metal phase, thus enhancing catalytic performance.

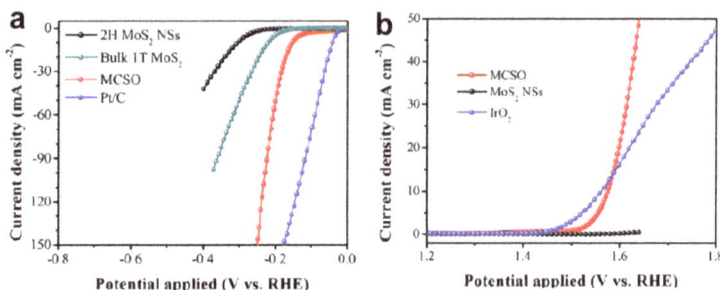

Figure 19. *(a) LSV curves of MCSO, MoS₂ NSs, and commercial Pt/C in 0.5 M H₂SO₄. (b) The LSV curves of MCSO in 1 M KOH. Reproduced with permission from ref [82]. Copyright 2018, Wiley-VCH.*

Li et al. [84] synthesized MoS_2 particle structures on reduced graphene oxide (RGO) by a one-step solvothermal reaction. By adding 22 mg of $(NH_4)_2MoS_4$ to 10 mg of graphene oxide (GO) dispersed in 10 mL of DMF, the $(NH_4)_2MoS_4$ precursor was reduced to MoS_2 nanoparticles, which were uniformly placed on RGO after reduction by hydrazine. The GO sheets acted as a substrate for nucleation and subsequent growth of MoS_2 nanoparticles, preventing the clustering of MoS_2 into 3D particles of varying sizes. The MoS_2/RGO hybrid material was tested for HER in 0.5 M H_2SO_4 and exhibited an initial overpotential of only 0.1 V.

Nickel sulfide is another highly interesting catalyst. NiS and Ni_3S_2 catalysts demonstrate strong stability in the OER, as well as low overpotentials in KOH electrolyte [85]. Ni_3S_2, a metal-rich nickel sulfide, exhibits superior electrocatalytic performance in OER, particularly when combined with other materials to form a heterogeneous nodular structure [86]. Luo et al. [87] synthesized NiS_2 hollow microspheres via a hydrothermal method using $Ni(NO_3)_2$ and $Na_2S_2O_3$ solutions as sources for Ni and S, respectively. Through annealing in a mixed atmosphere, the obtained NiS_2 hollow microspheres were further transformed into NiS porous hollow microspheres. The overpotentials of NiS_2 for the HER were 174 mV (acidic condition) and 148 mV (alkaline condition) at a current density of 10 mA cm^{-2}, while the overpotential of NiS for the OER was 320 mV, as indicated in **Figure 20a-c**. Yuan et al. [88] immersed a piece of Fe-Ni alloy foil in a mixture of Na_2S solution and ethanol. The reaction was conducted in an electric oven at 200 °C for 12 hours, resulting in the in-situ growth of Fe-doped Ni_3S_2 nanosheet arrays on the surface of the alloy foil. The obtained electrode exhibited high catalytic activity for the OER in a strong alkaline solution, with an overpotential of 282 mV at 10 mA cm^{-2}, as shown in **Figure 20d**.

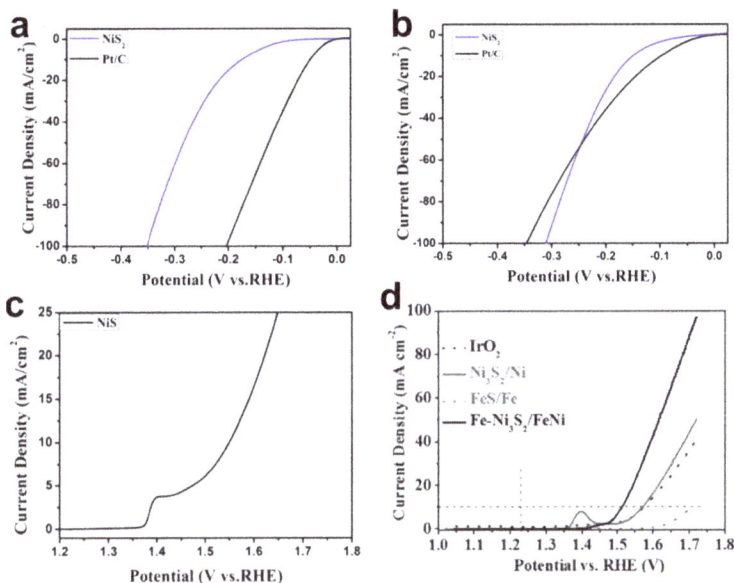

Figure 20. *(a) LSV curves of NiS₂ and 20% commercial Pt/C at a scan rate of 5 mV s⁻¹ for HER in 0.5 M H₂SO₄. (b) LSV curves of NiS₂ and 20% commercial Pt/C at a scan rate of 5 mV s⁻¹ for HER in 1.0 M KOH. (c) Polarization curve of NiS porous hollow microspheres at a scan rate of 2 mV s⁻¹ in 1 M KOH. Reproduced with permission from ref [87]. Copyright 2017, American Chemical Society. (d) LSV curves for Fe-Ni₃S₂/FeNi, Ni₃S₂/Ni, FeS/Fe, and IrO₂ in 1 M KOH solution. Reproduced with permission from ref [88]. Copyright 2017, Elsevier.*

7.   Transition Metal Selenides

Recently, transition metal chalcogenides, including iron (Fe), nickel (Ni), cobalt (Co), and molybdenum (Mo) selenides, have emerged as excellent catalysts for electrochemical water splitting. These materials possess remarkable properties, high stability, and cost-effectiveness. Transition metal selenides, in particular, have gained attention as a promising family of catalysts due to their unique characteristics compared to sulfides. The electronic structure of selenium (Se) consists of $4s^2 4p^4$, where the energy level of the empty $3d$ orbitals is in close proximity to the $3s$ and $3p$ orbitals. As a result, the $3d$ orbitals of Se atoms readily form covalent bonds with transition metal atoms [89]. This unique electronic structure imparts more metallic properties to transition metal selenides, facilitating electron transport and promoting reactions. Moreover, transition metal selenides offer several advantages, such as ease of preparation, excellent catalytic activity, and exceptional stability, making them highly desirable as catalysts for water electrolysis.

7.1. Single Metal Selenide Catalysts

Zhao et al. [90] conducted a study where they prepared $CoSe_2$ nanostrips with Co vacancies. The combined theoretical and experimental findings demonstrated that the improved performance of the catalyst was attributed to changes in the electronic state and a reduction in the binding energy of oxygen intermediates. In addition, ion doping has been explored as a strategy to enhance the catalytic performance of Co selenide. For instance, Zhang et al. [91] prepared a Mo-doped CoSe catalyst and investigated the source of its HER activity. The results indicated that Mo doping induced changes in the crystal and electronic structure of CoSe, leading to a shift in the electron density of the excited state toward the Fermi level, as depicted in **Figure 21a-c**. The synergistic effect between Mo and CoSe resulted in a high HER performance, with an overpotential of only 123.3 mV at 100 mA cm^{-2} in acidic conditions, as presented in **Figure 21d**.

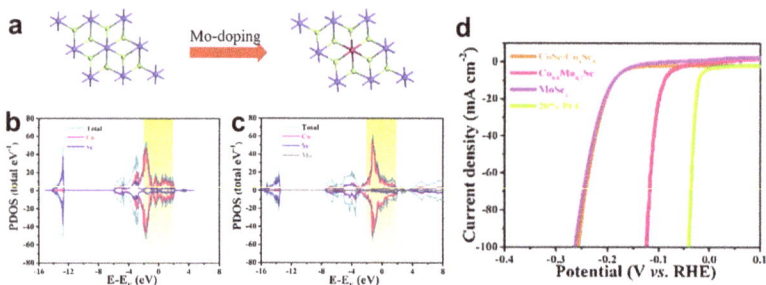

Figure 21. (a) The Mo-doping in CoSe. Calculated density of states for (b) CoSe and (c) $Co_{15}MoSe_{16}$. (d) HER polarization curves for $CoSe/Co_9Se_8$, $Co_{0.8}Mo_{0.2}Se$, $MoSe_2$ and 20% Pt/C in 0.5 M H_2SO_4. Reproduced with permission from ref [91]. Copyright 2020, Elsevier.

Moreover, non-stoichiometric Co selenides have also exhibited excellent catalytic performance. Jiang et al. [92] developed a nanoporous $Co_{0.85}Se$ catalyst loaded with single atomic Pt (Pt/np-$Co_{0.85}Se$), which demonstrated an initial overpotential for the HER that was nearly zero, surpassing commercial Pt/C catalysts, as illustrated in **Figure 22**.

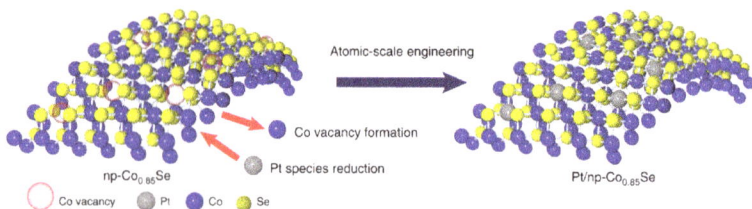

Figure 22. Schematic illustration of the fabrication of the (Pt/np-$Co_{0.85}Se$) catalyst. Reproduced with permission from ref [92]. Copyright 2019, Springer Nature.

In recent years, there has been increasing interest in Ni, Fe, and Mo selenides, with several promising results reported [93-97]. For example, Gao's group employed a hydrothermal method to prepare hierarchical arrays of Fe-doped NiSe nanorods/nanosheets loaded onto Ni films [98]. The resulting NiSe structure exhibited a combination of hexagonal and rhombohedral crystal structures. Notably, this catalyst demonstrated high and stable bifunctional electrocatalytic activity in alkaline media. Its overpotentials of 269 mV and 296 mV for the OER and HER, respectively, at a current density of up to 500 mA cm^{-2}, surpassed many of the previously reported Ni-based catalysts. Bonaccorso et al. [99] proposed an innovative concept by hybridizing the active components of HER and OER to design a bifunctional catalyst that can operate in both alkaline and acidic solutions. They designed hybrid structures consisting of flaky MoSe$_2$ and spherical Mo$_2$C loaded onto single-walled carbon nanotubes (SWCNTs). In addition, the electrochemical coupling between SWCNTs (as support) and MoSe$_2$:Mo$_2$C hybrids synergistically enhance both HER- and OER-activity of the native components, reaching small η_{10} in acidic and alkaline media (overpotentials of 0.049 and 0.089 V for HER in 0.5 M H$_2$SO$_4$ and 1 M KOH, respectively; overpotentials of 0.197 V and 0.241 V for OER in 0.5 M H$_2$SO$_4$, and 1 M KOH, respectively). The synergistic electrocatalytic effect among the three components is depicted in **Figure 23**.

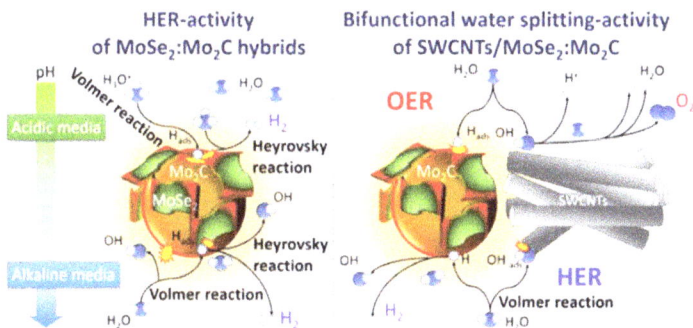

Figure 23. Illustration of the synergistic electrocatalytic effects in MoSe$_2$ flake/Mo$_2$C ball hybrids deposited on SWCNTs. Reproduced with permission from ref [99]. Copyright 2019, American Chemical Society.

7.2. Binary Metal Selenide Catalysts

In recent research on selenides, bimetallic selenides have emerged as a promising area, and some notable results have been reported. Yu and colleagues [100] prepared NiCoSe$_2$ through an electrodeposition method, as depicted in **Figure 24a**. High-resolution transmission electron microscopy (HR-TEM) images in **Figure 24b** exhibited distinct hexagonal lattice stripes corresponding to NiCoSe$_2$. Electrochemical experimental results in **Figure 24c** demonstrated that the bimetallic NiCoSe$_2$ catalyst exhibited a lower overpotential of 112.7 mV to produce the current density of 10 mA cm^{-2} and a smaller Tafel slope compared to CoSe, NiSe, and NiCo-OH.

Materials Research Forum LLC
https://doi.org/10.21741/9781644903070

Moreover, its hydrogen evolution catalytic performance surpassed that of the individual components.

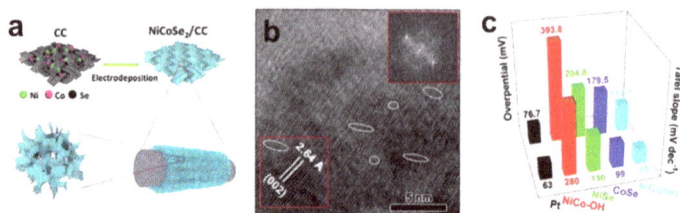

Figure 24. *(a) Scheme and (b) HR-TEM image of NiCoSe$_2$; (c) Overpotentials to generate the current density of 10 mA cm^{-2} and Tafel slopes. Reproduced with permission from ref [100]. Copyright 2018, the Royal Society of Chemistry.*

The electronic structure analysis of NiCoSe$_2$ revealed that Ni, Co, and Se elements synergistically contributed to the total density of states (TDOS) and that the overlapping *d* orbitals of Ni and Co indicated covalent interactions between the two metals. Notably, the PDOS analysis of NiCoSe$_2$ demonstrated that the dominant contributions to TDOS came from the d orbitals of Ni, Co, and Se, with significant involvement of the Se-*p* orbitals in the covalent interactions among the three elements, as shown in **Figure 25a-b**. This tuned electronic structure greatly enhanced the intrinsic electrocatalytic activity of NiCoSe$_2$. Additionally, the designed mixed-metal selenides exhibited high super-hydrophilicity, which facilitated the water adsorption process, as illustrated in **Figure 25c-d**.

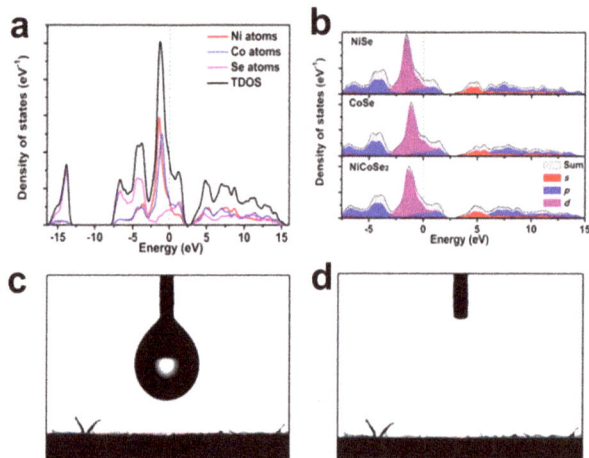

Figure 25. *(a) TDOS of NiCoSe$_2$. (b) Electronic DOS of NiSe, CoSe, and NiCoSe$_2$. Images (c) before and (d) after the droplet of water falls on NiCoSe$_2$/CC. Reproduced with permission from ref [100]. Copyright 2018, the Royal Society of Chemistry.*

8.   Conclusion and Prospects

In conclusion, the exploration of compound catalysts for HER has shown significant progress and holds promising potential. Transition metal carbides, nitrides, phosphides, oxides, sulfides, and selenides have emerged as excellent catalysts for electrochemical water splitting, benefiting from their unique electronic structures that enable enhanced metallic properties and efficient electron transport. The incorporation of transition metals like Fe, Ni, Co, and Mo into the structures of these compound catalysts has led to notable improvements in their catalytic performance and stability. The understanding of their electronic structures and covalent interactions has provided valuable insights into their enhanced electrocatalytic activity.

Future research in the field of compound catalysts for HER should focus on the exploration of novel materials, the optimization of composition and structure, and a deeper understanding of the fundamental mechanisms underlying their catalytic behavior. Efforts to enhance the durability and long-term stability of compound catalysts are crucial for their practical applications in large-scale hydrogen production and energy storage systems.

References

[1]   X. Peng, S. Xie, X. Wang, C. Pi, Z. Liu, B. Gao, L. Hu, W. Xiao, Energy-saving hydrogen production by the methanol oxidation reaction coupled with the hydrogen evolution reaction co-catalyzed by a phase separation induced heterostructure, J. Mater. Chem. A, 10 (2022) 20761-20769. https://doi.org/10.1039/D2TA02955C

[2]   M. Qiang, X. Zhang, H. Song, C. Pi, X. Wang, B. Gao, Y. Zheng, X. Peng, P.K. Chu, K. Huo, General synthesis of nanostructured Mo_2C electrocatalysts using a carbon template for electrocatalytic applications, Carbon, 197 (2022) 238-245. https://doi.org/10.1016/j.carbon.2022.06.016

[3]   X. Peng, Y. Yan, S. Xiong, Y. Miao, J. Wen, Z. Liu, B. Gao, L. Hu, P.K. Chu, Se-$NiSe_2$ hybrid nanosheet arrays with self-regulated elemental Se for efficient alkaline water splitting, J. Mater. Sci. Technol., 118 (2022) 136-143. https://doi.org/10.1016/j.jmst.2021.12.022

[4]   W. Li, D. Liu, N. Yang, J. Wang, M. Huang, L. Liu, X. Peng, Wang, Molybdenum diselenide-black phosphorus heterostructures for electrocatalytic hydrogen evolution, Appl. Surf. Sci., 467-468 (2019) 328-334. https://doi.org/10.1016/j.apsusc.2018.10.127

[5]   C. Huang, C. Pi, X. Zhang, K. Ding, P. Qin, J. Fu, X. Peng, In situ synthesis of MoP nanoflakes intercalated N-doped graphene nanobelts from MoO_3-amine hybrid for high-efficient hydrogen evolution reaction, Small, 14 (2018) 1800667. https://doi.org/10.1002/smll.201800667

[6]   C. Huang, X. Miao, C. Pi, B. Gao, X. Zhang, P. Qin, K. Huo, X. Peng, P.K. Chu, Mo_2C/VC heterojunction embedded in graphitic carbon network: An advanced electrocatalyst for hydrogen evolution, Nano Energy, 60 (2019) 520-526. https://doi.org/10.1016/j.nanoen.2019.03.088

[7]   X. Peng, L. Hu, L. Wang, X. Zhang, J. Fu, K. Huo, L.Y.S. Lee, K.-Y. Wong, P.K. Chu, Vanadium carbide nanoparticles encapsulated in graphitic carbon network nanosheets: A high-efficiency electrocatalyst for hydrogen evolution reaction, Nano Energy, 26 (2016) 603-609. https://doi.org/10.1016/j.nanoen.2016.06.020

[8]   X. Peng, X. Jin, N. Liu, P. Wang, Z. Liu, B. Gao, L. Hu, P.K. Chu, A high-performance electrocatalyst composed of nickel clusters encapsulated with a carbon network on TiN nanaowire arrays for the oxygen evolution reaction, Appl. Surf. Sci., 567 (2021) 150779. https://doi.org/10.1016/j.apsusc.2021.150779

[9]   J. Zhang, A.R. Woldu, X. Zhao, X. Peng, Y. Song, H. Xia, F. Lu, Plasmon-enhanced hydrogen evolution on Pt-anchored titanium nitride nanowire arrays, Appl. Surf. Sci., 598 (2022) 153745. https://doi.org/10.1016/j.apsusc.2022.153745

[10]   X. Wang, X. Zhang, Y. Xu, H. Song, X. Min, Z. Tang, C. Pi, Heterojunction Mo-based binary and ternary nitride catalysts with Pt-like activity for the hydrogen evolution reaction, Chem. Eng. J., 473 (2023) 144370. https://doi.org/10.1016/j.cej.2023.144370

[11]   X. Peng, A.M. Qasim, W. Jin, L. Wang, L. Hu, Y. Miao, Ni-doped amorphous iron phosphide nanoparticles on TiN nanowire arrays: An advanced alkaline hydrogen evolution electrocatalyst, Nano Energy, 53 (2018) 66-73. https://doi.org/10.1016/j.nanoen.2018.08.028

[12]   S. Feng, D. Li, H. Dong, S. Xie, Y. Miao, X. Zhang, Gao, Tailoring the Mo-N/Mo-O configuration in MoO_2/Mo_2N heterostructure for ampere-level current density hydrogen production, Appl. Catal. B-Environ., 342 (2024) 12451. https://doi.org/10.1016/j.apcatb.2023.123451

[13]   X. Peng, S. Xie, S. Xiong, R. Li, P. Wang, X. Zhang, Z. Liu, L. Hu, Ultralow-voltage hydrogen production and simultaneous Rhodamine B beneficiation in neutral wastewater, J. Energy Chem., 81 (2023) 574-582. https://doi.org/10.1016/j.jechem.2023.03.022

[14]   Z. Huang, A.R. Woldu, X. Peng, P. Chu, Remarkably boosted water oxidation activity and dynamic stability at large-current-density of $Ni(OH)_2$ nanosheet arrays by Fe ion association and underlying mechanism, Chem. Eng. J., 477 (2023) 147155. https://doi.org/10.1016/j.cej.2023.147155

[15]   A. Qayum, X. Peng, J. Yuan, Y. Qu, J. Zhou, Z. Huang, H. Xia, Z. Liu, D.Q. Tan, Highly durable and efficient $Ni-FeO_x/FeNi_3$ electrocatalysts synthesized by a facile in situ combustion-based method for overall water splitting with large current densities, ACS Appl. Mater. Interfaces, 14 (2022) 27842-27853. https://doi.org/10.1021/acsami.2c04562

[16]   M. Kuang, W. Huang, C. Hegde, W. Fang, X. Tan, C. Liu, J. Ma, Q. Yan, Interface engineering in transition metal carbides for electrocatalytic hydrogen generation and nitrogen fixation, Materials Horizons, 7 (2020) 32-53. https://doi.org/10.1039/C9MH01094G

[17]   Y. Liu, T.G. Kelly, J.G. Chen, W.E. Mustain, Metal carbides as alternative electrocatalyst supports, ACS Catal., 3 (2013) 1184-1194. https://doi.org/10.1021/cs4001249

[18] Y. Zhong, X. Xia, F. Shi, J. Zhan, J. Tu, H.J. Fan, Transition metal carbides and nitrides in energy storage and conversion, Adv. Sci., 3 (2016) 1500286. https://doi.org/10.1002/advs.201500286

[19] Q. Du, R. Zhao, T. Guo, L. Liu, X. Chen, J. Zhang, J. Du, J. Li, L. Mai, T. Asefa, Highly dispersed Mo_2C nanodots in carbon nanocages derived from Mo-based xerogel: Efficient electrocatalysts for hydrogen evolution, Small Methods, 5 (2021) 2100334. https://doi.org/10.1002/smtd.202100334

[20] K. Murugappan, E.M. Anderson, D. Teschner, T.E. Jones, K. Skorupska, Y. Román-Leshkov, Operando NAP-XPS unveils differences in MoO_3 and Mo_2C during hydrodeoxygenation, Nat. Catal., 1 (2018) 960-967. https://doi.org/10.1038/s41929-018-0171-9

[21] D. Reynard, B. Nagar, H. Girault, Photonic flash synthesis of Mo_2C/graphene electrocatalyst for the hydrogen evolution reaction, ACS Catal., 11 (2021) 5865-5872. https://doi.org/10.1021/acscatal.1c00770

[22] H. Huang, C. Yu, H. Huang, W. Guo, M. Zhang, X. Han, Q. Wei, S. Cui, X. Tan, J. Qiu, microwave-assisted ultrafast synthesis of molybdenum carbide nanoparticles grown on carbon matrix for efficient hydrogen evolution reaction, Small Methods, 3 (2019) 1900259. https://doi.org/10.1002/smtd.201900259

[23] W. An, J. Su, Z. Chen, B. Gao, X. Zhang, X. Peng, S. Peng, J. Fu, P.K. Chu, Low-temperature synthesis of mesoporous SiC Hollow spheres by magnesiothermic reduction, J. Am. Ceram. Soc., 99 (2016) 1859-1861. https://doi.org/10.1111/jace.14208

[24] D. Lancet, I. Pecht, Spectroscopic and immunochemical studies with nitrobenzoxadiazolealanine, a fluorescent dinitrophenyl analog, Biochemistry, 16 (1977) 5150-5157. https://doi.org/10.1021/bi00642a031

[25] J.S.J. Hargreaves, Heterogeneous catalysis with metal nitrides, Coord. Chem. Rev., 257 (2013) 2015-2031. https://doi.org/10.1016/j.ccr.2012.10.005

[26] S. Dong, X. Chen, X. Zhang, G. Cui, Nanostructured transition metal nitrides for energy storage and fuel cells, Coord. Chem. Rev., 257 (2013) 1946-1956. https://doi.org/10.1016/j.ccr.2012.12.012

[27] E. Furimsky, Metal carbides and nitrides as potential catalysts for hydroprocessing, Appl. Catal. A, 240 (2003) 1-28. https://doi.org/10.1016/S0926-860X(02)00428-3

[28] J. Xiao, Y. Xu, Y. Xia, J. Xi, S. Wang, Ultra-small Fe_2N nanocrystals embedded into mesoporous nitrogen-doped graphitic carbon spheres as a highly active, stable, and methanol-tolerant electrocatalyst for the oxygen reduction reaction, Nano Energy, 24 (2016) 121-129. https://doi.org/10.1016/j.nanoen.2016.04.026

[29] K. Schwarz, Band structure and chemical bonding in transition metal carbides and nitrides, Crit. Rev. Solid State, 13 (1987) 211-257. https://doi.org/10.1080/10408438708242178

[30]   D.J. Ham, J.S. Lee, Transition metal carbides and nitrides as electrode materials for low temperature fuel cells, Energies, 2 (2009) 873-899. https://doi.org/10.3390/en20400873

[31]   Z. Xue, J. Kang, D. Guo, C. Zhu, C. Li, X. Zhang, Y. Chen, Self-supported cobalt nitride porous nanowire arrays as bifunctional electrocatalyst for overall water splitting, Electrochim. Acta, 273 (2018) 229-238. https://doi.org/10.1016/j.electacta.2018.04.056

[32]   Z. Lv, M. Tahir, X. Lang, G. Yuan, L. Pan, X. Zhang, J.-J. Zou, Well-dispersed molybdenum nitrides on a nitrogen-doped carbon matrix for highly efficient hydrogen evolution in alkaline media, J. Mater. Chem. A, 5 (2017) 20932-20937. https://doi.org/10.1039/C7TA06981B

[33]   M. Sankar, N. Dimitratos, P.J. Miedziak, P.P. Wells, C.J. Kiely, G.J. Hutchings, Designing bimetallic catalysts for a green and sustainable future, Chem. Soc. Rev., 41 (2012) 8099-8139. https://doi.org/10.1039/c2cs35296f

[34]   S.H. Park, T.H. Jo, M.H. Lee, K. Kawashima, C.B. Mullins, H.-K. Lim, D.H. Youn, Highly active and stable nickel-molybdenum nitride (Ni_2Mo_3N) electrocatalyst for hydrogen evolution, J. Mater. Chem. A, 9 (2021) 4945-4951. https://doi.org/10.1039/D0TA10090K

[35]   H. Jin, X. Wang, C. Tang, A. Vasileff, L. Li, A. Slattery, S.-Z. Qiao, Stable and Highly Efficient Hydrogen Evolution from Seawater Enabled by an Unsaturated Nickel Surface Nitride, Adv. Mater., 33 (2021) 2007508. https://doi.org/10.1002/adma.202007508

[36]   Y. Yang, J. Yang, C. Kong, Q. Zhou, D. Qian, Y. Xiong, Z. Hu, Heterogeneous cobalt-iron phosphide nanosheets formed by in situ phosphating of hydroxide for efficient overall water splitting, J. Alloys Compd., 926 (2022) 166930. https://doi.org/10.1016/j.jallcom.2022.166930

[37]   L.f. Hong, R.t. Guo, Y. Yuan, X.y. Ji, Z.d. Lin, Z.s. Li, W.g. Pan, Recent progress of transition metal phosphides for photocatalytic hydrogen evolution, ChemSusChem, 14 (2021) 539-557. https://doi.org/10.1002/cssc.202002454

[38]   S.T. Oyama, T. Gott, H. Zhao, Y.-K. Lee, Transition metal phosphide hydroprocessing catalysts: A review, Catal. Today 143 (2009) 94-107. https://doi.org/10.1016/j.cattod.2008.09.019

[39]   L. Ji, J. Wang, X. Teng, T.J. Meyer, Z. Chen, CoP nanoframes as bifunctional electrocatalysts for efficient overall water splitting, ACS Catal., 10 (2019) 412-419. https://doi.org/10.1021/acscatal.9b03623

[40]   H. Du, R.-M. Kong, X. Guo, F. Qu, J. Li, Recent progress in transition metal phosphides with enhanced electrocatalysis for hydrogen evolution, Nanoscale, 10 (2018) 21617-21624. https://doi.org/10.1039/C8NR07891B

[41]   Y. Bai, L. Fang, H. Xu, X. Gu, H. Zhang, Y. Wang, Strengthened Synergistic Effect of Metallic M_xP_y (M= Co, Ni, and Cu) and Carbon Layer via Peapod-Like Architecture for Both Hydrogen and Oxygen Evolution Reactions, Small, 13 (2017) 1603718.

https://doi.org/10.1002/smll.201603718

[42] H.-M. Yang, C.-C. Weng, H.-Y. Wang, Z.-Y. Yuan, Transition metal nitride-based materials as efficient electrocatalysts: Design strategies and prospective applications, Coord. Chem. Rev., 496 (2023) 215410. https://doi.org/10.1016/j.ccr.2023.215410

[43] Z. Pu, T. Liu, I.S. Amiinu, R. Cheng, P. Wang, C. Zhang, P. Ji, W. Hu, J. Liu, S. Mu, Transition-metal phosphides: activity origin, energy-related electrocatalysis applications, and synthetic strategies, Adv. Funct. Mater., 30 (2020) 2004009. https://doi.org/10.1002/adfm.202004009

[44] Y. Shi, B. Zhang, Recent advances in transition metal phosphide nanomaterials: synthesis and applications in hydrogen evolution reaction, Chem. Soc. Rev., 45 (2016) 1529-1541. https://doi.org/10.1039/C5CS00434A

[45] S. Zhang, X. Zhang, Y. Rui, R. Wang, X. Li, Recent advances in non-precious metal electrocatalysts for pH-universal hydrogen evolution reaction, Green Energy Environ., 6 (2021) 458-478. https://doi.org/10.1016/j.gee.2020.10.013

[46] Y. Li, Z. Dong, L. Jiao, Multifunctional transition metal-based phosphides in energy-related electrocatalysis, Adv. Energy Mater., 10 (2020) 1902104. https://doi.org/10.1002/aenm.201902104

[47] Q. Dong, M. Li, M. Sun, F. Si, Q. Gao, X. Cai, Y. Xu, T. Yuan, S. Zhang, F. Peng, Phase-controllable growth Ni_xP_y modified CdS@ Ni_3S_2 Electrodes for efficient electrocatalytic and enhanced photoassisted electrocatalytic overall water splitting, small methods, 5 (2021) 2100878. https://doi.org/10.1002/smtd.202100878

[48] M. Liu, J. Li, Cobalt phosphide hollow polyhedron as efficient bifunctional electrocatalysts for the evolution reaction of hydrogen and oxygen, ACS Appl. Mater. Interfaces, 8 (2016) 2158-2165. https://doi.org/10.1021/acsami.5b10727

[49] Y. Pan, K. Sun, S. Liu, X. Cao, K. Wu, W.-C. Cheong, Z. Chen, Y. Wang, Y. Li, Y. Liu, Core-shell ZIF-8@ ZIF-67-derived CoP nanoparticle-embedded N-doped carbon nanotube hollow polyhedron for efficient overall water splitting, J. Am. Chem. Soc., 140 (2018) 2610-2618. https://doi.org/10.1021/jacs.7b12420

[50] D. Yoon, B. Seo, J. Lee, K.S. Nam, B. Kim, S. Park, H. Baik, S.H. Joo, K. Lee, Facet-controlled hollow Rh_2S_3 hexagonal nanoprisms as highly active and structurally robust catalysts toward hydrogen evolution reaction, Energy Environ. Sci., 9 (2016) 850-856. https://doi.org/10.1039/C5EE03456F

[51] E.J. Popczun, C.G. Read, C.W. Roske, N.S. Lewis, R.E. Schaak, Highly active electrocatalysis of the hydrogen evolution reaction by cobalt phosphide nanoparticles, Angew. Chem. Int. Ed., 53 (2014) 5427-5430. https://doi.org/10.1002/anie.201402646

[52] C. Wei, Q. Lu, J. Sun, F. Gao, Evolution of nickel sulfide hollow spheres through topotactic transformation, Nanoscale, 5 (2013) 12224-12230. https://doi.org/10.1039/c3nr03371f

[53] X. Yu, S. Zhang, C. Li, C. Zhu, Y. Chen, P. Gao, L. Qi, X. Zhang, Hollow CoP nanopaticle/N-doped graphene hybrids as highly active and stable bifunctional catalysts for full water splitting, Nanoscale, 8 (2016) 10902-10907. https://doi.org/10.1039/C6NR01867J

[54] M.H. Hansen, L.-A. Stern, L. Feng, J. Rossmeisl, X. Hu, Widely available active sites on Ni_2P for electrochemical hydrogen evolution-insights from first principles calculations, Phys. Chem. Chem. Phys., 17 (2015) 10823-10829. https://doi.org/10.1039/C5CP01065A

[55] Z. Chen, C.X. Kronawitter, B.E. Koel, Facet-dependent activity and stability of Co_3O_4 nanocrystals towards the oxygen evolution reaction, Phys. Chem. Chem. Phys., 17 (2015) 29387-29393. https://doi.org/10.1039/C5CP02876K

[56] L.-L. Feng, G. Yu, Y. Wu, G.-D. Li, H. Li, Y. Sun, T. Asefa, W. Chen, X. Zou, High-index faceted Ni_3S_2 nanosheet arrays as highly active and ultrastable electrocatalysts for water splitting, J. Am. Chem. Soc., 137 (2015) 14023-14026. https://doi.org/10.1021/jacs.5b08186

[57] Q. Qin, H. Jang, L. Chen, G. Nam, X. Liu, J. Cho, Low loading of Rh_xP and RuP on N, P codoped carbon as two trifunctional electrocatalysts for the oxygen and hydrogen electrode reactions, Adv. Energy Mater., 8 (2018) 1801478. https://doi.org/10.1002/aenm.201801478

[58] Z. Pu, I.S. Amiinu, Z. Kou, W. Li, S. Mu, RuP_2-based catalysts with platinum-like activity and higher durability for the hydrogen evolution reaction at all pH values, Angew. Chem. Int. Ed., 56 (2017) 11559-11564. https://doi.org/10.1002/anie.201704911

[59] J. Masud, T. Van Nguyen, N. Singh, E. McFarland, M. Ikenberry, K. Hohn, C.-J. Pan, B.-J. Hwang, A Rh_xS_y/C catalyst for the hydrogen oxidation and hydrogen evolution reactions in HBr, J. Electrochem. Soc., 162 (2015) F455. https://doi.org/10.1149/2.0901504jes

[60] Y. Li, N. Li, X. Yan, X. Li, R. Wang, X. Zhang, Watermelon-like $Rh_xS_y@C$ nanospheres: phase evolution and its influence on the electrocatalytic performance for oxygen reduction reaction, J. Mater. Sci., 52 (2017) 11402-11412. https://doi.org/10.1007/s10853-017-1294-0

[61] K. Wang, B. Huang, F. Lin, F. Lv, M. Luo, P. Zhou, Q. Liu, W. Zhang, C. Yang, Y. Tang, Wrinkled Rh_2P nanosheets as superior pH-universal electrocatalysts for hydrogen evolution catalysis, Adv. Energy Mater., 8 (2018) 1801891. https://doi.org/10.1002/aenm.201801891

[62] H. Duan, D. Li, Y. Tang, Y. He, S. Ji, R. Wang, H. Lv, P.P. Lopes, A.P. Paulikas, H. Li, High-performance Rh_2P electrocatalyst for efficient water splitting, J. Am. Chem. Soc., 139 (2017) 5494-5502. https://doi.org/10.1021/jacs.7b01376

[63] L. Huang, D. Chen, G. Luo, Y.-R. Lu, C. Chen, Y. Zou, C.-L. Dong, Y. Li, S. Wang, Zirconium-Regulation-Induced Bifunctionality in 3D Cobalt-Iron Oxide Nanosheets for Overall Water Splitting, Adv. Mater., 31 (2019) 1901439.

https://doi.org/10.1002/adma.201901439

[64]   S.H. Chang, N. Danilovic, K.-C. Chang, R. Subbaraman, A.P. Paulikas, D.D. Fong, M.J. Highland, P.M. Baldo, V.R. Stamenkovic, J.W. Freeland, J.A. Eastman, N.M. Markovic, Functional links between stability and reactivity of strontium ruthenate single crystals during oxygen evolution, Nat. Commun., 5 (2014) 4191. https://doi.org/10.1038/ncomms5191

[65]   T. Ling, T. Zhang, B. Ge, L. Han, L. Zheng, F. Lin, Z. Xu, W.-B. Hu, X.-W. Du, K. Davey, S.-Z. Qiao, Well-Dispersed Nickel- and Zinc-Tailored Electronic Structure of a Transition Metal Oxide for Highly Active Alkaline Hydrogen Evolution Reaction, Adv. Mater., 31 (2019) 1807771. https://doi.org/10.1002/adma.201807771

[66]   H. Zhou, X. Li, Y. Li, M. Zheng, H. Pang, Applications of M_xSe_y (M = Fe, Co, Ni) and Their Composites in Electrochemical Energy Storage and Conversion, Nano-Micro Lett., 11 (2019) 40. https://doi.org/10.1007/s40820-019-0272-2

[67]   N. Jiang, Q. Tang, M. Sheng, B. You, D.-e. Jiang, Y. Sun, Nickel sulfides for electrocatalytic hydrogen evolution under alkaline conditions: a case study of crystalline NiS, NiS_2, and Ni_3S_2 nanoparticles, Catal. Sci. Technol., 6 (2016) 1077-1084. https://doi.org/10.1039/C5CY01111F

[68]   J. Joo, T. Kim, J. Lee, S.I. Choi, K. Lee, Morphology-controlled metal sulfides and phosphides for electrochemical water splitting, Adv. Mater., 31 (2019) 1806682. https://doi.org/10.1002/adma.201806682

[69]   X.L. Yu-Lin Wu, Yong-Sheng Wei, Zhaoming Fu, Wenbo Wei, Xin-Tao Wu, Qi-Long Zhu,and Qiang Xu, Ordered Macroporous Superstructure of NitrogenDoped Nanoporous Carbon Implanted with Ultrafine Ru Nanoclusters for Efficient pH-Universal Hydrogen Evolution Reaction, Adv. Mater., 33 (2021) 2006965-2006976. https://doi.org/10.1002/adma.202006965

[70]   F. Jamal, A. Rafique, S. Moeen, J. Haider, W. Nabgan, A. Haider, M. Imran, G. Nazir, M. Alhassan, M. Ikram, Review of Metal Sulfide Nanostructures and their Applications, ACS Appl. Nano Mater., 6 (2023) 7077-7106. https://doi.org/10.1021/acsanm.3c00417

[71]   Y. Yang, H. Yao, Z. Yu, S.M. Islam, H. He, M. Yuan, Y. Yue, K. Xu, W. Hao, G. Sun, Hierarchical nanoassembly of $MoS_2/Co_9S_8/Ni_3S_2/Ni$ as a highly efficient electrocatalyst for overall water splitting in a wide pH range, J. Am. Chem. Soc., 141 (2019) 10417-10430. https://doi.org/10.1021/jacs.9b04492

[72]   Y. Guo, T. Park, J.W. Yi, J. Henzie, J. Kim, Z. Wang, B. Jiang, Y. Bando, Y. Sugahara, J. Tang, Y. Yamauchi, Nanoarchitectonics for Transition-Metal-Sulfide-Based Electrocatalysts for Water Splitting, Adv. Mater., 31 (2019) 1807134. https://doi.org/10.1002/adma.201807134

[73]   X.Y. Xingxing Zhu, Xingyou Lang, Jie Liu, Chandra-Veer Singh,Erhong Song, Yongfu Zhu, and Qing Jiang, Charge Self-Regulation of Metallic Heterostructure $Ni_2P@Co_9S_8$ for Alkaline Water Electrolysis with Ultralow Overpotential at Large Current Density,

Adv. Sci., 10 (2023) 2303682. https://doi.org/10.1002/advs.202303682

[74] S. Wang, L. Zhao, J. Li, X. Tian, X. Wu, L. Feng, High valence state of Ni and Mo synergism in NiS_2-MoS_2 hetero-nanorods catalyst with layered surface structure for urea electrocatalysis, J. Energy Chem., 66 (2022) 483-492. https://doi.org/10.1016/j.jechem.2021.08.042

[75] Z. Qin, Y. Chen, Z. Huang, J. Su, Z. Diao, L. Guo, Composition-Dependent Catalytic Activities of Noble-Metal-Free NiS/Ni_3S_4 for Hydrogen Evolution Reaction, J. Phys. Chem. C, 120 (2016) 14581-14589. https://doi.org/10.1021/acs.jpcc.6b05230

[76] Y. Zhang, F. Lu, L. Pan, Y. Xu, Y. Yang, Y. Bando, D. Golberg, J. Yao, X. Wang, Improved cycling stability of NiS_2 cathodes through designing a "kiwano" hollow structure, J. Mater. Chem. A, 6 (2018) 11978-11984. https://doi.org/10.1039/C8TA01551A

[77] C. Tsai, H. Li, S. Park, J. Park, H.S. Han, J.K. Nørskov, X. Zheng, F. Abild-Pedersen, Electrochemical generation of sulfur vacancies in the basal plane of MoS_2 for hydrogen evolution, Nat. Commun., 8 (2017) 15113. https://doi.org/10.1038/ncomms15113

[78] J. Zhang, W. Xiao, P. Xi, S. Xi, Y. Du, D. Gao, J. Ding, Activating and optimizing activity of CoS_2 for hydrogen evolution reaction through the synergic effect of N dopants and S vacancies, ACS Energy Lett., 2 (2017) 1022-1028. https://doi.org/10.1021/acsenergylett.7b00270

[79] T.F. Jaramillo, K.P. Jørgensen, J. Bonde, J.H. Nielsen, S. Horch, I. Chorkendorff, Identification of active edge sites for electrochemical H_2 evolution from MoS_2 nanocatalysts, Science, 317 (2007) 100-102. https://doi.org/10.1126/science.1141483

[80] X. Gan, L.Y.S. Lee, K.-y. Wong, T.W. Lo, K.H. Ho, D.Y. Lei, H. Zhao, 2H/1T phase transition of multilayer MoS_2 by electrochemical incorporation of S vacancies, ACS Appl. Energy Mater., 1 (2018) 4754-4765. https://doi.org/10.1021/acsaem.8b00875

[81] D. Merki, X. Hu, Recent developments of molybdenum and tungsten sulfides as hydrogen evolution catalysts, Energy Environ. Sci., 4 (2011) 3878-3888. https://doi.org/10.1039/c1ee01970h

[82] B. Shang, P. Ma, J. Fan, L. Jiao, Z. Liu, Z. Zhang, N. Chen, Z. Cheng, X. Cui, W. Zheng, Stabilized monolayer 1T MoS_2 embedded in CoOOH for highly efficient overall water splitting, Nanoscale, 10 (2018) 12330-12336. https://doi.org/10.1039/C8NR04218G

[83] F. Chen, D. Shi, M. Yang, H. Jiang, Y. Shao, S. Wang, B. Zhang, J. Shen, Y. Wu, X. Hao, Novel designed MnS-MoS2 heterostructure for fast and stable Li/Na storage: insights into the advanced mechanism attributed to phase engineering, Adv. Funct. Mater., 31 (2021) 2007132. https://doi.org/10.1002/adfm.202007132

[84] Y. Li, H. Wang, L. Xie, Y. Liang, G. Hong, H. Dai, MoS_2 nanoparticles grown on graphene: an advanced catalyst for the hydrogen evolution reaction, J. Am. Chem. Soc., 133 (2011) 7296-7299. https://doi.org/10.1021/ja201269b

[85]   P. Luo, F. Sun, J. Deng, H. Xu, H. Zhang, Y. Wang, Tree-like NiS-Ni$_3$S$_2$F heterostructure
        array and its application in oxygen evolution reaction, Acta Phys.-Chim. Sin, 34 (2018)
        1397-1404. https://doi.org/10.3866/PKU.WHXB201804022

[86]   Y. Yang, K. Zhang, H. Lin, X. Li, H.C. Chan, L. Yang, Q. Gao, MoS$_2$-Ni$_3$S$_2$
        heteronanorods as efficient and stable bifunctional electrocatalysts for overall water
        splitting, ACS Catal., 7 (2017) 2357-2366. https://doi.org/10.1021/acscatal.6b03192

[87]   P. Luo, H. Zhang, L. Liu, Y. Zhang, J. Deng, C. Xu, N. Hu, Y. Wang, Targeted synthesis
        of unique nickel sulfide (NiS, NiS$_2$) microarchitectures and the applications for the
        enhanced water splitting system, ACS Appl. Mater. Interfaces, 9 (2017) 2500-2508.
        https://doi.org/10.1021/acsami.6b13984

[88]   C.Z. Yuan, Z.T. Sun, Y.F. Jiang, Z.K. Yang, N. Jiang, Z.W. Zhao, U.Y. Qazi, W.H.
        Zhang, A.W. Xu, One-step In Situ growth of iron–nickel sulfide nanosheets on FeNi
        alloy foils: High-performance and self-supported electrodes for water oxidation, Small,
        13 (2017) 1604161. https://doi.org/10.1002/smll.201604161

[89]   K. Zhang, Y. Li, S. Deng, S. Shen, Y. Zhang, G. Pan, Q. Xiong, Q. Liu, X. Xia, X. Wang,
        J. Tu, Molybdenum Selenide Electrocatalysts for Electrochemical Hydrogen Evolution
        Reaction, ChemElectroChem, 6 (2019) 3530-3548.
        https://doi.org/10.1002/celc.201900448

[90]   Y. Dou, C.-T. He, L. Zhang, H. Yin, M. Al-Mamun, J. Ma, H. Zhao, Approaching the
        activity limit of CoSe$_2$ for oxygen evolution via Fe doping and Co vacancy, Nat.
        Commun., 11 (2020) 1664. https://doi.org/10.1038/s41467-020-15498-0

[91]   Y. Zhou, J. Zhang, H. Ren, Y. Pan, Y. Yan, F. Sun, X. Wang, S. Wang, J. Zhang, Mo
        doping induced metallic CoSe for enhanced electrocatalytic hydrogen evolution, Appl.
        Catal. B-Environ., 268 (2020) 118467. https://doi.org/10.1016/j.apcatb.2019.118467

[92]   K. Jiang, B. Liu, M. Luo, S. Ning, M. Peng, Y. Zhao, Y.-R. Lu, T.-S. Chan, F.M. de
        Groot, Y. Tan, Single platinum atoms embedded in nanoporous cobalt selenide as
        electrocatalyst for accelerating hydrogen evolution reaction, Nat. Commun., 10 (2019)
        1743. https://doi.org/10.1038/s41467-019-09765-y

[93]   L. Yang, Y. Deng, X. Zhang, H. Liu, W. Zhou, MoSe$_2$ nanosheet/MoO$_2$ nanobelt/carbon
        nanotube membrane as flexible and multifunctional electrodes for full water splitting in
        acidic electrolyte, Nanoscale, 10 (2018) 9268-9275.
        https://doi.org/10.1039/C8NR01572D

[94]   C. Panda, P.W. Menezes, C. Walter, S. Yao, M.E. Miehlich, V. Gutkin, K. Meyer, M.
        Driess, From a molecular 2Fe-2Se precursor to a highly efficient iron diselenide
        electrocatalyst for overall water splitting, Angew. Chem. Int. Ed., 56 (2017) 10506-
        10510. https://doi.org/10.1002/anie.201706196

[95]   R. Gao, H. Zhang, D. Yan, Iron diselenide nanoplatelets: Stable and efficient water-
        electrolysis catalysts, Nano Energy, 31 (2017) 90-95.
        https://doi.org/10.1016/j.nanoen.2016.11.021

[96] J. Yu, Q. Li, C.-Y. Xu, N. Chen, Y. Li, H. Liu, L. Zhen, V.P. Dravid, J. Wu, NiSe$_2$ pyramids deposited on N-doped graphene encapsulated Ni foam for high-performance water oxidation, J. Mater. Chem. A, 5 (2017) 3981-3986. https://doi.org/10.1039/C6TA10303K

[97] F. Ming, H. Liang, H. Shi, X. Xu, G. Mei, Z. Wang, MOF-derived Co-doped nickel selenide/C electrocatalysts supported on Ni foam for overall water splitting, J. Mater. Chem. A, 4 (2016) 15148-15155. https://doi.org/10.1039/C6TA06496E

[98] Z. Zou, X. Wang, J. Huang, Z. Wu, F. Gao, An Fe-doped nickel selenide nanorod/nanosheet hierarchical array for efficient overall water splitting, J. Mater. Chem. A, 7 (2019) 2233-2241. https://doi.org/10.1039/C8TA11072G

[99] L. Najafi, S. Bellani, R. Oropesa-Nuñez, M. Prato, B. Martín-García, R. Brescia, F. Bonaccorso, Carbon nanotube-supported MoSe$_2$ holey flake: Mo$_2$C ball hybrids for bifunctional pH-universal water splitting, ACS Nano, 13 (2019) 3162-3176. https://doi.org/10.1021/acsnano.8b08670

[100] J. Yu, Y. Tian, F. Zhou, M. Zhang, R. Chen, Q. Liu, J. Liu, C.-Y. Xu, J. Wang, Metallic and superhydrophilic nickel cobalt diselenide nanosheets electrodeposited on carbon cloth as a bifunctional electrocatalyst, J. Mater. Chem. A, 6 (2018) 17353-17360. https://doi.org/10.1039/C8TA04950E

Electrocatalytic Hydrogen Production: Catalysts and Applications   Materials Research Forum LLC
Materials Research Foundations **165** (2024)   https://doi.org/10.21741/9781644903070

CHAPTER 6

Composite Catalysts for Electrocatalytic Hydrogen Production

Xiang Peng*

Hubei Key Laboratory of Plasma Chemistry and Advanced Materials, Engineering Research Center of Phosphorus Resources Development and Utilization of Ministry of Education, School of Materials Science and Engineering, Wuhan Institute of Technology, Wuhan 430205, China

xpeng@wit.edu.cn

Abstract

Transition metal compounds have attracted considerable interest from researchers due to their cost and availability advantages. Consequently, catalysts based on transition metal compounds for hydrogen production in water electrolysis have been actively developed, playing a crucial role in promoting the industrialization of the hydrogen evolution reaction (HER). Recent reports have demonstrated significant progress in enhancing the hydrogen evolution efficiency of multicomponent substances through composite and structural design improvements. Researchers have successfully developed a range of composites by combining the strengths of different transition metal compounds, specifically for use in electrocatalytic HER.

Keywords

Composite Catalysts, Hydrogen Evolution, Metal/Metal Composites, Metal/Compound Composites, Compound/Compound Catalyst

1. Introduction

With the rapid depletion of fossil fuels and the increasing environmental problems they cause, there is a growing demand for the development of renewable energy sources [1]. Carbon-free hydrogen fuel has garnered significant attention worldwide as an ideal renewable energy source. It can be directly produced through an electrochemical water splitting reaction driven by clean electricity, without generating additional pollutants [2]. The efficiency of the hydrogen evolution reaction (HER) relies on the catalyst material's performance. Although the precious metal Pt is currently the most effective commercial catalyst, its high cost and limited reserves hinder its widespread industrial application [3]. Hence, there is an urgent need to develop inexpensive, high-performance, stable, and resource-rich catalysts for hydrogen production from electrolyzed water to advance the hydrogen economy.

Transition metal compounds have attracted considerable interest from researchers due to their cost and availability advantages [4, 5]. Consequently, catalysts based on transition metal compounds for hydrogen production in water electrolysis have been actively developed, playing a crucial role in promoting the industrialization of the HER. However, single-component materials often exhibit drawbacks, making it extremely challenging to achieve favorable outcomes in terms of both structure and performance. Recent reports have demonstrated significant progress in enhancing the hydrogen evolution efficiency of multicomponent substances through composite and structural design improvements [6]. Researchers have successfully developed a range of composites by combining the strengths of different transition metal compounds, specifically for use in electrocatalytic HER [7, 8]. The combination of transition metal compounds has led to exceptional electrochemical characteristics in these composites. This chapter focuses on the application of transition metal composites in the field of HER, encompassing metal/metal, metal/compound, and compound/compound composites.

2.   Metal/Metal Composites

When metals combine to form alloys, there is a redistribution of electrons due to variations in their chemical characteristics [9, 10]. This electron redistribution leads to a decrease in the energy barrier required for the formation of intermediates at the active site. As a result, alloying presents a viable approach to enhance electrocatalytic activity by introducing active sites that promote the dissociation of the Volmer step within the alloy. This process accelerates the adsorption and dissociation of water, ultimately increasing the rate of alkaline hydrogen evolution through dual-function synergy [11]. Alloys can be classified into three categories based on their composition and atomic organization: solid solution alloys with disordered atomic arrangements, intermetallic compounds with ordered atomic arrangements, and multi-component high-entropy alloys (HEAs) [12].

2.1.   Intermetallic Compounds

Intermetallic compounds exhibit distinct electronic and crystal structures that deviate from their constituent elements, resulting in exceptional surface adsorption and catalytic properties. Unlike metal solid solutions, intermetallic compounds possess an ordered atomic arrangement [13]. This atomic ordering and stable stoichiometry not only ensure the uniformity of active sites [14, 15], but also effectively isolate distinct active sites [16, 17]. Such isolation facilitates the optimization of intermediate adsorption through the ensemble effect, ultimately enhancing catalytic performance [18]. Given this inherent characteristic, intermetallic compounds play a crucial role in multiphase catalysts.

Sun et al. [19] described a highly effective approach for creating a monolithic design of self-supported Cu-Ni-Al ternary hybrid electrocatalysts with a three-dimensional (3D) interconnected nanoporous structure. This design enables excellent catalytic performance for the HER in an alkaline solution. The hybrid catalysts consist of intermetallic $Al_7Cu_4Ni@Cu_4Ni$ core/shell nanocrystals, which exhibit high electroactivity. These nanocrystals are anchored onto a 3D bicontinuous and bimodal nanoporous Cu skeleton (Bi-NP $Cu/Al_7Cu_4Ni@Cu_4Ni$) using a one-step chemical dealloying process. In this context, the former acts as extremely active sites for HER electrocatalysis, while the latter provides pathways for both electron and ion transportation through interconnected Cu ligaments and bimodal nanopore channels.

The fabrication process of Bi-NP Cu/Al$_7$Cu$_4$Ni@Cu$_4$Ni hybrid catalysts involves simple alloying and dealloying processes. These processes allow for the adjustment of the nanoporous structures by controlling the initial microstructure of precursor alloy ribbons made of Cu$_{20-x}$Ni$_x$Al$_{80}$ (with x values of 0, 1.54, 2.00, and 2.86 at%) and the diffusion of Cu atoms at the solid-electrolyte interface. The precursor ribbons are initially created through a vacuum melt-spinning process. These ribbons are made from alloy ingots consisting of pure Cu and Al, with the option of adding small amounts of Ni. The microstructures and component distribution in the ribbons are modified by adjusting the Ni/Cu ratio, leading to the formation of a stable intermetallic compound known as Al$_7$Cu$_4$Ni.

The scanning electron microscopy (SEM) images in **Figure 1a** of the Bi-NPCu/Al$_7$Cu$_4$Ni@Cu$_4$Ni catalyst, which was fabricated by chemically dealloying Cu$_{18.46}$Ni$_{1.54}$Al$_{80}$, reveal a consistent and interconnected nanoporous structure. This structure consists of metallic ligaments and nanopore channels with two distinct sizes: approximately 200 nm and 10 nm. The wider channels are formed due to the rapid dissolution of α-Al, while the smaller nanopores are created during the subsequent dealloying of the intermetallic CuAl$_2$ phase. Through the complete removal of α-Al and CuAl$_2$, a Bi-NP Cu framework is formed. This exposes the existing Al$_7$Cu$_4$Ni nanocrystals to the electrolyte, resulting in only surface etching. This etching leads to the formation of a Cu$_4$Ni(110) surface alloy, which then grows on the Al$_7$Cu$_4$Ni(114) planes, as indicated in **Figure 1b**.

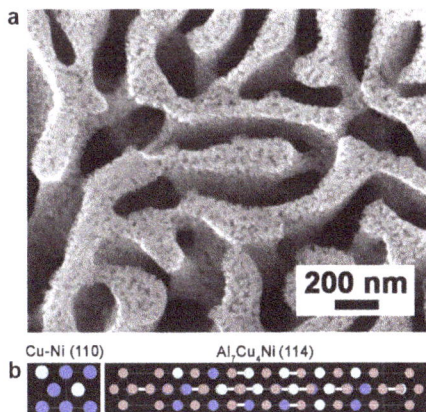

Figure 1. (a) Typical top-view SEM image of Bi-NP Cu$_{12}$Ni$_1$Al$_{2.6}$ catalyst with bimodal nanoporous structures consisting of micrometer-and nanometer-scaled pore channels. (b) Atomic illustrations of crystalline structure for both Cu$_4$Ni alloy and Al$_7$Cu$_4$Ni intermetallic compound. Blue, gray, and magenta balls denote Cu, Ni, and Al atoms, respectively. Reproduced with permission from ref [19]. Copyright 2018, Wiley-VCH.

The electroactive Al$_7$Cu$_4$Ni@Cu$_4$Ni nanocrystals seamlessly merge with the conductive Bi-NP Cu skeleton through metallic bonding. The Al$_7$Cu$_4$Ni@Cu$_4$Ni maintains its rhombohedral phase as the intermetallic Al_7Cu_4Ni core, while the tetragonal $CuAl_2$ undergoes a transformation to the cubic Cu. These findings are confirmed through the analysis of X-ray diffraction patterns (XRD) (**Figure 2a**). The distinct diffraction peaks observed in the patterns correspond to the (101), (018), (110), (0216), and (300) planes of rhombohedral Al_7Cu_4Ni (JCPDS card No. 28-0016), along with the diffraction peaks of cubic Cu (JCPDS card No. 04-0836). The Bi-NP $Cu/Al_7Cu_4Ni@Cu_4Ni$ hybrid catalysts demonstrate a catalytic current density of 10 mA cm^{-2} with a low overpotential of 139 mV in an alkaline solution, as shown in **Figure 2b**.

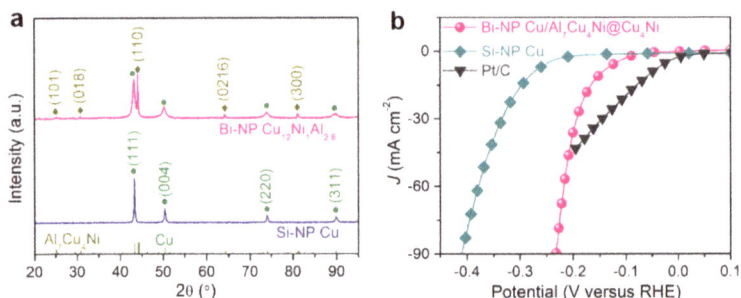

Figure 2. *(a) XRD patterns of $Cu_{12}Ni_1Al_{2.6}$. (b) Polarization curves of Bi-NP $Cu/Al_7Cu_4Ni@Cu_4Ni$. Reproduced with permission from ref [19]. Copyright 2018, Wiley-VCH.*

2.2.   Solid Solution Alloys

The electronic structure of solid solutions, which are randomly mixed at the atomic level, has been extensively investigated. By modifying the composition of constituent atoms, their electrical properties can be manipulated. Solid solutions, similar to metal compounds, offer significant potential as they enable interactions between different metal species, thereby enhancing catalytic activity and durability. Unlike intermetallic compounds, solid solutions can undergo continuous and arbitrary changes in their composition ratio. Furthermore, while solid solutions maintain the lattice structure of the solvent atoms, alloying introduces lattice distortion, which can be advantageous in finely adjusting the adsorption of H* and enhancing the HER performance [20].

Nairan et al. presented a method for synthesizing a nanowire array electrode of NiMo solid solution using an aqueous solution [21]. The resulting electrode demonstrated exceptional performance in HER. The NiMo alloy is uniformly dispersed throughout the Ni nanowires through a one-step growing procedure. The deposition of the NiMo alloy onto Ni nanowires provides exceptional electrical conductivity, a high concentration of catalytic active sites, and superior capabilities in charge/mass transport and hydrogen bubble formation. The technique of growing NiMo solid solution nanowire arrays on Ti foil involved deposition and was facilitated by the use of a magnetic field. The deposition process, as shown in **Figure 3a**, utilized an aqueous solution containing $NiCl_2\cdot 6H_2O$, $Na_3C_6H_5O_7\cdot 2H_2O$, and $Na_2MoO_4\cdot 2H_2O$ in different weight ratios of Ni:Mo salts, specifically 95:05, 90:10, and 80:20 wt%. The high-resolution transmission electron microscopy (HR-TEM) image in **Figure 3b** exhibits the formation of a NiMo alloy, accompanied by a significant amount of lattice misfit or surface defects. However, these flawed attributes are rarely observable in the HR-TEM image of NiMo samples with varying levels of Mo incorporation (1.60, 0%) prepared at 80 °C. These findings indicate that the degree of disorder can be regulated by adjusting the synthesis temperature, which can also influence the catalytic activity of HER.

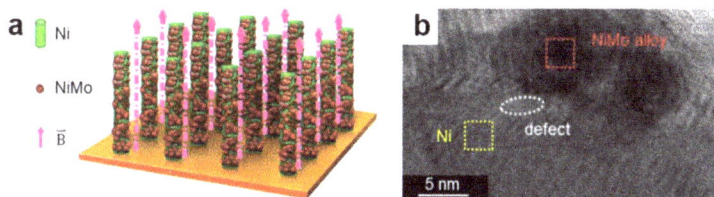

Figure 3. (a) Schematic illustration of fabrication of NiMo nanowire arrays through magnetic field assisted growth process, where dotted lines represent the magnetic field lines. (b) HR-TEM image of NiMo-65. Reproduced with permission from ref [21]. Copyright 2019, Wiley-VCH.

The crystalline nature of the solid solution nanowire arrays in their as-prepared state was confirmed through XRD analysis. **Figure 4a** depicts the XRD patterns obtained for NiMo electrode samples with varying Mo concentration, synthesized at a temperature of 80 °C. The diffraction peaks observed at $2\theta \approx 44.6°$, $52.0°$, and $76.6°$ correspond to the (111), (200), and (220) facets of nickel, respectively (JCPDS card No. 04-0850). Notably, the MoNi$_4$ alloy does not exhibit a distinct peak, which can be attributed to the relatively low Mo content determined by inductive coupled plasma (ICP) spectroscopy. The measured Mo concentration in NiMo nanowire arrays was found to be 1.60, 2.63, and 2.85 atomic percent (at%) for Ni and Mo precursor ratios of 95:05, 90:10, and 80:20, respectively. Nevertheless, the diffraction peaks displayed a shift towards smaller 2θ values, indicating lattice expansion resulting from the successful incorporation of Mo into Ni.

The enhanced HER performance of NiMo-65 in HER can be attributed to several factors, including the vertical alignment of nanowire arrays, high surface curvature, abundant edges/step sites, surface imperfections, and optimized Mo concentration. Among these characteristics, the optimized integration of Mo plays a crucial role in modifying the electronic structure and optimizing the hydrogen adsorption energy in the NiMo alloy. This is achieved by introducing strain through lattice mismatch and forming the MoNi$_4$ phase. Additionally, the lattice expansion (d-band shift) of surface atoms leads to a significant enhancement in the intrinsic activity of metallic catalysts. The NiMo alloy, with its MoNi$_4$ phase and surface defective sites, acts as active electrocatalytic sites. These findings indicate that the exceptional HER activity stems from the combined influence of both the phase composition and surface imperfections. The NiMo alloy exhibits remarkable catalytic activity in HER, characterized by remarkably low overpotentials of 17 and 98 mV at the current density of 10 and 400 mA cm^{-2}, respectively, in an alkaline environment (**Figure 4b**). These overpotentials are superior to those of the advanced catalysts based on non-noble metals and are even comparable to platinum-based electrodes.

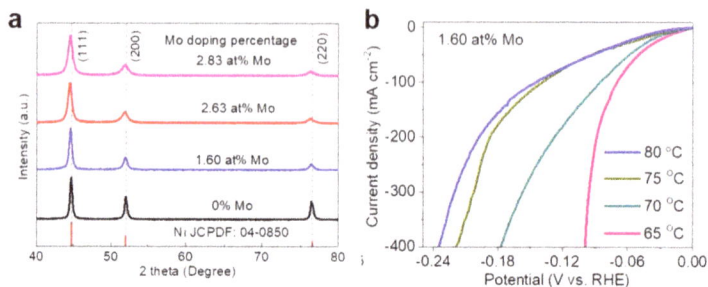

Figure 4. *(a) XRD patterns of NiMo-based electrodes prepared at 80 °C with different Mo content. (b) LSV curves of NiMo electrode (1.60 at% Mo) synthesized at different temperatures. Reproduced with permission from ref [21]. Copyright 2019, Wiley-VCH.*

2.3.   High-Entropy Alloys

High-entropy alloys (HEAs) belong to a significant class of high-entropy materials (HEMs). Unlike traditional alloys, high-entropy alloys typically consist of five or more alloying elements, with varying atomic ratios among the constituents [22, 23]. They are formed through the synergistic interaction of multiple fundamental elements, rather than relying on a single element. When combined through alloying, these components exhibit exceptional stability and form intricate crystalline solid solution phases. HEAs possess remarkable characteristics compared to traditional alloys due to their unique disordered atomic arrangement and elevated mixed entropy. These attributes include exceptional strength, resistance to fatigue and fractures [24], wear resistance [25], corrosion resistance [26], thermal stability [27], and more [28]. The high entropy effect, lattice distortion, diffusion retardation, and cocktail effect of HEAs provide them with superiority over single-material alloys, making them excellent candidates for developing nanostructured catalytic materials [29-31]. Since the initial discovery of multicomponent high-entropy alloys in 2004 [22], numerous sophisticated techniques have been developed for their production. These methods include electrosynthesis [32], carbon thermal shock [17], fast moving bed pyrolysis [33], and ultrasonication-assisted wet chemistry procedures [34]. Given the presence of multiple metals, dealloying is a prevalent technique used to create porous materials, thereby increasing the exposure of active surface areas.

Yao et al. [35] made a significant discovery by introducing a new type of electrode composed of a monolithic nanoporous CuAlNiMoFe multielemental alloy (MEA). This electrode features a surface with an in-situ-formed high-entropy CuNiMoFe alloy, which is securely attached to a hierarchical Cu skeleton. It has demonstrated great potential for highly efficient electrocatalysis of HER in nonacidic media. The surface high-entropy alloy, formed in the body-centered cubic (bcc) CuAl-NiMoFe precipitates, consists of different Cu, Ni, Mo, and Fe atoms. It serves as both a catalyst to enhance water dissociation and as a site for H* adsorption/desorption. The hierarchical nanoporous Cu skeleton acts as a framework for efficient electron transfer and facilitates the movement of ions and molecules through interconnected Cu ligaments and large channels with a lamella structure.

The monolithic nanoporous MEA hybrid electrodes are fabricated using a simple and scalable method involving alloying and dealloying. To synthesize the precursor materials, Cu (99.9%) and Al (99.9%) are combined using the arc-melting technique, with or without the inclusion of Ni (99.9%), Mo (99.9%), and/or Fe (99.9%), in an argon environment. The resulting mixture is then cooled using a furnace. Nanoporous MEA sheets with a thickness of approximately 400 μm are produced by chemically dealloying the alloy precursor slices in a N_2-purged 6 M KOH aqueous electrolyte at room temperature until no gas is being produced.

The XRD patterns of the as-dealloyed $(Cu_{8/w}Ni_{x/w}Mo_{y/w}Fe_{z/w})_{17.7}Al_{82.3}$ (x, y, z = 1) alloy (**Figure 5a**) clearly show characteristic peaks at 2θ = 30.86°, 44.24°, 64.25°, and 81.31°. These peaks correspond to the (300), (330), (600), and (633) planes of the body-centered cubic (bcc)-Al_4Cu_9 phase in space group P43/m. Additionally, there are peaks attributed to the face-centered cubic (fcc)-Cu skeleton (JCPDS card No. 04-0836). The bcc diffraction peaks of the Al_4Cu_9 phase show a slight decrease in angles and a significant reduction in intensity compared to the typical line patterns of intermetallic Al_4Cu_9 (JCPDS card No. 24-0003), which can be attributed to the presence of Ni, Mo, and Fe atoms with a strong high-entropy effect.

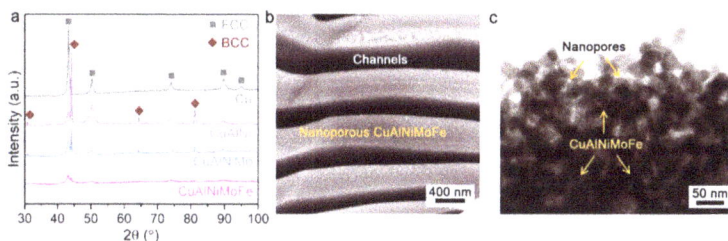

Figure 5. *(a) XRD patterns of as-dealloyed nanoporous Cu, CuAlNi, CuAlNiMo, and CuAlNiMoFe catalysts. (b) Typical cross-sectional SEM image of nanoporous CuAlNiMoFe catalyst. (c) Low-magnification STEM image of nanoporous CuAlNiMoFe with uniform small nanopores. Reproduced with permission from ref [35]. Copyright 2020, Wiley-VCH.*

A SEM image of a hierarchical nanoporous CuAlNiMoFe electrode is shown in **Figure 5b**. The image reveals large channels with a width of approximately 200 nm, as well as nanoporous CuAlNiMoFe lamellas with a width of around 400 nm. These lamellas consist of uniform interpenetrative nanopores and interconnected metallic ligaments, with a characteristic length as small as approximately 10 nm (**Figure 5c**). The distinctive bimodal porous structure of this material facilitates efficient electron transmission through the interconnected Cu ligaments. Additionally, it provides excellent accessibility to the electroactive sites on the surface of the high-entropy CuNiMoFe alloy due to the presence of large lamellar channels and small nanopores [36].

The density functional theory (DFT) calculations have been conducted to investigate the electrochemical properties of the CuAlNiMoFe electrode. The results demonstrate that the energy barrier for water dissociation decreases from 0.91 eV on the CuNiFe surfaces to 0.52 eV on the CuNi-MoFe surface when Mo is included (**Figure 6a**). This inclusion of Mo plays a crucial role in facilitating the efficient dissociation of H_2O into H* and HO* by promoting hydrogen and hydroxyl adsorption on Ni(Fe) and Mo, respectively. The CuAlNiMoFe catalyst, with a monolithic nanoporous structure, exhibits impressive electrocatalytic performance. It achieves current densities of approximately 1840 and 100 mA cm^{-2} at -240 mV in 1 M KOH (**Figure 6b**). This result surpasses the performance of lots of reported advanced nonprecious electrocatalytic materials.

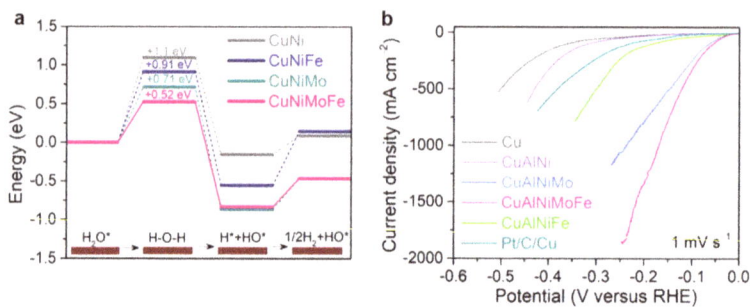

Figure 6. (a) Energy diagram of water dissociation on surface CuNi, CuNiFe, CuNiMo, and CuNiMoFe alloys at different stages of the reaction. (b) Polarization curves for nanoporous Cu, CuAlNi, CuAlNiMo, CuAlNiMoFe and CuAlNiFe catalysts. Reproduced with permission from ref [35]. Copyright 2020, Wiley-VCH.

3. Metal/Compound Composites

The production of hetero-nanostructures consisting of metal or metal compounds has emerged as a promising approach for developing cost-effective and high-performing electrocatalysts. Incorporating metal compounds such as sulfides, phosphides, oxides, and carbides [37-39] can effectively modify the energy levels of the metal catalyst, maximizing its efficiency and reducing the reliance on expensive noble metal species, ultimately lowering the overall cost. The hetero-interface between the metal and metal compounds in the catalyst plays a crucial role, as modifying each surface can have a synergistic influence on the catalytic process [40].

Wu et al. presented a catalyst composed of Ni and MoN nanoparticles on amorphous MoN nanorods, which exhibited exceptional performance in sustaining high-current-density HER [41]. The catalyst was prepared by initially depositing a self-sustaining NiMoO$_4$•xH$_2$O precursor on copper foam (CF) through a water bath reaction at 90 °C for 8 hours. It was then transformed into Ni-MoN through a subsequent ammonia reduction process at 400 °C for 2 hours. The SEM image depicts a configuration of vertically oriented nanorods with nanoparticles present on their surfaces. TEM images in **Figure 7a-b** show that the nanoparticles are incorporated inside the nanorod matrix and have diameters ranging from several to tens of nanometers. The hierarchical nanorod-nanoparticle structure with significant surface roughness offers two advantages: it provides a large number of active sites and prevents catalyst agglomeration during the HER process. This structure improves both the catalytic activity and durability of the catalyst.

TEM analysis of the crystalline nanoparticles shows precise measurement of lattice fringes at the center of each nanoparticle, with interplanar spacings of 0.176 nm (**Figure 7c-d**), corresponding to the (200) planes of metallic Ni. The nanoparticle edges (**Figure 7c** and **e**) exhibit a lattice fringe with an interplanar spacing of 0.186 nm, specifically assigned to the (202) plane of MoN. HR-TEM images clearly reveal the borders between the amorphous nanorod matrix and the crystalline nanoparticles, as well as the boundaries between the metallic Ni and MoN phases. The presence of these boundaries and flaws in multiple dimensions can increase the number of active sites, thereby enhancing the catalytic reaction.

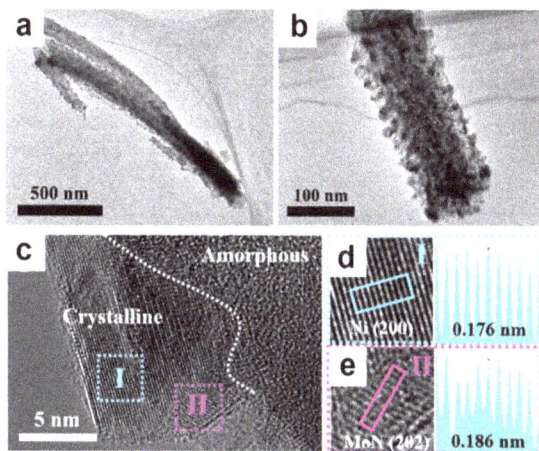

Figure 7. *(a,b) TEM images and (c-e) HR-TEM images of Ni-MoN. Reproduced with permission from ref [41]. Copyright 2022, Wiley-VCH.*

The charge density discrepancies observed in **Figure 8a-b** indicate the presence of electron accumulation at the heterointerface, suggesting a significant exchange of charges and electronic adjustment between Ni and MoN. In contrast to the electronic structure of MoN, when the electronic structure in Ni-MoN is modulated, there is an accumulation of charge density on both the N and Mo sites. The consecutive arrangement of the density of states (**Figure 8c**) near the Fermi level indicates that Ni-MoN is in a metallic state, which is advantageous for achieving high electronic conductivity.

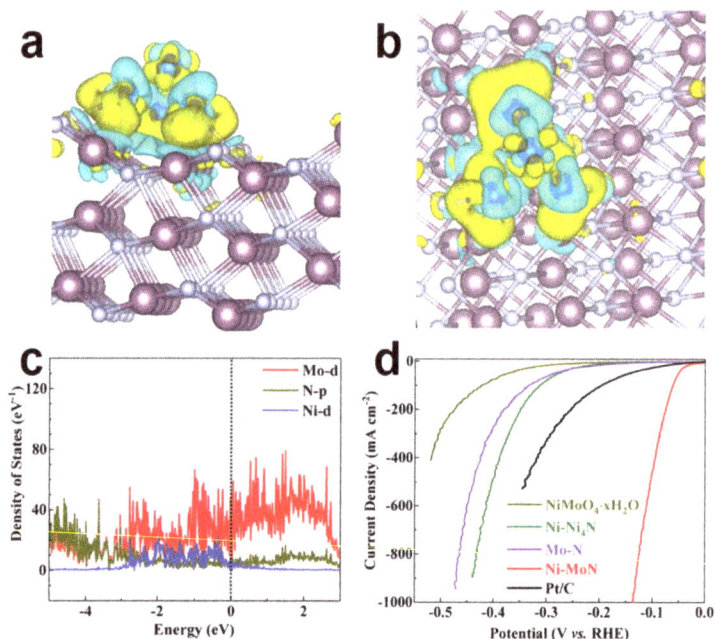

Figure 8. *(a) Side and (b) top views of charge density differences in Ni-MoN. Yellow and cyan regions represent electron accumulation and depletion, respectively. (c) DOS calculated for Mo, N, and Ni in Ni-MoN. The black dotted line indicates the Fermi level. (d) LSV curves. Reproduced with permission from ref [41]. Copyright 2022, Wiley-VCH.*

The Ni-MoN material possesses a significant surface area and several multidimensional boundaries/defects, which expose a large number of active sites. Its mesoporous structure and hydrophilic surface contribute to efficient electrolyte diffusion and rapid release of gas bubbles during the catalytic process. Theoretical calculations suggest that when Ni and MoN come into contact, there is a transfer of charge and redistribution of electrons. Consequently, the Ni-MoN interface exhibits metallic behavior with high conductivity. The presence of Mo sites in Ni-MoN significantly enhances the kinetics of the sluggish alkaline HER by increasing water adsorption and dissociation capacity. As a result, the Ni-MoN catalyst requires minimal overpotentials, specifically 61 and 136 mV (**Figure 8d**), to achieve current densities of 100 and 1000 mA cm^{-2}, respectively, in 1 M KOH solution.

Xiao et al. [42] developed a bifunctional catalyst composed of a metallic hierarchical Co/MoNi heterostructure decorated with self-supported monocrystalline CoNiMoO$_x$ nanorods grown on Ni foam for the HER and hydrogen oxidation reaction (HzOR). The synthesis method for the CoNiMo/CoNiMoO$_x$ catalyst involves a simple hydrothermal process followed by calcination in an H$_2$/Ar environment, as depicted in **Figure 9a**. This approach results in the formation of a metallic hierarchical Co/MoNi heterostructure on self-supported monocrystalline $CoNiMoO_x$ nanorods modified with oxygen vacancies. The purpose of this structure is to enable highly efficient electrolysis of HzOR and HER.

Figure 9b clearly illustrates the distinct boundaries between the $CoNiMoO_x$ nanorod and the loaded nanoclusters. The $CoNiMoO_x$ nanorod, located in the specified region, exhibits a characteristic monocrystalline nature, as observed in the inset dot-like selected area electron diffraction (SAED) pattern after the calcination process. The monocrystalline form of the material provides exceptional stability as it lacks anisotropic stress induced by polycrystalline species. Furthermore, the HR-TEM image reveals important structural details. The gap between lattice fringes, measuring 0.351 nm (**Figure 9c**), corresponds to the (102) plane of $CoMoO_3$ in the $CoNiMoO_x$ nanorod region. In the nanocluster region, the HR-TEM image shows lattice fringe spacing of 0.206 nm, which corresponds to the (133) plane of MoNi. Additionally, lattice fringe spacings of 0.217 nm and 0.191 nm, with an intersection angle of 45°, are attributed to the (100) and (101) planes of Co.

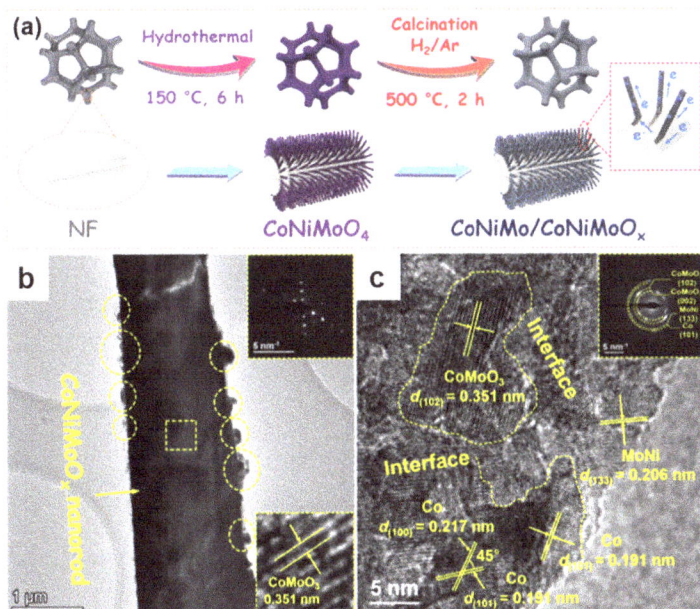

Figure 9. (a) Schematic illustration of the synthesis process. (b) TEM image and SAED pattern for the $CoNiMoO_x$ nanorod. (c) HR-TEM image and SAED pattern of the nanocluster. Reproduced with permission from ref [42]. Copyright 2023, the Royal Society of Chemistry.

Figure 10a displays the high-resolution X-ray photoelectron spectroscopy (XPS) spectra of Co-$2p$ in the $CoNiMo/CoNiMoO_x$ catalyst. The spectra exhibit two sets of spin-orbit doublets of

Co-2p, along with two satellite peaks. The peaks observed at binding energies of 780.9 and 784.0 eV can be attributed to the Co^{3+} and Co^{2+} states, respectively. Conversely, the peak observed at a binding energy of 777.8 eV corresponds to the metallic Co^0 state. Similarly, the Ni-2p spectrum in **Figure 10b** can be fitted into two pairs of spin–orbit doublets. The peak at binding energy of 852.4 eV belongs to the metallic Ni^0 state and the other peaks at binding energies of 855.7 and 858.2 eV are respectively attributed to the Ni^{2+} and Ni^{3+} states. The peaks observed at binding energies of 229.1 and 230.2 eV in **Figure 10c** represent the Mo^{4+} and Mo^{5+} states, respectively. These peaks indicate the presence of oxygen vacancies in the $CoNiMoO_x$ nanorods that were formed during the calcination process. Additionally, the highest point of the metallic Mo^0 state can be found at a binding energy of 227.9 eV.

The presence of zero-valence metallic states observed in the aforementioned findings indicates the successful formation of the Co/MoNi heterostructure on oxygen vacancies-modified $CoNiMoO_x$ nanorods. As depicted in **Figure 10d**, the $CoNiMo/CoNiMoO_x$ material demonstrates outstanding performance in HER. It achieves current densities of 100 and 300 mA cm^{-2} with remarkably low applied potentials of only -82 and -146 mV (*vs.* RHE) in a 1.0 M NaOH solution, respectively. Notably, the exceptional HER performance of the catalyst effectively counters the interference caused by a high concentration of chloride ions (Cl^-).

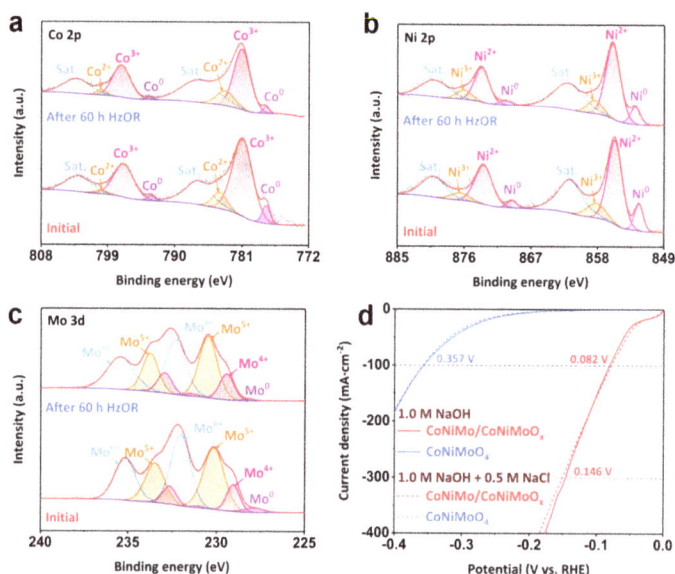

Figure 10. *High-resolution XPS spectrum of (a) Co-2p, (b) Ni-2p, and (c) Mo-3d for $CoNiMo/CoNiMoO_x$ and $CoNiMoO_4$. (d) Polarization curves of NF, $CoNiMoO_4$, and $CoNiMo/CoNiMoO_x$ in solutions of 1.0 M NaOH with/without 0.5 M NaCl. Reproduced with permission from ref [42]. Copyright 2023, the Royal Society of Chemistry.*

4.   Compound/Compound Composites

The development of economical and efficient water electrolytic catalysts is crucial for large-scale production of green hydrogen. To meet the requirements of industrial conditions, it is necessary to address the challenge of achieving both activity and stability at high current densities. One approach to tackle this challenge is by creating a wider range of active sites. Composite catalysts play a significant role in introducing multiple active sites, which can enhance the overall catalytic performance [43].

4.1.   Crystal-Amorphous Composites

Compared to crystalline materials, amorphous materials offer several advantages. These advantages include [44-46] : (1) Sufficient active sites: Amorphous materials possess unsaturated coordination sites and random-oriented suspension bonds, which enhance the adsorption of reactants and provide more active sites for catalytic reactions; (2) Enhanced reaction kinetics: The highly defective structure and disordered atomic arrangement in amorphous materials facilitate ion diffusion and electron transfer, leading to accelerated reaction kinetics; (3) Self-restructuring capability: The flexible and transportable nature of amorphous materials allows for self-restructuring during electrocatalysis, enabling good reaction adaptability and improved catalytic performance. However, amorphous materials also have inherent drawbacks, such as high cation solubility, poor electrical conductivity, and low stability [47]. To overcome these limitations, constructing crystal-amorphous composite materials has emerged as an effective strategy to develop catalysts with high activity and stability.

Guo et al. [48] proposed a simple and mild method to fabricate crystalline-amorphous metal phosphate (MPi) and NiS composites on nickel foam (NF) for hydroelectrolysis applications. The resulting catalyst exhibited excellent performance, with FePi-NiS/NF achieving an ultralow overpotential of 345 mV for the oxygen evolution reaction (OER) and NiCoPi-NiS/NF achieving an overpotential of 223 mV for the HER at a current density of 1000 mA cm^{-2}. Under high current density conditions, the composite material can undergo in situ conversion into highly active and target species. The unique pore structure of the composite, interlinked by nanosheets, provides abundant catalytic sites and efficient channels for bubble diffusion.

When tested under industrial conditions (6 M KOH, 70 $^\circ$C), the assembled electrode achieved a current density of 1000 mA cm^{-2} with only 1.712 V voltage. Upon scaling up the electrode to a larger size (area: $\approx$ 25 cm^2), it exhibited excellent performance for alkaline electrolysis using anion exchange membranes (AEMWE). Under industrial electrolysis conditions (1 M KOH, 50 $^\circ$C), the electrode achieved a high current of 12.5 A at a voltage of 1.87 V, with an energy efficiency of up to 79.2% and excellent durability for up to 30 hours. It outperformed the standard Pt/C/NF||IrO$_2$/NF electrode, resulting in energy savings of 0.215 kWh Nm^{-3}. In addition to crystal-amorphous composites, carbon nanotube composites and molybdenum matrix composites are also introduced as promising strategies for designing and constructing stable and active catalysts for various applications.

4.2.   Carbon-based Composites

Numerous studies have demonstrated the catalytic activity of transition metal oxides, particularly cobalt-based materials, for the HER [49, 50]. Cobalt-containing electrocatalysts exhibit excellent HER catalytic activity due to their unique external electronic structure [51]. For example, Dai et al. [51] synthesized Co/CoO anchored to a nitrogen-doped carbon composite, achieving an overpotential of 152 mV (at 10 mA cm^{-2}) in the HER. Similarly, Li et al. [52] developed an amorphous CoFe Oxide@2D black phosphorus material, demonstrating an overpotential of 88 mV (at 10 mA cm^{-2}) in the HER. Shu et al. [53] reported a carbon material doped with Co/P (Co/P/C), which exhibited an HER overpotential of 240 mV (at 10 mA cm^{-2}). Wang et al. [54] constructed a Ni-doped layered nanosheet of Co_3S_4 derived from a metal-organic framework, showing an overpotential of 173 mV at the current density of 10 mA cm^{-2} in HER. Feng et al. [55] synthesized a Co-doped VS_2 composite on NF, achieving an HER overpotential of only 164.5 mV at a current density of 10 mA cm^{-2}. These results establish cobalt as a core element in the construction of excellent electrocatalysts for the HER.

In cobalt compound catalysts, the cobalt antimony oxide has demonstrated good electrochemical activity, particularly in OER performance [56, 57]. However, the electrochemical hydrogen evolution properties of cobalt antimony oxide are still not fully elucidated. Additionally, efficient electron conduction capacity is crucial to enhance the efficiency of the HER, as it occurs under electrochemical conditions [58, 59]. Semiconductor catalysts, often used as HER catalysts, frequently exhibit unsatisfactory electron transfer performance. To address this issue and achieve better HER reaction performance, combining HER catalysts with materials possessing excellent conductive properties such as nickel foam [60, 61], carbon cloth [62, 63], carbon felt [64, 65], and carbon nanotubes [66, 67] has proven effective.

For example, Ding et al. [59] synthesized a cobalt phosphide/carbon nanotube composite, which exhibited significantly higher HER activity compared to the original cobalt phosphide. Wang et al. [68] coated carbon nanotubes with covalent metal porphyrin polymers, resulting in a significant decrease in HER overpotential. Li et al. [69] prepared a Ni_3P-Ni heterostructure and carbon nanotube composite electrode, demonstrating excellent HER catalytic efficiency. Liu et al. [70] supported Ir-Ni/NiO nanoparticles on carbon nanotubes, achieving lower HER overpotential. Moreover, Peng et al. [71] synthesized a layered nanosheet structure composed of separated vanadium carbide nanoparticles wrapped in a highly conductive mesoporous graphitic carbon network (VC-NS) through hydrothermal and low temperature magnesia thermal reactions. This structure exhibited a large specific surface area, minimal overpotential, fast proton discharge kinetics, excellent durability, and efficient HER activity. These results provide evidence that combining active electrocatalysts with carbon-based materials is beneficial for improving the efficiency of electrochemical hydrogen evolution.

4.3.   Mo-based Composites

In their pioneering work, Wang et al. [72] presented a groundbreaking method for fabricating biphasic nitride nanoribbons consisting of Mo_2N and $Ni_{0.2}Mo_{0.8}N$. Their approach involved intercalating MoO_3 nanoribbons (NBs) with Ni^{2+} nitride. The resulting $Mo_2N/Ni_{0.2}Mo_{0.8}N$ catalyst exhibited exceptional stability and remarkable performance in the HER. It displayed an impressively low overpotential of only 26 mV (at 10 mA cm^{-2}) and a Tafel slope of 31 mV dec^{-1}, both in alkaline electrolyte and simulated seawater, rivalling or even surpassing the benchmark Pt/C catalyst. The outstanding alkaline HER properties of the Mo$_2$N/Ni$_{0.2}$Mo$_{0.8}$N catalyst can be attributed to its unique heterostructure, which boasts adjustable content and a robust interface that effectively prevents metallic nickel segregation and structural collapse during the nitriding process. Through a combination of DFT calculations and experimental investigations, Wang et al. unveiled that the Mo$_2$N/Ni$_{0.2}$Mo$_{0.8}$N interface, characterized by strong electron interactions, optimizes the adsorption and desorption of hydrogen molecules. This results in the creation of moderately weak bonding metal sites with positive ΔG_{H*} values (representing the Gibbs free energy change for the hydrogen adsorption step), which act as highly efficient catalytic centers. Consequently, the HER kinetics are significantly accelerated, leading to improved overall HER activity.

In addition, Feng et al. [73] employed a programmed in situ nitriding process to fabricate MoO$_2$/Mo$_2$N heterostructures, as illustrated in **Figure 11a**. They employed XPS to study the changes in Mo-N/Mo-O content and subsequently regulate the coordination of Mo atoms within the heterostructures, as shown in **Figure. 11b**. Notably, their findings, illustrated in **Figure 11c**, underscored the effectiveness of coordinating Mo atoms as a strategy for optimizing the electronic configuration and expediting the dynamics of the HER.

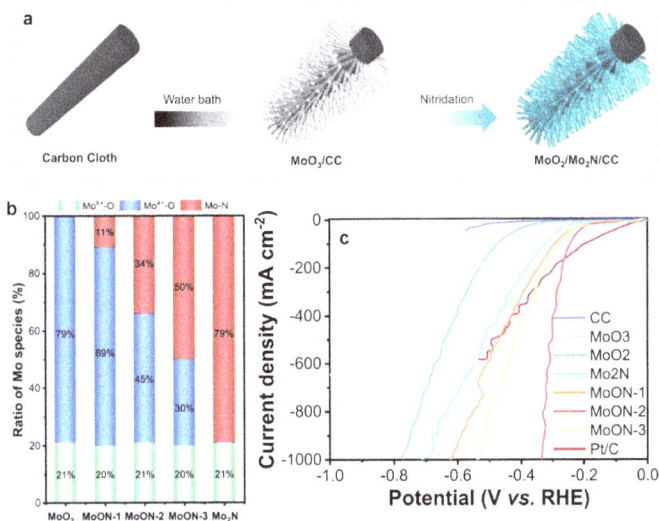

Figure 11. (a) Schematic showing the fabrication of the MoO$_2$/Mo$_2$N electrocatalysts. (b) Ratios of Mo species in MoO$_2$, MoON-1, MoON-2, MoON-3, and Mo2N. (c) LSV curves of MoO$_2$, MoON-1, MoON-2, MoON-3, and Mo$_2$N in 0.5 M H$_2$SO$_4$. Reproduced with permission from ref [73]. Copyright 2023, Elsevier.

Molybdenum selenide ($MoSe_2$) based nanocomposites have gained significant attention as highly promising electrochemical catalysts for hydroxyl ion reactions. These nanocomposites exhibit exceptional catalytic activity, long-term stability, and cost-effectiveness. The incorporation of $MoSe_2$ with other composite materials leads to a synergistic effect that further enhances the electrocatalytic performance. Zhou et al. [74] designed a $MoSe_2$-NiSe heterostructure, which demonstrated a low overpotential of 210 mV and a Tafel slope of 56 mV dec^{-1} at 10 mA cm^{-2} (**Figure 12a-c**). The vertical nanohybrids with well-aligned bands facilitate the transfer of electrons from metallic NiSe to semiconductive $MoSe_2$, enabling electronic modulation and improved HER activity.

In another study, Mu et al. [75] synthesized a layered nanosheet assembled $MoSe_2$/$CoSe_2$ microcage using a one-step hydrothermal method, resulting in a highly efficient HER catalyst. Generally, during the hydrothermal reaction, $CoSe_2$ tends to form nanoparticles while $MoSe_2$ tends to form nanosheets. However, the $MoSe_2$/$CoSe_2$ composites with varying ratios of molybdenum and cobalt form unique layered nanosheet assembly tubular microcage morphologies. The formation of these novel nanostructures can be attributed to the different crystallization and competitive nucleation and growth mechanisms of $MoSe_2$ and $CoSe_2$. The hydrothermal process ensures the complete mixing of $CoSe_2$ and $MoSe_2$ at the nanoscale level, and the synergistic effect of $CoSe_2$ and $MoSe_2$ is induced by defects. It was discovered that controlling the Co/Mo ratio allows for the adjustable structure and opens possibilities for preparing Mo-Co matrix composites.

Electrochemical tests provided compelling evidence of the synergistic effect, as the $MoSe_2$/$CoSe_2$ microcage exhibited lower overpotential, higher current density, and a smaller Tafel slope compared to the individual $MoSe_2$ and $CoSe_2$. Surprisingly, the sample with the highest cobalt content did not exhibit the best HER behavior. Instead, samples with a moderate Mo/Co molar ratio demonstrated higher HER electrocatalytic activity, with an initial overpotential of 110 mV and a Tafel slope of 73 mV dec^{-1}. These results highlight the influence of the microcage morphology assembled by layered nanosheets and the synergistic effect induced by defect formation on the enhanced HER activity. However, the competitive nucleation and growth mechanisms require further investigation, and a deeper understanding of the synergistic effects between these composites is necessary.

Ren et al. [76] also reported a synergistic effect between $MoSe_2$ and MoS_2. They synthesized $MoSe_2$@MoS_2 nanostructures through a two-step hydrothermal method, which provided additional active sites and exhibiting synergistic effects in catalyzing hydrogen evolution. The synthesized $MoSe_2$@MoS_2 nuclear/shell nanostructures exhibited a more complex and porous morphology compared to the original $MoSe_2$ nanosheets, indicating that the co-growth of $MoSe_2$ and MoS_2 prevented aggregation and provided more active sites. XPS analysis revealed a strong electronic interaction between the MoS_2 core and $MoSe_2$ shell, as indicated by the shifted peaks in the S-$2p$ spectra from 161.50 eV and 162.62 eV for original MoS_2 to 161.67 eV and 162.78 eV for $MoSe_2$@MoS_2 composites. This strong electron interaction between the core and shell resulted in a synergistic effect during the reaction, leading to enhanced HER activity, a lower initial overpotential of 161 mV, a smaller Tafel slope of 60 mV dec^{-1}, and improved stability.

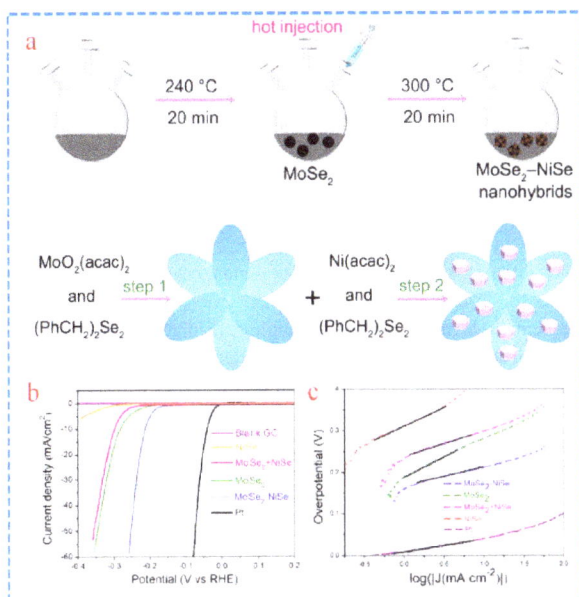

Figure 12. (a) Illustration of the synthesis of MoSe₂-NiSe nanohybrids. (b) Polarization curves for HER on a blank glassy carbon (GC) electrode and modified GC electrodes comprising MoSe₂-NiSe nanohybrids, pure MoSe₂, pure NiSe, physically mixed MoSe₂ and NiSe (denoted as MoSe₂+NiSe), and Pt. (c) Tafel plots. Reproduced with permission from ref [74]. Copyright 2016, American Chemical Society.

5.   Conclusion and Prospects

Transition metal compounds have emerged as effective catalyst materials for the HER due to their favorable properties such as low cost, good electrical conductivity, and high stability. Researchers have made significant progress in developing various composite materials based on transition metal compounds for HER.   This chapter aims to classify and discuss the characteristics and electrocatalytic activities of different types of composites, including metal/metal, metal/compound, and compound/compound composites.   These transition metal composites offer a promising avenue for achieving sustainable and cost-effective hydrogen production and utilization.

To design efficient transition metal composite catalysts, several key issues need to be better understood:

(i) Surface Atomic Arrangement and Defects: HER primarily occurs at the catalyst's surface, where the atomic arrangement and defects have a dominant role in catalytic activity during water splitting. Gaining a profound understanding of the evolution of active sites and the HER mechanism is crucial for optimizing the catalyst's performance in hydrogen evolution.

(ii) Electrolyte Polarization: HER is typically conducted in alkaline or acidic electrolytes, which can polarize the surface of transition metal compound catalysts at high electrolyte concentrations. This polarization process can significantly impact the stability of catalysts. Therefore, comprehensively understanding the polarization phenomena occurring on the catalyst's surface is essential for enhancing their long-term stability.

(iii) Synergistic Effects in Composites: Composites combining transition metal compounds with carbon materials or other transition metal compounds often exhibit higher electrocatalytic activity compared to single compounds. This enhancement is attributed to synergistic effects between the components. However, the underlying mechanisms behind these synergistic effects are still not fully understood and require further investigation.

References

[1]   S.O. Ganiyu, C.A. Martínez-Huitle, The use of renewable energies driving electrochemical technologies for environmental applications, Curr. Opin. Electroche., 22 (2020) 211-220. https://doi.org/10.1016/j.coelec.2020.07.007

[2]   S. Chu, Y. Cui, N. Liu, The path towards sustainable energy, Nat. Mater., 16 (2017) 16-22. https://doi.org/10.1038/nmat4834

[3]   T.-H. Yang, J. Ahn, S. Shi, P. Wang, R. Gao, D. Qin, Noble-metal nanoframes and their catalytic applications, Chem. Rev., 121 (2021) 796-833. https://doi.org/10.1021/acs.chemrev.0c00940

[4]   C.-X. Zhao, H.-F. Wang, B.-Q. Li, Q. Zhang, Multianion transition metal compounds: synthesis, regulation, and electrocatalytic applications, Acc. Mater. Res., 2 (2021) 1082-1092. https://doi.org/10.1021/accountsmr.1c00136

[5]   J. Hu, C. Zhang, X. Meng, H. Lin, C. Hu, X. Long, S. Yang, Hydrogen evolution electrocatalysis with binary-nonmetal transition metal compounds, J. Mater. Chem. A, 5 (2017) 5995-6012. https://doi.org/10.1039/C7TA00743D

[6]   R. Bose, V.R. Jothi, K. Karuppasamy, A. Alfantazi, S.C. Yi, High performance multicomponent bifunctional catalysts for overall water splitting, J. Mater. Chem. A, 8 (2020) 13795-13805. https://doi.org/10.1039/D0TA02697B

[7]   W. Guo, Q.V. Le, H.H. Do, A. Hasani, M. Tekalgne, S.-R. Bae, T.H. Lee, H.W. Jang, S.H. Ahn, S.Y. Kim, Ni_3Se_4@$MoSe_2$ composites for hydrogen evolution reaction, Appl. Sci., 9 (2019) 5035. https://doi.org/10.3390/app9235035

[8]   R. Saini, F. Naaz, A.H. Bashal, A.H. Pandit, U. Farooq, Recent advances in nitrogen-doped graphene-based heterostructures and composites: mechanism and active sites for electrochemical ORR and HER, Green Chem., 26 (2024) 57-102. https://doi.org/10.1039/D3GC03576J

[9]   C. Cai, K. Liu, Y. Zhu, P. Li, Q. Wang, B. Liu, S. Chen, H. Li, L. Zhu, H. Li, J. Fu, Y.
        Chen, E. Pensa, J. Hu, Y.-R. Lu, T.-S. Chan, E. Cortés, M. Liu, Optimizing Hydrogen
        Binding on Ru Sites with RuCo Alloy Nanosheets for Efficient Alkaline Hydrogen
        Evolution, Angew. Chem. Int. Ed., 61 (2022) e202113664.
        https://doi.org/10.1002/anie.202113664

[10]   D. Lancet, I. Pecht, Spectroscopic and immunochemical studies with
        nitrobenzoxadiazolealanine, a fluorescent dinitrophenyl analog, Biochemistry, 16 (1977)
        5150-5157. https://doi.org/10.1021/bi00642a031

[11]   X. Mu, J. Gu, F. Feng, Z. Xiao, C. Chen, S. Liu, S. Mu, RuRh Bimetallene Nanoring as
        High-efficiency pH-Universal Catalyst for Hydrogen Evolution Reaction, Adv. Sci., 8
        (2021) 2002341. https://doi.org/10.1002/advs.202002341

[12]   Y. Nakaya, S. Furukawa, Catalysis of Alloys: Classification, Principles, and Design for a
        Variety of Materials and Reactions, Chem. Rev., 123 (2023) 5859-5947.
        https://doi.org/10.1021/acs.chemrev.2c00356

[13]   L. Rößner, H. Schwarz, I. Veremchuk, R. Zerdoumi, T. Seyller, M. Armbrüster,
        Challenging the Durability of Intermetallic Mo-Ni Compounds in the Hydrogen
        Evolution Reaction, ACS Appl. Mater. Interfaces, 13 (2021) 23616-23626.
        https://doi.org/10.1021/acsami.1c02169

[14]   B.P. Williams, Z. Qi, W. Huang, C.-K. Tsung, The impact of synthetic method on the
        catalytic application of intermetallic nanoparticles, Nanoscale, 12 (2020) 18545-18562.
        https://doi.org/10.1039/D0NR04699J

[15]   Z. Pu, T. Liu, G. Zhang, Z. Chen, D.-S. Li, N. Chen, W. Chen, Z. Chen, S. Sun, General
        Synthesis of Transition-Metal-Based Carbon-Group Intermetallic Catalysts for Efficient
        Electrocatalytic Hydrogen Evolution in Wide pH Range, Adv. Energy Mater., 12 (2022)
        2200293. https://doi.org/10.1002/aenm.202200293

[16]   W. Xiao, W. Lei, M. Gong, H.L. Xin, D. Wang, Recent Advances of Structurally Ordered
        Intermetallic Nanoparticles for Electrocatalysis, ACS Catal., 8 (2018) 3237-3256.
        https://doi.org/10.1021/acscatal.7b04420

[17]   Y. Wang, W. Zheng, B. Wang, L. Ling, R. Zhang, The effects of doping metal type and
        ratio on the catalytic performance of C_2H_2 semi-hydrogenation over the intermetallic
        compound-containing Pd catalysts, Chem. Eng. Sci. , 229 (2021) 116131.
        https://doi.org/10.1016/j.ces.2020.116131`

[18]   D. Yuan, L. Cai, T. Xie, H. Liao, W. Hu, Selective hydrogenation of acetylene on Cu–Pd
        intermetallic compounds and Pd atoms substituted Cu(111) surfaces, Phys. Chem. Chem.
        Phys., 23 (2021) 8653-8660. https://doi.org/10.1039/D0CP05285J

[19]   J.-S. Sun, Z. Wen, L.-P. Han, Z.-W. Chen, X.-Y. Lang, Q. Jiang, Nonprecious
        Intermetallic Al_7Cu_4Ni Nanocrystals Seamlessly Integrated in Freestanding Bimodal
        Nanoporous Copper for Efficient Hydrogen Evolution Catalysis, Adv. Funct. Mater., 28
        (2018) 1706127. https://doi.org/10.1002/adfm.201706127

[20] M. Shao, J.H. Odell, A. Peles, D. Su, The role of transition metals in the catalytic activity of Pt alloys: quantification of strain and ligand effects, Chem. Commun., 50 (2014) 2173-2176. https://doi.org/10.1039/c3cc47341d

[21] A. Nairan, P. Zou, C. Liang, J. Liu, D. Wu, P. Liu, C. Yang, NiMo Solid Solution Nanowire Array Electrodes for Highly Efficient Hydrogen Evolution Reaction, Adv. Funct. Mater., 29 (2019) 1903747. https://doi.org/10.1002/adfm.201903747

[22] P.K. Huang, J.W. Yeh, T.T. Shun, S.K. Chen, Multi-Principal-Element Alloys with Improved Oxidation and Wear Resistance for Thermal Spray Coating, Adv. Eng. Mater., 6 (2004) 74-78. https://doi.org/10.1002/adem.200300507

[23] D.B. Miracle, O.N. Senkov, A critical review of high entropy alloys and related concepts, Acta Mater., 122 (2017) 448-511. https://doi.org/10.1016/j.actamat.2016.08.081

[24] M.A. Hemphill, T. Yuan, G.Y. Wang, J.W. Yeh, C.W. Tsai, A. Chuang, P.K. Liaw, Fatigue behavior of $Al_{0.5}CoCrCuFeNi$ high entropy alloys, Acta Mater., 60 (2012) 5723-5734. https://doi.org/10.1016/j.actamat.2012.06.046

[25] M.-H. Chuang, M.-H. Tsai, W.-R. Wang, S.-J. Lin, J.-W. Yeh, Microstructure and wear behavior of $Al_xCo_{1.5}CrFeNi_{1.5}Ti_y$ high-entropy alloys, Acta Mater., 59 (2011) 6308-6317. https://doi.org/10.1016/j.actamat.2011.06.041

[26] Y. Shi, B. Yang, X. Xie, J. Brechtl, K.A. Dahmen, P.K. Liaw, Corrosion of Al xCoCrFeNi high-entropy alloys: Al-content and potential scan-rate dependent pitting behavior, Corros. Sci., 119 (2017) 33-45. https://doi.org/10.1016/j.corsci.2017.02.019

[27] Y.D. Wu, Y.H. Cai, T. Wang, J.J. Si, J. Zhu, Y.D. Wang, X.D. Hui, A refractory $Hf_{25}Nb_{25}Ti_{25}Zr_{25}$ high-entropy alloy with excellent structural stability and tensile properties, Mater. Lett., 130 (2014) 277-280. https://doi.org/10.1016/j.matlet.2014.05.134

[28] X. Chang, M. Zeng, K. Liu, L. Fu, Phase Engineering of High-Entropy Alloys, Adv. Mater., 32 (2020) 1907226. https://doi.org/10.1002/adma.201907226

[29] S. Han, C. He, Q. Yun, M. Li, W. Chen, W. Cao, Q. Lu, Pd-based intermetallic nanocrystals: From precise synthesis to electrocatalytic applications in fuel cells, Coord. Chem. Rev., 445 (2021) 214085. https://doi.org/10.1016/j.ccr.2021.214085

[30] D. Feng, Y. Dong, P. Nie, L. Zhang, Z.-A. Qiao, CoNiCuMgZn high entropy alloy nanoparticles embedded onto graphene sheets via anchoring and alloying strategy as efficient electrocatalysts for hydrogen evolution reaction, Chem. Eng. J., 430 (2022) 132883. https://doi.org/10.1016/j.cej.2021.132883

[31] Y. Pan, J.-X. Liu, T.-Z. Tu, W. Wang, G.-J. Zhang, High-entropy oxides for catalysis: A diamond in the rough, Chem. Eng. J., 451 (2023) 138659. https://doi.org/10.1016/j.cej.2022.138659

[32] F. Waag, Y. Li, A.R. Ziefuß, E. Bertin, M. Kamp, V. Duppel, G. Marzun, L. Kienle, S. Barcikowski, B. Gökce, Kinetically-controlled laser-synthesis of colloidal high-entropy alloy nanoparticles, RSC Adv., 9 (2019) 18547-18558.

https://doi.org/10.1039/C9RA03254A

[33] S. Gao, S. Hao, Z. Huang, Y. Yuan, S. Han, L. Lei, X. Zhang, R. Shahbazian-Yassar, J. Lu, Synthesis of high-entropy alloy nanoparticles on supports by the fast moving bed pyrolysis, Nat. Commun., 11 (2020) 2016. https://doi.org/10.1038/s41467-020-15934-1

[34] M. Liu, Z. Zhang, F. Okejiri, S. Yang, S. Zhou, S. Dai, Entropy-Maximized Synthesis of Multimetallic Nanoparticle Catalysts via a Ultrasonication-Assisted Wet Chemistry Method under Ambient Conditions, Adv. Mater. Interfaces, 6 (2019) 1900015. https://doi.org/10.1002/admi.201900015

[35] R.-Q. Yao, Y.-T. Zhou, H. Shi, W.-B. Wan, Q.-H. Zhang, L. Gu, Y.-F. Zhu, Z. Wen, X.-Y. Lang, Q. Jiang, Nanoporous Surface High-Entropy Alloys as Highly Efficient Multisite Electrocatalysts for Nonacidic Hydrogen Evolution Reaction, Adv. Funct. Mater., 31 (2021) 2009613. https://doi.org/10.1002/adfm.202009613

[36] C. Zhu, D. Du, A. Eychmüller, Y. Lin, Engineering Ordered and Nonordered Porous Noble Metal Nanostructures: Synthesis, Assembly, and Their Applications in Electrochemistry, Chem. Rev., 115 (2015) 8896-8943. https://doi.org/10.1021/acs.chemrev.5b00255

[37] Q. Shao, P. Wang, X. Huang, Opportunities and Challenges of Interface Engineering in Bimetallic Nanostructure for Enhanced Electrocatalysis, Adv. Funct. Mater., 29 (2019) 1806419. https://doi.org/10.1002/adfm.201806419

[38] X. Liu, M.C. Hersam, Interface Characterization and Control of 2D Materials and Heterostructures, Adv. Mater., 30 (2018) 1801586. https://doi.org/10.1002/adma.201801586

[39] Y. Yang, M. Luo, W. Zhang, Y. Sun, X. Chen, S. Guo, Metal surface and interface energy electrocatalysis: fundamentals, performance engineering, and opportunities, Chem, 4 (2018) 2054-2083. https://doi.org/10.1016/j.chempr.2018.05.019

[40] T. Kwon, M. Jun, J. Joo, K. Lee, Nanoscale hetero-interfaces between metals and metal compounds for electrocatalytic applications, J. Mater. Chem. A, 7 (2019) 5090-5110. https://doi.org/10.1039/C8TA09494B

[41] L. Wu, F. Zhang, S. Song, M. Ning, Q. Zhu, J. Zhou, G. Gao, Z. Chen, Q. Zhou, X. Xing, T. Tong, Y. Yao, J. Bao, L. Yu, S. Chen, Z. Ren, Efficient Alkaline Water/Seawater Hydrogen Evolution by a Nanorod-Nanoparticle-Structured Ni-MoN Catalyst with Fast Water-Dissociation Kinetics, Adv. Mater., 34 (2022) 2201774. https://doi.org/10.1002/adma.202201774

[42] Z. Xiao, J. Wang, H. Lu, Y. Qian, Q. Zhang, A. Tang, H. Yang, Hierarchical Co/MoNi heterostructure grown on monocrystalline CoNiMoO$_x$ nanorods with robust bifunctionality for hydrazine oxidation-assisted energy-saving hydrogen evolution, J. Mater. Chem. A, 11 (2023) 15749-15759. https://doi.org/10.1039/D3TA02930A

[43] A. Li, X. Chang, Z. Huang, C. Li, Y. Wei, L. Zhang, T. Wang, J. Gong, Thin

heterojunctions and spatially separated cocatalysts to simultaneously reduce bulk and surface recombination in photocatalysts, Angew. Chem. Int. Ed., 55 (2016) 13734-13738. https://doi.org/10.1002/ange.201605666

[44]   S. Shen, Z. Wang, Z. Lin, K. Song, Q. Zhang, F. Meng, L. Gu, W. Zhong, Crystalline-amorphous interfaces coupling of CoSe$_2$/CoP with optimized d-band center and boosted electrocatalytic hydrogen evolution, Adv. Mater., 34 (2022) 2110631. https://doi.org/10.1002/adma.202110631

[45]   B. Zhou, R. Gao, J.J. Zou, H. Yang, Surface design strategy of catalysts for water electrolysis, Small, 18 (2022) 2202336. https://doi.org/10.1002/smll.202202336

[46]   Y. Zhang, F. Gao, D. Wang, Z. Li, X. Wang, C. Wang, K. Zhang, Y. Du, Amorphous/crystalline heterostructure transition-metal-based catalysts for high-performance water splitting, Coord. Chem. Rev., 475 (2023) 214916. https://doi.org/10.1016/j.ccr.2022.214916

[47]   T.I. Singh, A. Maibam, D.C. Cha, S. Yoo, R. Babarao, S.U. Lee, S. Lee, High-alkaline water-splitting activity of mesoporous 3D heterostructures: an amorphous-shell@crystalline-core nano-assembly of Co-Ni-phosphate ultrathin-nanosheets and V-doped cobalt-nitride nanowires, Adv. Sci., 9 (2022) 2201311. https://doi.org/10.1002/advs.202201311

[48]   L. Guo, J. Xie, S. Chen, Z. He, Y. Liu, C. Shi, R. Gao, L. Pan, Z.-F. Huang, X. Zhang, Self-supported crystalline-amorphous composites of metal phosphate and NiS for high-performance water electrolysis under industrial conditions, Appl. Catal. B-Environ., 340 (2024) 123252. https://doi.org/10.1016/j.apcatb.2023.123252

[49]   W. Zhang, L. Cui, J. Liu, Recent advances in cobalt-based electrocatalysts for hydrogen and oxygen evolution reactions, J. Alloys Compd., 821 (2020) 153542. https://doi.org/10.1016/j.jallcom.2019.153542

[50]   Y. Hua, X. Li, C. Chen, H. Pang, Cobalt based metal-organic frameworks and their derivatives for electrochemical energy conversion and storage, Chem. Eng. J., 370 (2019) 37-59. https://doi.org/10.1016/j.cej.2019.03.163

[51]   X. Peng, X. Jin, B. Gao, Z. Liu, P.K. Chu, Strategies to improve cobalt-based electrocatalysts for electrochemical water splitting, J. Catal., 398 (2021) 54-66. https://doi.org/10.1016/j.jcat.2021.04.003

[52]   X. Li, L. Xiao, L. Zhou, Q. Xu, J. Weng, J. Xu, B. Liu, Adaptive bifunctional electrocatalyst of amorphous CoFe oxide@2D black phosphorus for overall water splitting, Angew. Chem. Int. Ed., 59(2020) 21106-21113. https://doi.org/10.1002/ange.202008514

[53]   Y. Shu, K. Sasaki, Y. Fujimoto, K. Miyake, Y. Uchida, S. Tanaka, N. Nishiyama, Electrochemical hydrogen evolution reaction over Co/P doped carbon derived from triethyl phosphite-deposited 2D nanosheets of Co/Al layered double hydroxides, Int. J. Hydrogen Energ., 47 (2022) 10638-10645. https://doi.org/10.1016/j.ijhydene.2022.01.120

[54] J. Wang, Y. Wang, Z. Yao, Z. Jiang, Metal–organic framework-derived Ni doped Co_3S_4 hierarchical nanosheets as a monolithic electrocatalyst for highly efficient hydrogen evolution reaction in alkaline solution, Chin. J. Chem. Eng., 42 (2022) 380-388. https://doi.org/10.1016/j.cjche.2021.02.010

[55] T. Feng, C. Ouyang, Z. Zhan, T. Lei, P. Yin, Cobalt doping VS_2 on nickel foam as a high efficient electrocatalyst for hydrogen evolution reaction, Int. J. Hydrogen Energ., 47 (2022) 10646-10653. https://doi.org/10.1016/j.ijhydene.2022.01.132

[56] K. Ham, S. Hong, S. Kang, K. Cho, J. Lee, Extensive active-site formation in trirutile $CoSb_2O_6$ by oxygen vacancy for oxygen evolution reaction in anion exchange membrane water splitting, ACS Energy Lett., 6 (2021) 364-370. https://doi.org/10.1021/acsenergylett.0c02359

[57] T.A. Evans, K.-S. Choi, Electrochemical synthesis and investigation of stoichiometric, phase-pure $CoSb_2O_6$ and $MnSb_2O_6$ electrodes for the oxygen evolution reaction in acidic media, ACS Appl. Energy Mater., 3 (2020) 5563-5571. https://doi.org/10.1021/acsaem.0c00526

[58] E. Cao, Z. Chen, H. Wu, P. Yu, Y. Wang, F. Xiao, S. Chen, S. Du, Y. Xie, Y. Wu, Boron-induced electronic-structure reformation of CoP nanoparticles drives enhanced pH-universal hydrogen evolution, Angew. Chem. Int. Ed., 59 (2020) 4154-4160. https://doi.org/10.1002/anie.201915254

[59] Z. Ding, H. Yu, X. Liu, N. He, X. Chen, H. Li, M. Wang, Y. Yamauchi, X. Xu, M.A. Amin, Prussian blue analogue derived cobalt-nickel phosphide/carbon nanotube composite as electrocatalyst for efficient and stable hydrogen evolution reaction in wide-pH environment, J. Colloid Interf. Sci., 616 (2022) 210-220. https://doi.org/10.1016/j.jcis.2022.02.039

[60] H. Gao, J. Zang, Y. Wang, S. Zhou, P. Tian, S. Song, X. Tian, W. Li, One-step preparation of cobalt-doped $NiS@MoS_2$ core-shell nanorods as bifunctional electrocatalyst for overall water splitting, Electrochim. Acta, 377 (2021) 138051. https://doi.org/10.1016/j.electacta.2021.138051

[61] D. Liu, R. Tong, Y. Qu, Q. Zhu, X. Zhong, M. Fang, K.H. Lo, F. Zhang, Y. Ye, Y. Tang, Highly improved electrocatalytic activity of NiSx: Effects of Cr-doping and phase transition, Appl. Catal. B-Environ., 267 (2020) 118721. https://doi.org/10.1016/j.apcatb.2020.118721

[62] T. Shaker, H. Mehdipour, A.Z. Moshfegh, Low loaded MoS_2/Carbon cloth as a highly efficient electrocatalyst for hydrogen evolution reaction, Int. J. Hydrogen Energ., 47 (2022) 1579-1588. https://doi.org/10.1016/j.ijhydene.2021.10.136

[63] P.-W. Yu, S. Elmas, T. Roman, X. Pan, Y. Yin, C.T. Gibson, G.G. Andersson, M.R. Andersson, Highly active platinum single-atom catalyst grafted onto 3D carbon cloth support for the electrocatalytic hydrogen evolution reaction, Appl. Surf. Sci., 595 (2022) 153480. https://doi.org/10.1016/j.apsusc.2022.153480

[64] E.A. Şahin, Preparation of CoAg modified carbon felt electrodes for alkaline water electrolysis, Russ. J. Phys. Chem. A 96 (2022) 664-672. https://doi.org/10.1134/S0036024422030104

[65] M.B. Koca, G.G. Çelik, G. Kardaş, B. Yazıcı, NiGa modified carbon-felt cathode for hydrogen production, Int. J. Hydrogen Energ., 44 (2019) 14157-14163. https://doi.org/10.1016/j.ijhydene.2018.09.031

[66] W. Xu, H. Huang, X. Wu, Y. Yuan, Y. Liu, Z. Wang, D. Zhang, Y. Qin, J. Lai, L. Wang, Mn-doped Ru/RuO₂ nanoclusters@CNT with strong metal-support interaction for efficient water splitting in acidic media, Compos. Part B-Eng., 242 (2022) 110013. https://doi.org/10.1016/j.compositesb.2022.110013

[67] B. Liu, B. He, H.-Q. Peng, Y. Zhao, J. Cheng, J. Xia, J. Shen, T.-W. Ng, X. Meng, C.-S. Lee, W. Zhang, Unconventional nickel nitride enriched with nitrogen vacancies as a high-efficiency electrocatalyst for hydrogen evolution, Adv. Sci., 5 (2018) 1800406. https://doi.org/10.1002/advs.201800406

[68] Y. Wang, D. Song, J. Li, Q. Shi, J. Zhao, Y. Hu, F. Zeng, N. Wang, Covalent metalloporphyrin polymer coated on carbon nanotubes as bifunctional electrocatalysts for water splitting, Inorg. Chem., 61 (2022) 10198-10204. https://doi.org/10.1021/acs.inorgchem.2c01415

[69] D. Li, Z.-F. Zhang, Z.-Y. Yang, W.-Y. Wu, M.-H. Zhang, T.-R. Yang, Q.-S. Zhang, J.-Y. Xie, Ni₃P-Ni heterostructure electrocatalyst for alkaline hydrogen evolution, J. Alloys Compd., 921 (2022) 166204. https://doi.org/10.1016/j.jallcom.2022.166204

[70] J. Liu, Z. Wang, D. Zhang, Y. Qin, J. Xiong, J. Lai, L. Wang, Systematic engineering on Ni-based nanocatalysts effectively promote hydrogen evolution reaction, Small, 18 (2022) 2108072. https://doi.org/10.1002/smll.202108072

[71] X. Peng, L. Hu, L. Wang, X. Zhang, J. Fu, K. Huo, L.Y.S. Lee, K.-Y. Wong, P.K. Chu, Vanadium carbide nanoparticles encapsulated in graphitic carbon network nanosheets: A high-efficiency electrocatalyst for hydrogen evolution reaction, Nano Energy, 26 (2016) 603-609. https://doi.org/10.1016/j.nanoen.2016.06.020

[72] X. Wang, X. Zhang, Y. Xu, H. Song, X. Min, Z. Tang, C. Pi, J. Li, B. Gao, Y. Zheng, X. Peng, P.K. Chu, K. Huo, Heterojunction Mo-based binary and ternary nitride catalysts with Pt-like activity for the hydrogen evolution reaction, Chem. Eng. J., 470 (2023) 144370. https://doi.org/10.1016/j.cej.2023.144370

[73] S. Feng, D. Li, H. Dong, S. Xie, Y. Miao, X. Zhang, B. Gao, P.K. Chu, X. Peng, Tailoring the Mo-N/Mo-O configuration in MoO₂/Mo₂N heterostructure for ampere-level current density hydrogen production, Appl. Catal. B-Environ., 342 (2024) 123451. https://doi.org/10.1016/j.apcatb.2023.123451

[74] X. Zhou, Y. Liu, H. Ju, B. Pan, J. Zhu, T. Ding, C. Wang, Q. Yang, Design and epitaxial growth of MoSe₂-NiSe vertical heteronanostructures with electronic modulation for enhanced hydrogen evolution reaction, Chem. Mater., 28 (2016) 1838-1846.

https://doi.org/10.1021/acs.chemmater.5b05006

[75]   C. Mu, H. Qi, Y. Song, Z. Liu, L. Ji, J. Deng, Y. Liao, F. Scarpa, One-pot synthesis of
       nanosheet-assembled hierarchical $MoSe_2/CoSe_2$ microcages for the enhanced
       performance of electrocatalytic hydrogen evolution, RSC Adv., 6 (2016) 23-30.
       https://doi.org/10.1039/C5RA21638A

[76]   X. Ren, Q. Wei, P. Ren, Y. Wang, R. Chen, Synthesis of flower-like $MoSe_2@MoS_2$
       nanocomposites as the high efficient water splitting electrocatalyst, Mater. Lett., 231
       (2018) 213-216. https://doi.org/10.1016/j.matlet.2018.08.049

Materials Research Forum LLC
https://doi.org/10.21741/9781644903070

CHAPTER 7

Surface Structure Modulation of the Catalysts

Xiang Peng*

Hubei Key Laboratory of Plasma Chemistry and Advanced Materials, Engineering Research Center of Phosphorus Resources Development and Utilization of Ministry of Education, School of Materials Science and Engineering, Wuhan Institute of Technology, Wuhan 430205, China

xpeng@wit.edu.cn

Abstract

The electrocatalytic performance of the catalysts is influenced by various factors, including morphology, chemical state, electronic structure, and environment. Electrocatalysis is well-known as a surface interface reaction process. Consequently, researchers have focused on enhancing the surface atom utilization of catalysts by manipulating their morphology. Constructing catalysts with different sizes (bulk and nanoparticles) and dimensions increases their specific surface area and improves their non-intrinsic catalytic activity. Metal valence and internal electronic structure also exert significant influence on the catalytic process, subsequently affecting material performance. Therefore, combining non-intrinsic and intrinsic performance optimization through strategies like morphology, elemental valence, and electronic structure modulation can enhance the performance of electrocatalysts, resulting in high activity, stability, and selectivity.

Keywords

Surface Structure Modulation, Hydrogen Evolution, Geometrical Morphology, Chemical State, Electronic Structure

1. Introduction

Hydrogen (H_2) offers several advantages, including zero carbon emissions, a high combustion value of 143 kJ g^{-1}, and environmentally friendly combustion products. Consequently, it is considered one of the most promising clean energy sources. To produce "green hydrogen", water electrolysis technology has been developed in conjunction with renewable energy sources. This technology aims to achieve zero carbon emissions during the hydrogen production process. The water electrolysis process typically consists of the hydrogen evolution reaction (HER) and the oxygen evolution reaction (OER). Developing high-performance catalysts for HER and OER is crucial for reducing the actual potential of electrocatalytic water decomposition and improving energy efficiency. At present, water electrolysis relies heavily on precious metal-based materials, such as Pt-based HER catalysts and Ir/Ru-based OER catalysts. However, the high cost and limited availability of these precious metals hinder their large-scale application. Alternatively, research into low-cost transition metal-based catalysts, including hydroxides/hydroxyl oxides, phosphates, oxides, nitrides, and carbides, has made significant progress. Nevertheless, their performance falls short of platinum group metals, making it challenging to meet the requirements of large-scale use.

In practice, the electrocatalytic performance of the catalysts is influenced by various factors, including morphology, chemical state, electronic structure, and environment. Electrocatalysis is well-known as a surface interface reaction process. Consequently, researchers have focused on enhancing the surface atom utilization of catalysts by manipulating their morphology. Constructing catalysts with different sizes (bulk and nanoparticles) and dimensions, such as one-dimension (1D) [1-6], two-dimension (2D) [7-10], and three-dimension (3D), increases their specific surface area and improves their non-intrinsic catalytic activity. As testing methods have advanced, researchers have discovered that performance is not solely determined by morphology. Metal valence and internal electronic structure also exert significant influence on the catalytic process, subsequently affecting material performance [11, 12]. Therefore, combining non-intrinsic and intrinsic performance optimization through strategies like morphology, elemental valence, and electronic structure modulation can enhance the performance of electrocatalysts, resulting in high activity, stability, and selectivity. This chapter provides an overview of modification strategies, including morphology regulation, chemical state regulation, and electronic structure regulation.

2. Geometrical Morphology Regulation

Morphology refers to the shape, size, and geometric structure of the catalyst surface. By manipulating the morphology of a catalyst, it is possible to increase the specific surface area and active site density, thereby enhancing its catalytic activity. There are various methods available for morphology modulation, including crystal surface manipulation, nanoparticle synthesis, and morphology control. For instance, by selecting a specific crystal surface or synthesis method, catalysts with higher hydrogen precipitation activity and selectivity can be obtained.

Both the morphology and dimensions of catalysts play a crucial role in their performance. Generally, smaller-sized materials with reduced dimensions have larger specific surface areas. This increased surface area exposes more unsaturated active sites, facilitating the adsorption of H^+ and H_2O and promoting surface reactions, thereby exhibiting higher catalytic activity.

2.1. Zero-Dimensional Structures

Zero-dimensional (0D) materials, such as nanoparticles or quantum dots, exhibit unique properties that make them particularly suitable for electrocatalytic hydrogen production. These materials possess a high surface-to-volume ratio, which means that a large number of active sites are available for catalytic reactions. This increased surface area provides more opportunities for HER to occur, leading to enhanced catalytic activity. Additionally, 0D materials often exhibit size-dependent properties, including electronic structure and quantum confinement effects. These size effects can significantly influence the catalytic performance of the material. For example, the electronic structure of nanoparticles can be tailored by controlling their size, resulting in optimized energy levels and enhanced charge transfer kinetics during the HER.

Carbon dots (CDs), a 0D carbon nanomaterial, have garnered significant interest due to their unique properties. These include abundant inherent vacancies, high surface area, fast electron transfer, high stability, and seamless integration with other nanomaterials, making them ideal for designing new electrocatalysts in conjunction with Ru. CDs, with their plentiful intrinsic vacancies and functional groups, can stabilize the Ru-CDs structure through anchoring and coordination. Furthermore, the confinement effect exerted by CDs can enhance the stability of Ru nanoparticles by impeding their agglomeration. By combining Ru with vacancy-rich CDs (Ru@CDs), a ruthenium/vacancy-rich carbon dot (Ru@CDs) electrocatalyst was synthesized , as illustrated in **Figure 1a-b** [13]. This material not only introduces vacancies skillfully into the Ru composite structure but also effectively mitigates the instability and agglomeration of Ru nanoparticles, leading to a significant improvement in their electrocatalytic activity. The Ru@CDs catalyst demonstrates outstanding performance, achieving an overpotential of 30 mV at 10 mA cm^{-2} and exhibiting excellent stability over 10,000 cycles. Remarkably, these results are comparable to, or even surpass, those of commercial Pt/C catalysts and Ru-based catalysts reported in the literature.

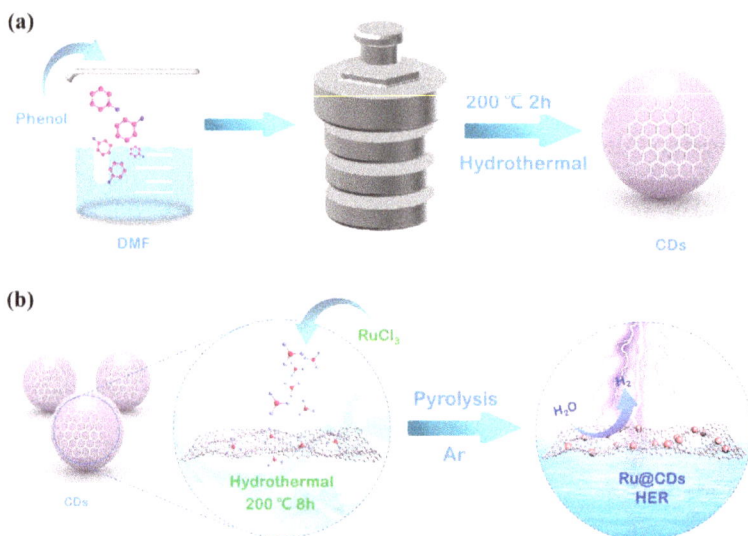

Figure 1. *Schematic illustration for the synthesis of (a) CDs and (b) Ru@CDs. Reproduced with permission from ref [13]. Copyright 2021, Wiley-VCH.*

2.2. One-Dimensional Structure

The 1D structure possesses a high aspect ratio and a large specific surface area, enabling efficient directional electron transport and providing ample surface reaction sites, thereby enhancing the performance of the HER. In a study by Peng et al. [14], 1D nanowires with a NiSe/MoSe$_2$ heterointerface on carbon cloth (NMS/CC) were prepared, serving as co-catalysts for HER and methanol oxidation reaction (MOR). The scanning electron microscope (SEM) images of the $NiMoO_4$ precursor, as shown in **Figure 2a-b**, display a smooth surface with uniform nanowire arrays covering the carbon fibers. After selenization, as depicted in **Figure 2c-d,** the array structure remained unchanged, but the nanowires became rough, and the nanoparticles on the surface provided additional active sites for electrocatalysis. The $NiSe/MoSe_2$ catalyst exhibited excellent MOR characteristics, requiring lower potential and Tafel slope than those of the OER, while maintaining high efficiency and stability. The energy-efficient hydrogen production via the combined MOR/HER configuration could be powered by a solar cell with an output voltage of 1.5 V. These findings demonstrate the potential application of the MOR/HER strategy for zero-carbon emission energy production, offering valuable insights into the coordination of electrosynthesis and electrocatalysis.

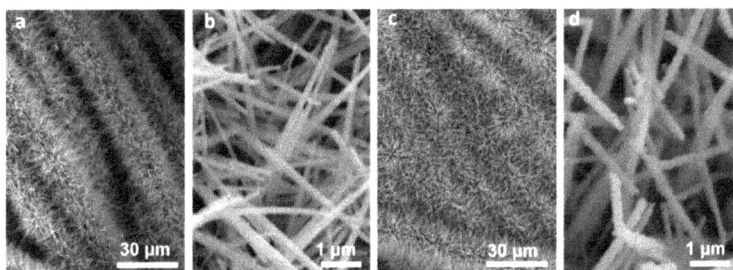

Figure 2. SEM images of the catalysts: (a-b) $NiMoO_4/CC$, (c-d) $NiSe/MoSe_2/CC$. Reproduced with permission from ref [14]. Copyright 2022, the Royal Society of Chemistry.

In another study by Yang's group [15], they synthesized 1D hollow Fe-CoP catalysts, as depicted in **Figure 3a**. These catalysts not only provided a high electrochemically active specific surface area of 14 mF cm^{-2} but also reduced the charge transport distance, facilitating rapid electron transport. The Fe-CoP catalyst exhibited excellent HER activity in both alkaline and acidic solutions, with overpotentials of 80 mV and 49 mV, respectively, to generate a current density of 10 mA cm^{-2}, as shown in **Figure 3b-c**.

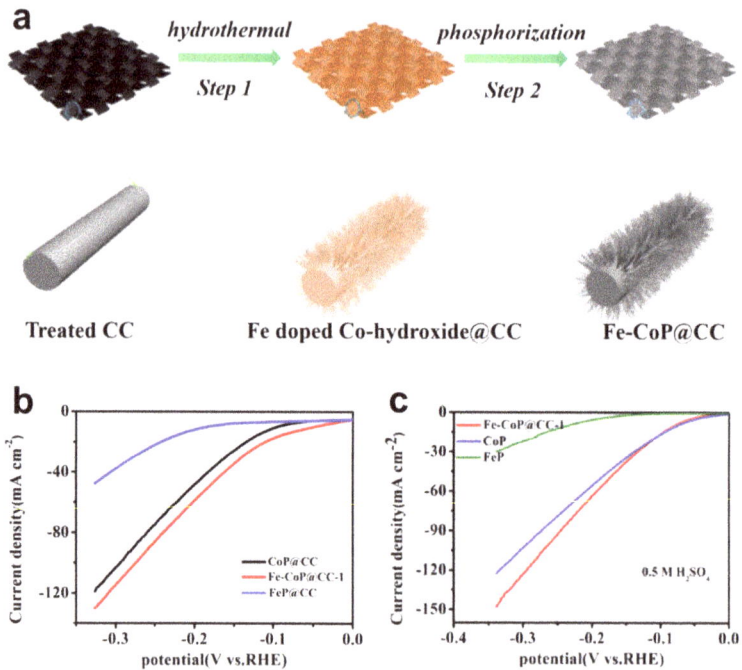

Figure 3. *(a) Schematic illustration of the formation of Fe-CoP@CC. The HER polarization curves in (b) 1 M KOH and (c) 0.5 M H₂SO₄. Reproduced with permission from ref [15]. Copyright 2021, Elsevier.*

2.3. Two-Dimensional Structure

The 2D structure refers to materials with a thickness of less than 100 nm. This thin structure facilitates rapid electron transport to the surface for subsequent reactions. Furthermore, the larger plane size of 2D materials provides more adsorption sites, thereby enhancing reaction activity. Additionally, these catalysts exhibit good stability, with minimal changes in morphology even after extended periods of continuous testing. In a study by Li et al. [16], $MoSe_2/Ni_{0.85}Se$ (NMS) catalysts were prepared using a two-step hydrothermal method. **Figure 4a-b** shows that $NiMoO_4$ possesses a noticeable nanosheet structure. After a 12-hour hydrothermal reaction at 180 °C, the layered spheres of nanosheet structure were preserved, as depicted in **Figure 4c**. The open structure between the spherical nanosheets provides a large exposed surface area for active sites and allows electrolyte penetration into the interior.

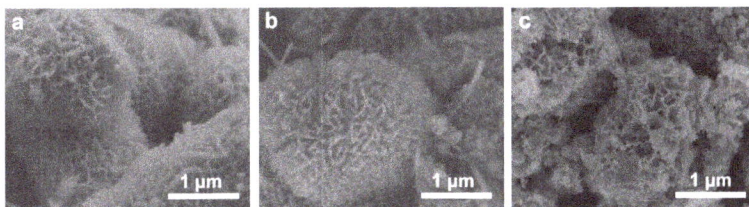

Figure 4. *SEM images of (a) NiMoO$_4$ ·xH$_2$O, (b) NiMoO$_4$, and (c) NMS-3. Reproduced with permission from ref [16]. Copyright 2024, Elsevier.*

Nanosheets with rough and porous surfaces tend to expose more active sites, thereby enhancing catalytic activity. The performance of these catalysts and the relationship between morphology and performance have been explored. For example, as shown in **Figure 5a-c**, Yang et al. [17] deposited NiCoP@FeNi LDH layered nanosheets on nickel foam by hydrothermal phosphating electrodeposition process. For the HER and OER, NiCoP@FeNi LDH/NF requires an overpotential of only 195 and 230 mV respectively to reach 1000 mA cm^{-2}. The layered structure has good electron transport ability, and the porous-large pore structure improves the gas release rate, thus improving the hydrogen production ability of high current density. NiCoP@FeNi LDH/NF has electrochemically active surface area (ECSA) of 17.68 and 45.34 mF cm^{-2} for HER and OER, meaning that the layered electrodes exposed more active sites.

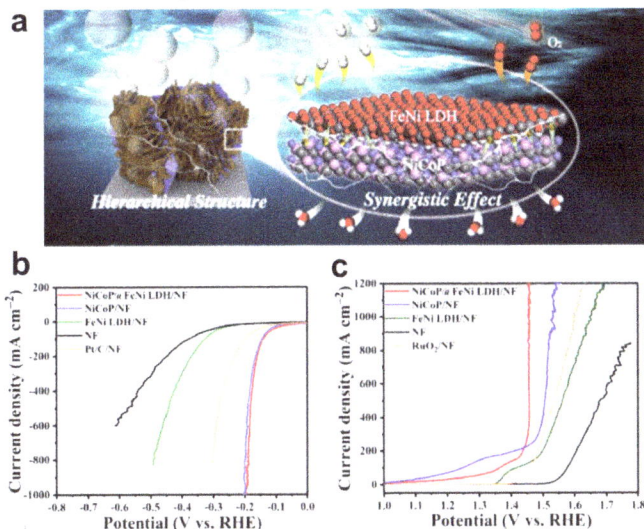

Figure 5. *(a) Schematic illustration of the formation of NiCoP@FeNi LDH. LSV curves of (b) HER and (c) OER. Reproduced with permission from ref [17]. Copyright 2023, Elsevier.*

Xie et al. [18] prepared vertical NiO nanosheets (NiO/CC) on CC using a solution method, followed by annealing to stabilize the nanostructure and improve adhesion to the substrate. Subsequently, the product underwent ultrasonic treatment in a $FeSO_4$ solution to introduce iron atoms into the nickel oxide nanoparticle. **Figure 6a** demonstrates the nanosheet structure, which remains intact after annealing, as shown in **Figure 6b**. **Figure 6c** confirms the preservation of the nanosheet structure after the incorporation of iron elements. The vertically aligned nanosheet array on CC allows the exposure of active sites and facilitates electrolyte penetration, providing favorable conditions for electrocatalytic activity. As a result, an overpotential of merely 288 mV is required for a current density of 100 mA cm^{-2} by the Fe/NiO/CC electrocatalyst and a small Tafel slope of 72.6 mV dec^{-1} is observed in the alkaline electrolyte.

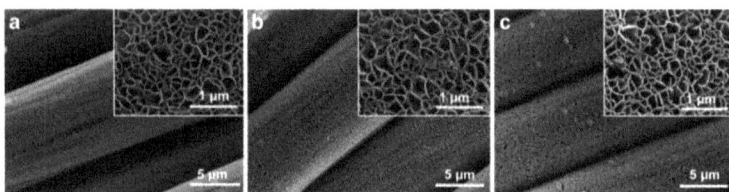

Figure 6. *(a) Ni-hydroxide/CC, (b) NiO/CC, and (c) Fe/NiO/CC with the insets showing the high-resolution SEM images. Reproduced with permission from ref [18]. Copyright 2022, Elsevier.*

In a study by Zhao et al. [19], a near-monolayer of NiCoP with a thickness of 0.98 nm was synthesized. This quasi-monolayer structure provides abundant intermediate adsorption sites and exhibits good hydrogen precipitation performance in an alkaline medium. The catalyst shows an overpotential of 84 mV to produce a current density of 10 mA cm^{-2}, with no decay in performance after 10,000 cyclic scans and demonstrates stability and durability comparable to that of RuO_2.

2.4.   Three-Dimensional Structures

The catalytic effect of 3D nanostructures presents a promising avenue for the development of high-efficiency catalysts [20]. The essence of this morphology lies in the selective adjustment of the catalyst's surface, which exposes more catalytically active sites, thereby enhancing activity, performance, or modulation response in catalytic reactions. The surface chemisorption properties of nanostructured catalysts [21] play a crucial role in their catalytic activity, stability, and selectivity [22]. In particular, the 3D nanowire array structure offers abundant active sites for efficient gas release [23], leading to a substantial improvement in the catalytic activity of the HER.

In recent years, three-dimensional graphene has emerged as a focus of research [24], obtained by assembling 2D graphene through appropriate methods. It not only inherits the excellent properties of 2D 4graphene but also possesses remarkable structural features such as a three-dimensional porous structure and a large surface area [25]. Scholars in the field of electrocatalysis have widely employed 3D graphene-supported cobalt-based catalysts due to their low overpotential and excellent electrocatalytic activity [26].

For instance, as depicted in **Figure 7**, Tang et al. [26] successfully synthesized three-dimensional graphene-supported cobalt metaphosphate ($Co(PO_3)_2$-3D RGO) through a one-step hydrothermal low-temperature synthesis. Graphene has been extensively utilized as a catalytic carrier in the HER field, with three-dimensional graphene (**Figure 7a-c**) offering advantages such as excellent electrical conductivity, abundant pore structure, large specific surface area, and stable physical and chemical properties. The $Co(PO_3)_2$-3D RGO material exhibits a favorable three-dimensional porous structure and a large specific surface area (**Figure 7d-f**). Consequently, the resulting electrocatalyst demonstrates outstanding electrochemical activity in the HER in a 0.5 M H_2SO_4 solution. It achieves an overpotential of only 176 mV and a Tafel slope of 63 mV dec^{-1}. Furthermore, the electrocatalyst exhibits excellent hydrogen evolution activity in a 1 M KOH solution, with an overpotential of 158 mV and a Tafel slope of 88 mV dec^{-1} at a current density of 10 mA cm^{-2}. $Co(PO_3)_2$-3D RGO remains catalytically active for at least 10 hours in both acidic and basic conditions.

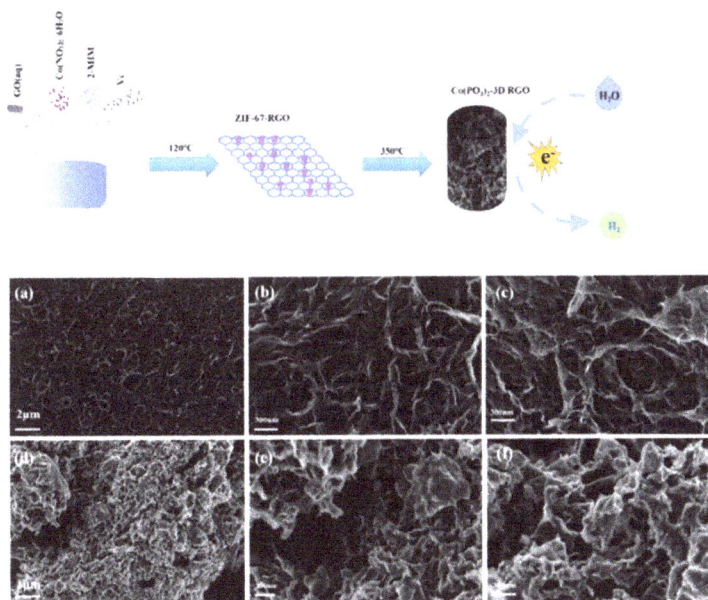

Figure 7. *Synthesis process and (a–c) SEM images of 3D RGO, (d–f) SEM images of $Co(PO_3)_2$-3D RGO. Reproduced with permission from ref [26]. Copyright 2022, the Royal Society of Chemistry.*

In addition, Meng et al. [27] prepared a novel three-dimensional (3D) flower-like WP$_2$ nanowire arrays, as illustrated in **Figure 8a-d**, which helps electrolyte penetration and has a low cell voltage of 1.65 V.

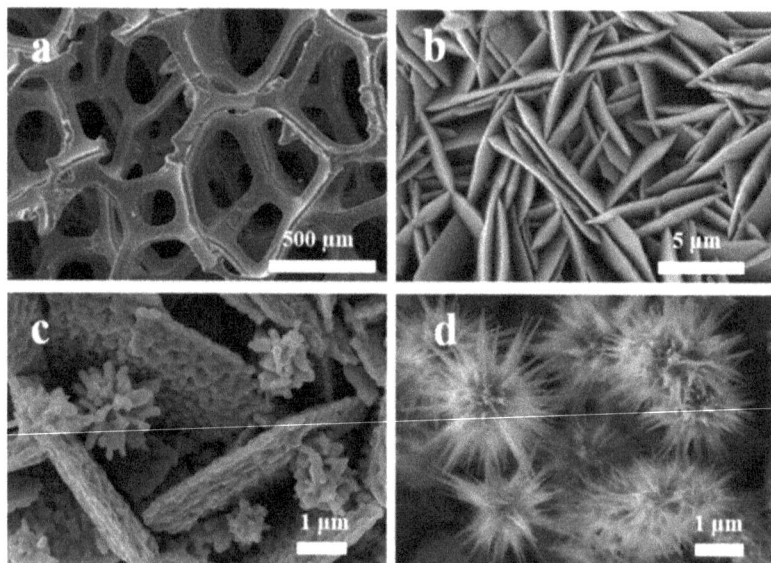

Figure 8. *SEM images taken from (a) bare Ni foam, (b) WO$_x$ precursor before 450 °C calcination, (c) tungsten trioxide after 450 °C calcination, and (d) the 3D nanostructured WP$_2$ NW/NF. Reproduced with permission from ref [27]. Copyright 2021, Elsevier.*

2.5.   Other Structures

Due to the adjustable pore structure of metal-organic frameworks (MOFs) and the coordination of metal centers with different organic ligands in different environments, various unique morphologies and structures can be generated. MOFs have been extensively studied as a template/precursor for the preparation of transition metal sulfur group compounds-based composites, which play vital roles in electrocatalysis [28, 29]. For instance, Park et al. [30] elegantly transformed ZIF-67 into N-doped carbon-coated CoSe$_2$ nanorods and then neatly blanketed them into 3D porous nanocages constructed from components, as illustrated in **Figure 9**. This combination yielded a CoSe$_2$@NC-NR/CNT merge with a unique structure boasting ample active sites and conductive pathways for active electrons, facilitating swift electron transport. In addition, the resilient and robust backbone of carbon nanotubes upholds the integrity of the composite, resulting in outstanding electrochemical stability. CoSe$_2$@NC-NR/CNT microspheres exhibit a significantly higher current density value of 23.5 mA cm^{-2} at the overpotential of 200 mV compared to that of CoSe$_2$@NC polyhedrons, which achieve 9.6 mA cm^{-2} at η = 200 mV. These findings indicate that the CoSe$_2$@NC nanorods contribute to the

HER activity, while the porous CNT microspheres provide a conductive pathway for electron transfer during electrochemical reactions. The combination of $CoSe_2@NC$-NR and CNT backbones leads to a synergistic effect, resulting in an enhanced electrocatalytic activity for HER.

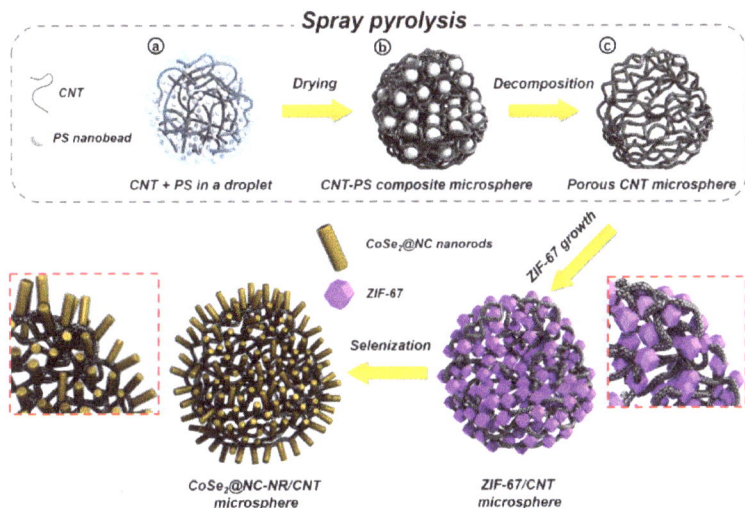

Figure 9. Synthesis process and morphology diagrams of $CoSe_2@NC$-NR/CNT. Reproduced with permission from ref [30]. Copyright 2018, American Chemical Society.

Chen et al. [31] decorated $MoSe_2$ nanosheets on hollow $CoSe_2$ nanocube surfaces, producing a robust $CoSe_2@MoSe_2$ catalyst, as shown in **Figure 10**. These hollow nanocube structures offer plenty of active sites for accelerating electron transfer and bubble release. Intriguingly, their internal voids enhance active site density and shorten diffusion distances for intermediate adsorption and reactions, delivering high stability and simplified charge transport in the porous shell. As a result, the overpotential of the $CoSe_2@MoSe_2$ catalyst at the current density of 10 $mA\ cm^{-2}$ was 183 mV, remarkably smaller than that of the $CoSe_2$ catalyst at 220 mV.

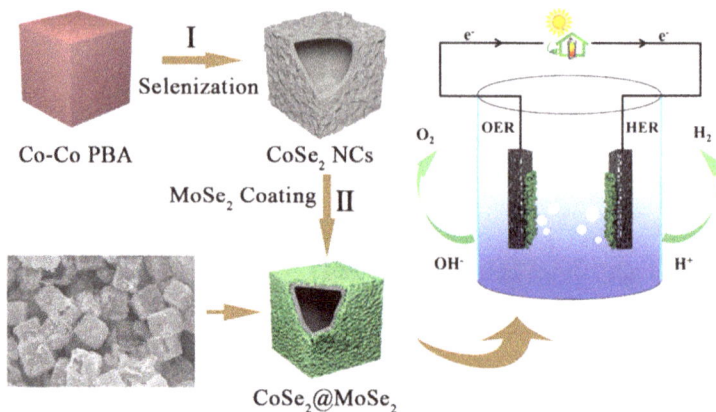

Figure 10. Synthesis process and morphology diagrams of $CoSe_2@MoSe_2$. Reproduced with permission from ref [31]. Copyright 2020, the Royal Society of Chemistry.

Huang et al. [32] synthesized $NiSe_2@NC$ core-shell nano octahedra protected by N-doped carbon shells via a well-adjusted MOF template, as illustrated in **Figure 11**. The optimized core-shell $NiSe_2@NC$ nano-octahedron with pyridinic-N content amounts to 63.4% of the total nitrogen content, termed as $NiSe_2@NC$-PZ, and shows the highest HER activity with a low overpotential of 162 mV at the current density of 10 mA·cm^{-2} in alkaline media. This is because the N-rich PZ ligand transforms into an N-doped carbon layer covering the $NiSe_2$ surface, increasing the number of active sites and facilitating electron transport.

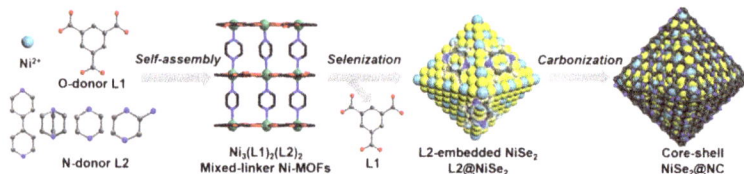

Figure 11. Synthesis process of $NiSe_2@NC$-PZ. Reproduced with permission from ref [32]. Copyright 2020, Elsevier.

Recently, Yu et al. [33] innovatively transformed spinel-type $NiCo_2O_4$ into porous nanosheets of monoclinic $NiCo_2Se_4$ (NCS). Its porous nanostructure and heterogeneous atom doping create a vast number of active sites. Consequently, NCS demonstrated promising OER catalytic behavior with an overpotential of 295 mV to produce the current density of 10 mA cm^{-2}, alongside a potential of 1.68 V when decomposing total water, as shown in **Figure 12a**. By adjusting the morphologies and structures of catalysts, increasing unsaturated active sites and reducing electron transport distance can significantly boost their electrochemical activity, as illustrated in **Figure 12b**.

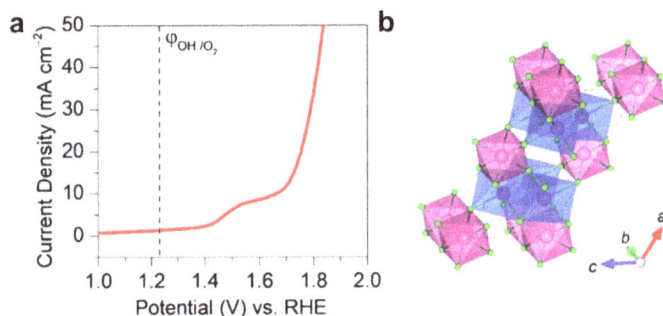

Figure 12. *(a) 95% iR-corrected polarization curves of NiCo₂Se₄ holey nanosheets for overall water splitting in a two-electrode configuration. (b) Crystal structure of NiCo₂Se₄. Reproduced with permission from ref [33]. Copyright 2017, American Chemical Society.*

3. Chemical State Modulation

Valence signifies the oxidation or reduction state of the metal or metal oxide on the catalyst's surface. Modulating this state can alter catalyst-reactant interactions and its activity. Common techniques include redox reactions, oxidation and reduction treatments. For example, hydrogen precipitation activity can be improved by depositing a metal oxide layer.

Active site valence configuration is vital in regulating the electronic structure of non-precious metal electrocatalysts. Nickel-, cobalt-, or iron-based oxides demonstrate desirable characteristics including d-band electrons similar to noble metals, high corrosion resistance, and abundant metastable valence states [34]. Interest lies in structurally flexible, mixed-valence spinel-type oxides (e.g., MFe_2O_4, M=Co, Ni, Cu, etc.) rather than common monooxides. These structures possess 64 tetrahedral and 32 octahedral sites coordinated to oxygen, and 16 octahedral sites and 8 tetrahedral sites occupied by metal cations, facilitating cation migration and redistribution, an effective means to enhance electrochemical properties through cation modulation. For example, partial substitution of $CoFe_2O_4$ with Cr, Ni and/or Mn was successfully prepared, confirming the effects of cation distribution, oxygen vacancies, lattice crystals, bond lengths, etc. on the electrocatalytic activity [35-37].

In addition, Li et al. [16] leveraged electronic interactions between Ni and Mo atoms through in situ phase separation of $Ni_{0.85}Se$ and $MoSe_2$ from $NiMoO_4$ to optimize electronic structure, supported by the peaks shift of both Ni and Mo, as shown in **Figure 13a-b**. The simultaneous selective reduction tunes Ni and Mo oxidation states (**Figure 13c**), which supports better water molecule adsorption and hydrogen desorption. The optimized catalyst required an overpotential of 124 mV to generate a current density of 10 mA cm⁻², a small Tafel slope of 63 mV dec⁻¹ in alkaline electrolytes, Faraday's efficiency of 98.9% in hydrogen production, as well as excellent long-term stability for 170 hours.

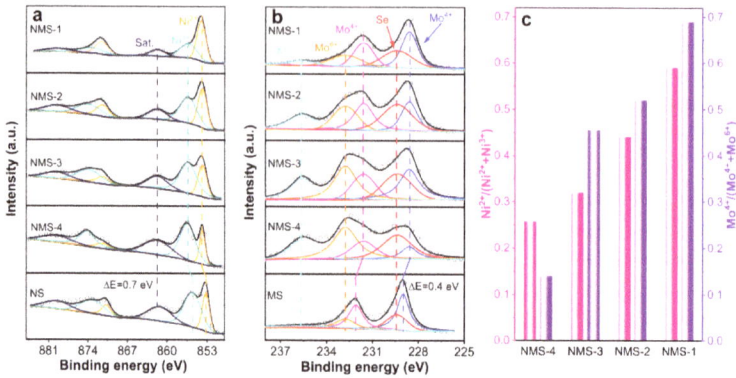

Figure 13. High-resolution XPS spectra of (a) Ni-2p and (b) Mo-3d; (c) calculated ratios of Ni^{2+} / $(Ni^{2+} + Ni^{3+})$ and Mo^{4+} / $(Mo^{4+} + Mo^{6+})$ of the NMS catalysts. Reproduced with permission from ref [16]. Copyright 2024, Elsevier.

Valence modulation offers further flexibility in tuning electronic structure. It optimizes adsorption state and charge transfer properties by regulating $3d$ energy and electron density, thereby enhancing intrinsic activity [35]. However, synchronizing the valence states of different elements to regulate them into reverse valence states remains a significant challenge in integrated synthesis environments. Wang et al. [38] pioneer an anti-spinel nickel-cobalt-iron oxide nanocube superstructure with both preferential Co^{2+} and Ni^{3+} valence modulation ($Ni^{III}Co^{II}Fe$-O@NF, NF stands for nickel foam) through in situ topochemical transformations. $Ni^{III}Co^{II}Fe$-O@NF outperforms the NiCoFe oxide in catalyzing water splitting. As a bifunctional catalyst, $Ni^{III}Co^{II}Fe$-O@NF achieves a current density of 10 mA cm^{-2} in 1 M KOH at a low cell voltage of 1.455 V, as depicted in **Figure 14**. The density functional theory (DFT) calculations indicate that favorable Co^{2+} and Ni^{3+} synergistically reduce the free energy for hydrogen adsorption (ΔG_{H*}) reduction in HER.

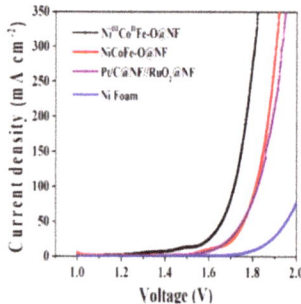

Figure 14. Overall water splitting LSV curves of the electrolyzers with the as-prepared catalysts. Reproduced with permission from ref [38]. Copyright 2021, Elsevier.

Valence regulation of nonmetallic elements selenium and ionic selenium in selenides by hydrothermal temperature, designing Se-NiSe$_2$ hybrid nanosheets with a self-adjusting ratio of ionic Se (I-Se) and elemental Se (E-Se) on carbon cloth through solution synthesis and hydrothermal treatment [8]. Experimental and theoretical studies explored the effects of I-Se/E-Se ratio on the electrocatalytic properties of HER and OER. As shown in **Figure 15a**, the optimized bifunctional catalyst achieves a current density of 10 mA cm^{-2} with an overpotential of 133 mV in 1.0 M KOH for HER, and an overpotential of 350 mV to produce the current density of 100 mA cm^{-2} for OER. According to DFT calculations in **Figure 15b**, electronic environment and hydrogen/water adsorption/desorption free energy, promote the electrocatalytic water separation process. Optimal I-Se/E-Se ratio illustrated in **Figure 15c** improves catalytic activity and reaction kinetics, balancing I-Se and E-Se interactions for appropriate Se-H binding and active site exposure to achieve exceptional electrocatalytic activity.

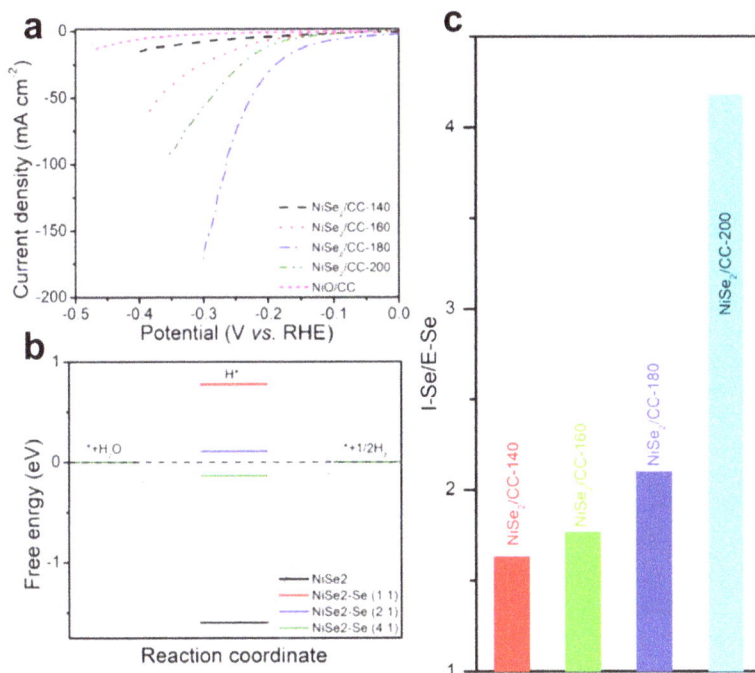

Figure 15. *(a) Polarization curves, (b) Gibbs free energy diagram for alkaline HER on the Se sites of NiSe$_2$ for different I-Se/E-Se ratios (*representing the clean surface), and (c) Calculated ratios of I-Se/E-Se (1) NiSe$_2$/CC-140, (2) NiSe$_2$/CC-160, (3) NiSe$_2$/CC-180, (4) NiSe$_2$/CC-200. Reproduced with permission from ref [18]. Copyright 2022, Elsevier.*

4.   Electronic Structure Optimization

Spatial electronic configuration alludes to the spatial distribution of electron occupancy levels across the metallic centers situated at the catalyst interface. By strategically modifying this configuration, we can affect the rates and extent of electron exchange between the catalyst and the chemical constituents involved in the reaction. Multiple techniques exist for regulating electronic structure, including incorporating electron donors or accepting molecules and altering the immediate surroundings and surface area interfaces. Consider, for instance, the introduction of electron donors or modifications to the surrounding coordination environment which can alter the electron concentration and molecular makeup of the catalyst, thereby impacting its capacity to absorb and subsequently dissociate hydrogen.

At the heart of electrochemical reactions lies the interplay between the catalytically active deposits located on the electrocatalyst surface and the reactive entities. The electronic profile of these catalysts, typically comprising transition metals, plays a crucial role in facilitating or hampering the binding and activation of these reactive substances onto the surface, thereby setting inherent limits to the efficiency of the electrocatalysts. When systematically evaluating the hydrogen evolution properties of catalysts from a thermodynamics standpoint, we come across four primary parameters: hydrogen adsorption energy (ΔG_H), water adsorption energy (ΔG_{H2O}), water dissociation energy, and hydroxyl adsorption energy.

Interestingly, ΔG_H emerges as an essential guidepost in the caravan towards efficient HER catalysts due to its integral involvement across various hydrogen evolution pathways. In particular, when the value of ΔG_H falls closer to zero, that specific alloy tends to exhibit the most exceptional hydrogen evolution activity [39, 40]. On the contrary, a negative value implies higher surface hydrogen atom adsorption potential, supporting the Volmer stage. However, an excessively high absolute value of $|\Delta G_H|$ propels unfavorable conditions for the succeeding Tafel and Heyrovsky stages. Similarly, when ΔG_H assumes a positive value, the immobilization of H^+ onto the material's surface as adsorbed hydrogens (H_{ads}) becomes challenging, posing obstacles to the entire HER procedure. Hence, scientists have ingeniously employed certain strategy optimization measures to refine the electronic structure of the catalytic active sites to fabricate catalysts exhibiting superior efficiency regiments.

4.1.   Heteroatom Doping

Heteroatom or radical integration can introduce lattice irregularities, subsequently altering local electron distribution and causing lattice deformations, vacancies, and phase transformations, consequently providing numerous new catalytic sites [41-43]. Significant improvements in electrical conductivity and catalytic activity arise with various doping. These modifications effectively modify the substrate's capacity for reaction intermediates, consequently enhancing electrochemical efficiency [44]. It is imperative to note that excessive heteroatom incorporation may result in significant decreases in HER efficiency due to potential structural deterioration.

Doping procedures are typically carried out by introducing necessary heteroatoms directly into raw materials, subsequently subjecting them to hydrothermal/solvothermal treatments, chemical vapor deposition (CVD), electrodeposition, or annealing under diverse atmospheric conditions.

The use of plasmas or microwaves for doping has also proven effective. For instance, Peng et al. [45] constructed a hierarchically structured composite consisting of Ni-doped amorphous FeP nanoparticles, porous TiN nanowires, and graphitic carbon fibers (Ni-FeP/TiN/CC). Ni was incorporated into the FeP/TiN/CC matrix through plasma ion implantation (**Figure 16a**), executed using a high-energy metal ion implanter utilizing Ni as the cathodic arc source. The material's electronic structure undergoes modifications upon Ni ion penetration, simultaneously yielding an amorphous surface (**Figure 16b**). This surface assists in accelerating charge transfer via its crystalline TiN core and augmenting electrochemical activity. Remarkable HER performance is attributed to the synergistic effect of Ni and Fe atoms within the Ni-doped FeP particles, the active amorphous surface, and the conductive nanowire support matrix, resulting in an increased number of active sites and hence enhanced charge transfer efficiency and hindering catalyst migration and aggregation. The Ni-FeP/TiN/CC catalyst offers superior HER performance, including an overpotential of 75 mV for a cathodic current density of 10 mA cm^{-2} and a Tafel slope comparable to commercial Pt/C catalysts (**Figure 16c**).

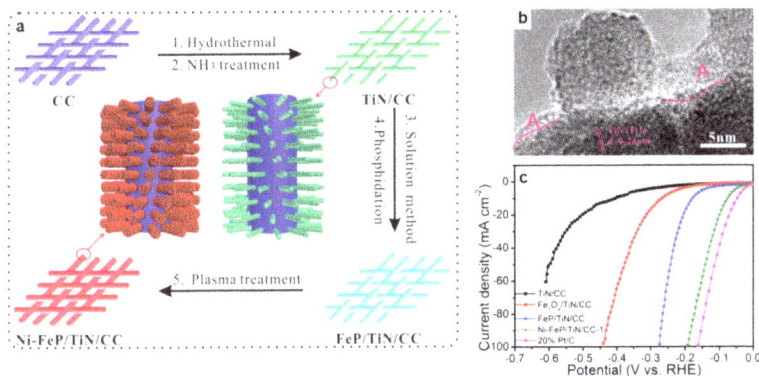

Figure 16. *(a) Schematic illustration of the preparation of the hierarchical Ni-FeP/TiN/CC electrocatalyst. (b) Ni-FeP/TiN/CC with symbol A showing the amorphous TiN surface. (c) Polarization curves. Reproduced with permission from ref [45]. Copyright 2018, Elsevier.*

Zhou et al. [46] introduced an innovative hydrothermal-impregnation-pyrolysis method to incorporate W into Co nanoparticles enveloped by nitrogen-doped carbon (CoW@N-C) on nickel foam (NF) as efficient HER catalysts. The vertically aligned and uniform CoW nanoneedles are anchored onto the surface of NF utilizing a conventional hydrothermal technique. The X-ray diffraction (XRD) peak of Co (111) migrates towards the lower diffraction angle, indicating that the inclusion of W elevates the lattice spacing of Co due to its larger atomic radius (130 pm) compared with that of Co (116 pm). The electronic constitution of Co@N-C, which is conducive to charge transfer, reaction kinetics, active site exposure, and intrinsic activity augmentation, may be improved by the inserted W atoms. Metal atoms (Cr, Mo, or Ce) were effectively incorporated into the Co@N-C system utilizing this procedure to generate CoM@N-C hybrids, which exhibited considerably enhanced HER activity compared to Co@N-

C. This exhibited the versatility of the approach. As predicted, the inclusion of these metals can enhance the electronic constitution of Co@N-C and accelerate the HER.

DFT computations have been executed to elucidate the reason behind the enhancement of HER performance through the assimilation of tungsten into the material. At variance with Co@N-C, CoW@N-C manifests a lower water adsorption free energy of −0.43 eV (**Figure 17a**). This delineates that the addition of tungsten fortifies the material's capacity to bind water, potentially enhancing the Volmer step. Tungsten proves beneficial to the HER procedure as its ΔG_H of -0.107 eV aligns closer to zero compared to −0.163 eV for Co@N-C. The electronic density of states (DOS) of Co@N-C and CoW@N-C were also computed (**Figure 17b**). Compared to Co@N-C, the CoW@N-C possesses an increased number of electronic states adjacent to the Fermi level, typically considered beneficial for the charge transfer of the material. Importantly, the incorporation of tungsten augments the distribution of electronic states near the Fermi level, propelling electron transfer during the HER progression. Notably, the Co atoms in CoW@N-C exhibit discernable electronic states near the Fermi level, evincing that the incorporation of tungsten amplifies Co's capability for electron transportation and expedites the migration of electrons from Co to the N-doped C layer.

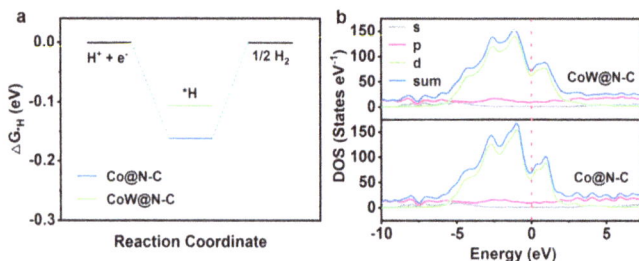

Figure 17. (a) The calculated free energies of H adsorption on Co@N-C and CoW@N-C. (b) DOS of Co@N-C and CoW@N-C. The Fermi level is shifted to zero. Reproduced with permission from ref [46]. Copyright 2022, the Royal Society of Chemistry.

The XPS analysis was employed to scrutinize the HER augmentation mechanism. The C=N bond's binding energy within the C-1s spectrum for Co@N-C (**Figure 18a**) registers at 285.4 eV, yet, it displays a positive shift in CoW@N-C. This signifies that tungsten doping exerts influence on the electron distribution between the C and N atoms. Moreover, after tungsten doping, the binding energy of M-N$_x$ in the N-1s spectrum for Co@N-C (**Figure 18b**) is concurrently negatively shifted, standing at 399.1 eV. Tungsten doping induces a positive translocation in the binding energy of the Co-N$_x$ bond from 782.1 eV to 782.5 eV (**Figure 18c**), congruent with the negative shift of the N peaks. This denotes that an increased number of electrons are transported from Co atoms to N atoms in CoW@N-C compared to that in Co@N-C. The negative shift of the Co(0) peak proposes that additional electrons may be transposed from the W atoms to the Co atoms due to the tungsten addition. According to the XPS results, tungsten can be incorporated into the Co@N-C system to revise its electronic structure and foster the transport of interface charges from metallic cobalt to the surface's N-doped carbon layer.

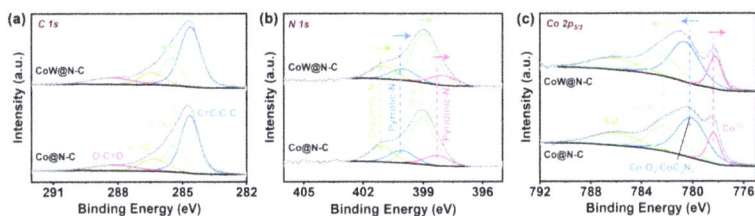

Figure 18. *High-resolution XPS spectra of (a) C 1s, (b) N-1s, and (c) Co-2p$_{3.2}$ for Co@N-C and CoW@N-C. Reproduced with permission from ref [46]. Copyright 2022, the Royal Society of Chemistry.*

A potential mechanism for enhancing the HER was diagrammatically illustrated in **Figure 19**. The HER would take place on the carbon surface in contact with an alkaline electrolyte solution, which provides the electrons required by the HER. We can speculate that in the absence of tungsten doping, "x" (x > 0) H_2O molecules would interact with "x" electrons on the carbon surface at a certain stage during electrolysis to generate "x/2" H_2 and "x" OH^-. However, following tungsten doping, some interfacial charges would migrate from metallic cobalt to the nitrogen-enriched carbon layer on the surface, augmenting the number of electrons in the carbon layer and augmenting the number of electrons accessible for electrolysis to coordinate with H_2O molecules. Expressed differently, the carbon surface may furnish "x + y" (y > 0) electrons under the identical applied potential, which would subsequently react with "x + y" H_2O molecules to generate "(x + y)/2" H_2 and "x + y" OH^-. Therefore, the suitable W-doped material would result in a superior current density compared to the original one under the same applied potential. The attained $Co_{1.5}W_{0.5}$-py-500@NF exhibits exceptional HER performance, attaining current densities of 25 and 100 mA cm^{-2} in 1.0 M KOH, necessitating remarkably low overpotentials of 55 and 100 mV, respectively.

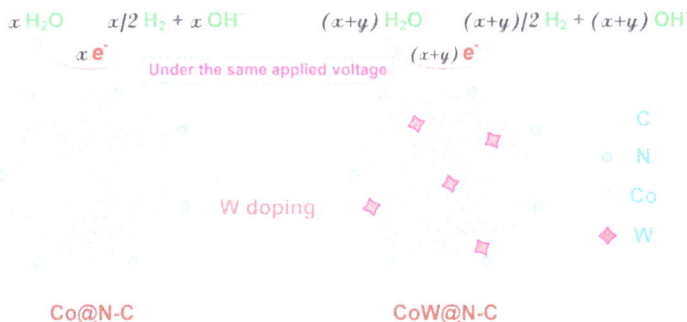

Figure 19. *Schematic diagram of the HER enhancement mechanism. Reproduced with permission from ref [46]. Copyright 2022, the Royal Society of Chemistry.*

4.2. Vacancy

Anionic and cationic vacancies can enhance intrinsic activity in catalysts by modulating surface charge distributions while maintaining the original lattice structure [47-50]. Various methods exist to generate vacancies such as annealing in diverse atmospheres, reduction, ion exchange, liquid phase etching, hydrothermal/solvothermal procedures, and plasma processing. These vacations or defects modify the local atomic structure and coordination number, facilitating the pooling of surface charges in a specific direction and an augmentation in the quantity of electrocatalytic active centers [51]. Primarily, vacancies modify the local electron distribution by introducing defects and inflicting structural deformation in the catalysts. Furthermore, modifying the atomic arrangement adjacent to vacancies induces a modification in the electronic configuration, culminating in a harmonized binding energy of H* [50].

Exploitation of oxygen vacancy engineering showcases promise towards improvement of the HER attributes. This is due to the absence of oxygen sites facilitating quick electron transfer, reducing the energy barrier for the genesis of intermediates, decomposing water molecules, absorbing OH entities, and optimizing the ΔG_H [50, 52, 53]. Karmakar et al. [54] synthesized vacancy-enriched $NiMoO_4$ precipitation by subjecting the initially cultivated $NiMoO_4$ nanorods, grown on nickel foam, to treatment with $NaBH_4$. This treatment instigated the formation of internal oxygen vacancies within the $NiMoO_4$ structure, recognized as $NiMoO4(V_O)$. Electron paramagnetic resonance (EPR) spectra were obtained for both $NiMoO_4$ and $NiMoO4(V_O)$ to corroborate the existence of oxygen vacancy, as illustrated in **Figure 20**. Application of $NaBH_4$ results in a palpable surge in structural imperfections when compared to the untreated $NiMoO_4$. The noteworthy elevation in the maximum intensity at $g = 1.99$ implies that the structural deformities emanate from the generation of oxygen vacancies. The inception of oxygen vacancies within the lattice frequently engenders the evolution of a functional electronic structure that fosters sustainable electrocatalytic activity.

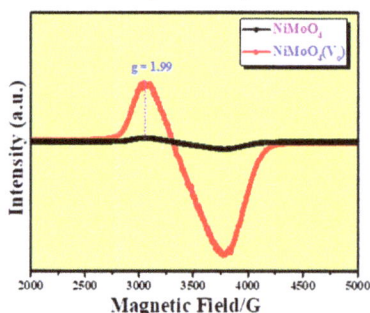

Figure 20. *EPR spectra of $NiMoO_4$ and $NiMoO4(V_O)$ at room temperature. Reproduced with permission from ref [54]. Copyright 2021, the Royal Society of Chemistry.*

It is reported that the electrochemical performance of transition metal compounds is significantly influenced by the electron occupancy of the crystal field, segmented into e_g orbitals. A catalyst possessing e_g orbital occupancy close to unity would manifest the highest degree of catalytic activity. Upon comparing the crystal field splitting electronic configurations of Ni^{2+} ($t_{2g}^6 e_g^2$) and Ni^{3+} ($t_{2g}^6 e_g^1$), apparent that the latter has a superior probability of exhibiting electron occupancy close to unity. Consequently, Ni^{3+} disparities more electrocatalytic activity. As aforementioned, the advent of oxygen vacancies resulted in an insufficiency of electrons in the Ni^{2+} sites, consequently promoting the oxidation of Ni^{2+} to Ni^{3+}. The oxidation of Ni^{2+} imparts nearly complete occupancy of the e_g orbital, thereby enhancing the electrocatalytic property with reduced overpotentials.

Beyond the electrical effect, the bond strength of the metal oxide fabricated during the reaction is another paramount aspect. The most efficacious production of metal oxide bonds transpires when the hydroxide ion assaults with balanced strength, neither overly robust nor excessively weak. Within the context of molecular orbital theory, the strength of specific bonds is chiefly determined by the electronic density present in the antibonding orbitals. Nonstoichiometric oxides, which harbour an abundance of oxygen vacancies, can generate a novel electronic state designated as a band gap state (BGS). These states are induced by the accumulation of numerous unpaired d electrons at elevated energies, situated above the Fermi levels. The augmentation of energy within the d bands prompts the antibonding states to ascend above the Fermi level (E_F) with substantial energy increments as depicted in **Figure 21**. This leads to the partial occupancy of antibonding states, indirectly facilitating the establishment of metallic bonds with optimal bond strength.

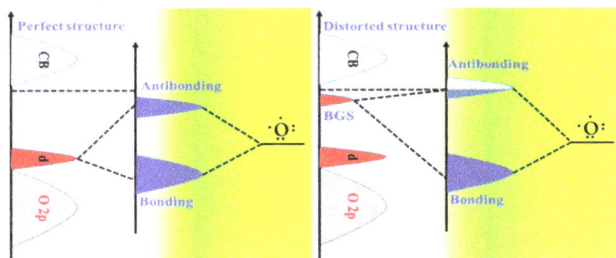

Figure 21. *Schematic representation of molecular orbital and band structure, for the origination of high electrocatalytic activity of NiMoO4(V$_O$) through vacancy formation. Reproduced with permission from ref [54]. Copyright 2021, the Royal Society of Chemistry.*

To encapsulate, the generation of oxygen vacancies confers substantial benefits as it promotes the ingress of electroactive agents and diminishes the Fermi-level energy by maintaining antibonding states at higher energy levels. This efficient utilization of creating vacancies established advantageous electrical configuration and redox properties of metal ions under lesser applied overpotentials, whilst also promoting the bonding capacity of oxygen with abundant electroactive sites, culminating in further enhancements of electrocatalytic activity. When NiMoO4(V$_O$) nanorods with vacancies were employed for OER and HER under alkaline conditions, they necessitated only 220 and 255 mV overpotential, respectively, at a current density of 50 mA cm^{-2}, as shown in **Figure 22**.

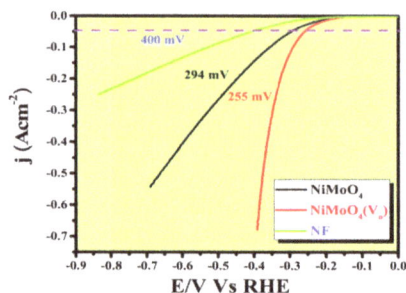

Figure 22. *LSV curve of rod-like NiMoO4(V$_O$) and NiMoO$_4$. Reproduced with permission from ref [54]. Copyright 2021, the Royal Society of Chemistry.*

Dual vacancies (DV) play crucial roles in preserving the equilibrium of hydrogen adsorption energy and enabling electron transfer, which in turn stimulates the process of HER. Liu et al. [55] developed MnO$_2$ ultrathin nanosheets that were doped with dual vacancies (DV-MnO$_2$) utilizing a well-defined chemical approach. A system containing lithium and ethylenediamine (Li-EDA) was utilized to incorporate DV into MnO$_2$ nanosheets. In this procedure, Li (a potent electron donor) interacted with the metal sites, inducing the formation of oxygen vacancies (V$_O$). Simultaneously, the Mn atoms were extracted by EDA chelating molecules under a metastable state, leaving behind manganese vacancies (V$_{Mn}$). High-resolution transmission electron microscopy (HR-TEM) image unveils atomic-scale modifications such as lattice distortion, medium-range atom disorder, and numerous defects following Li-EDA treatment, as illustrated in **Figure 23a**. The atomic-resolution scanning transmission electron microscopy (STEM) images of DV-MnO$_2$ reveal the absence of Mn and O atoms, with numerous atomically disseminated Mn (white ring) and O (yellow ring) vacancies observed across the nanosheets, as depicted in **Figure 23b-c**.

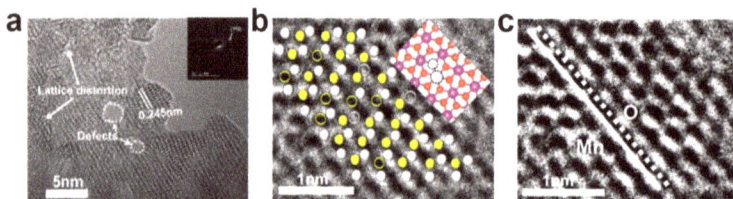

Figure 23. *(a) HR-TEM image of DV-MnO$_2$. (b,c) Atomic-resolution STEM images of (110) plane belonging to DV-MnO$_2$. The disks show oxygen (yellow) and manganese (white) atoms, while hollow disks represent oxygen (V$_O$) and manganese (V$_{Mn}$) vacancies. Reproduced with permission from ref [55]. Copyright 2021, Wiley-VCH.*

X-ray absorption fine structure spectroscopy (XAFS) was utilized to scrutinize the complex particulars and substantiate the formation of vacancies. The R space plot of MnO$_2$ presents the anticipated ranges for the first shell Mn-O and second shell Mn-Mn, which are approximately 1.53 Å and 2.56 Å, respectively, as illustrated in **Figure 24**. The intensity of these two coordination shells persistently mirrors the local symmetry in the bulk phase. The intensity of these two coordination shells closely matches that of the standard, indicating the exceptional integrity of the crystal structure of MnO_2. Remarkably, the strength of the Mn-Mn bond degrades substantially in Mn^VO_2, while the Mn-O bond demonstrates a marginal shift. This implies the absence of adjacent Mn atoms. Therefore, V_{Mn} precipitates anatomic coordination defects, analogous to prior discoveries. A divergent pattern is discerned in MnO^V_2 due to the partial fracture of Mn-O, implying the presence of V_O and the potential scarcity of V_{Mn}.

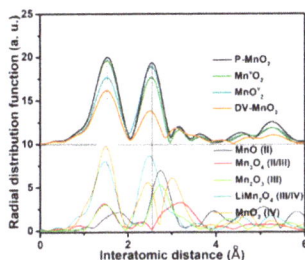

Figure 24. EXAFS $k^2\chi(k)$ Fourier transform spectra of P-MnO$_2$, Mn^{V}O$_2$, MnO$^V_2$, DV-MnO$_2$, and references. Reproduced with permission from ref [55]. Copyright 2021, Wiley-VCH.

Within DV-MnO$_2$, there exists a substantial diminution in the magnitude of the two coordinative bonds when compared to MnO$_2$. These alterations are attributed to the fact that DV mitigates the local symmetry coordinating environment and culminates in the distortion of atomic configurations. The amalgamation of anion-oxygen and cation-manganese vacancies in manganese dioxide nanosheets generates atypical local lattice deformation and electronic modulation. The advent of cation vacancies, specifically V_{Mn}, escalates conductivity, whilst anion vacancies, specifically V_O, augment the active centers by optimizing local electronic configurations. These amendments amplify the active sites accessible to reaction, enhance conductivity, facilitate the dissociation of water, and encourage the adsorption and desorption of intermediates. Consequently, the kinetics of the HER and OER are accelerated, with overpotentials of 260 and 59 mV at a current density of 10 mA cm^{-2}, respectively.

4.3. Phase Transition

Phase transition denotes the process of transition amongst disparate crystal phases, from crystalline to amorphous states, or from conductor to semiconducting states [50, 56, 57]. These transitions yield alterations in stability, activity, and conductivity. Phase transition can be potentiated electrochemically or through the application of heat and near-infrared radiation.

Transition metal dichalcogenides (TMDs) commonly exist in two distinct phases known as the 2H phase and the 1T phase [58, 59]. The MoS$_2$ compound in its trigonal prismatic form (2H phase) habitually behaves as a semiconductor, with a band gap in the range of 1.3 to 2 eV. Conversely, the octahedral 1T phase of MoS$_2$ displays metallic attributes and offers a

Electrocatalytic Hydrogen Production: Catalysts and Applications   Materials Research Forum LLC
Materials Research Foundations **165** (2024)   https://doi.org/10.21741/9781644903070

substantially reduced barrier to charge conductance. The improved HER performance of 1T phase-transition TMDs is pivotal due to the similarity in Gibbs free energy of the adsorption and desorption processes of the intermediate H* and that of platinum [60].

For example, Gao et al. [61] utilized magnetron sputtering to apply myriad MoS$_2$ coatings incorporating different metal heteroatoms as dopants. Both experimental and theoretical studies indicate that the exchange of an additional electron between the dopants and Mo atoms is a significant factor in the transformation from the 2H phase to the 1T phase. The magnetron sputtering methodology involves bombarding the MoS$_2$ target surface with the argon ion glow discharge emanated by the magnetron sputtering apparatus. The ions have ample kinetic energy to dislodge the atoms within the lattice. The unbound atoms possess the potential to nucleate on the surface of the substrate. MoS$_2$ can be co-sputtered with numerous metal targets (Ag, Al, Au, Cr, Hf, Cu, Ta, and Zr), resulting in atomic doping or atom displacement inside the MoS$_2$ film. The optimized 2H-MoS$_2$ structure presents a hexagonal lattice with minimal distinction between neighboring sites, due to the amplified signal triggered by the intersection of two sulfur atoms alongside the trajectory of the electron beam.

In contrast, in the optimal 1T-MoS$_2$ structures, the intensity disparity between the two Mo sites is considerably reduced, with the S atoms in the upper and lower layers organized in a staggered manner, culminating in the formation of a triangular lattice. The crystal field theory provides insights into the transition observed in doped MoS$_2$, illustrated in **Figure 25**. The 4d orbitals of 2H-MoS$_2$ (D_{3h}) undergo elaboration into d_{z2}, $d_{x2-y2,xy}$, and $d_{xz,yz}$. In pristine MoS$_2$, two electrons have a propensity for occupying the d_{z2} orbital and exhibit semiconductor-like characteristics. The 4d orbitals of Mo in the 1T-phase exhibit a separation into two unique energy strata: the lower energy level encompasses the $d_{xy,yz,xz}$ orbitals (t_{2g}), whereas the higher energy level comprises the $d_{z2,x2-y2}$ orbitals (e_g). They are positioned between the lowest and medium levels of D_{3h}, and between the highest and middle levels of D_{3h}, respectively. The donor atoms (Cu, Al, Au, and Ag) inject additional electrons into the Mo sites that are substituted by dopants.

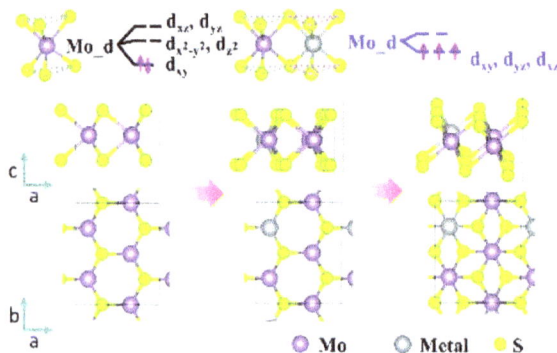

Figure 25. Splitting of the Mo 4d orbitals for the 2H (D_{3h}) and 1T (O_h) phases before and after replacing a Mo atom with a donor atom. Reproduced with permission from ref [61]. Copyright 2021, the Royal Society of Chemistry.

In the $2H$-phase of MoS_2, the additional electrons reside in the orbitals with greater energy, leading to a notable energy alteration of more than 1 eV. Contrarily, in the $1T$ phase, the additional electrons are located in the unoccupied $d_{xy, yz, xz}$ orbitals (lowest) with a slight energy recalibration. Consequently, the disposition of the additional electrons results in a phase transformation from $2H$ to $1T$. Moreover, the $1T$ phase of MoS_2 exhibits a thermoneutral ΔG_{H^*} value that surpasses that of the $2H$ phase of MoS_2. To encapsulate, the inclusion of additional electrons, resulting from potent dopants, instigates the phase transition of MoS_2. The $1T$-MoS_2 material exhibits exceptional HER capabilities, characterized by a minimal overpotential value. The Cu-MoS_2 electrode exhibited an overpotential of 71.6 mV to achieve a current density of 10 mA cm^{-2}.

4.4.   Strain

The lattice strain, whether compressive or tensile, can alter the surface electronic structure by refining the dispersal of surface atoms and their bond lengths [62, 63]. Such strains may be classified into inherent and acquired strains, based on the various techniques utilized to induce lattice strain. Intrinsic stress is naturally generated during the material fabrication process, such as hetero-atom substitution, lattice vacancy, phase transition, lattice mismatch, and geometry effect [64]. Conversely, acquired strain is instigated by an external intervention. The modifications in the energy band structure and atomic structure can result in an augmentation of electrocatalytic efficiency. It is noteworthy that modifying the obtained strain is more versatile than the intrinsic strain [65].

Li et al. [66] fabricated the Ni_3Fe alloy embedded in Ni_3FeN ultrathin nanosheets through thermal ammonolysis treatment of cation-deficient monolayered NiFe layered double hydroxides (LDH). The process of generating defect-rich Ni_3FeN/Ni_3Fe (d-Ni_3FeN/Ni_3Fe) nanosheets involved subjecting Ni_3FeAl-LDH to alkali-etching treatment, catalyzing the formation of cation vacancies known as Ni_3Fe-LDH-V_{Al}. Subsequently, d-Ni_3FeN/Ni_3Fe was synthesized by exposing Ni_3Fe-LDH-V_{Al} to thermal ammonolysis at a temperature of 500 °C for 2 hours. The thermal ammonolysis process triggers cation vacancies and nitrogen doping, disrupting the initial lattice configuration. Furthermore, the generation of the Ni_3Fe alloy during the nitriding process may result in strain formation in proximate microstructures within incredibly thin nanosheets. This strain is favorable for the formation of numerous dislocations and boundaries in thin nanosheets. The d-Ni_3FeN/Ni_3Fe exhibit notable lattice defects, including boundaries between distinct planes and dislocations resulting from lattice deformation. These lattice flaws are deemed as critical active sites due to their controlled electronic architectures.

The Ni K-edge Fourier transform-extended X-ray absorption fine structure (FT-EXAFS) curve presented in **Figure 26a** of d-Ni_3FeN/Ni_3Fe shows a peak at approximately 1.6 Å, which is induced by the atomic junction between Ni and N. Another peak in the range of 2.1-2.5 Å is attributed to the atomic interplay between Ni and Fe (or Ni and Ni). The d-Ni_3FeN/Ni_3Fe material presents a diminished intensity of the Ni-Fe peak and a decreased interatomic distance compared to Ni_3FeN. This indicates that the material has undergone considerable structural disruption due to the presence of numerous defects and alloy strains, resulting in a deficiency of atomic coordination between Ni and Fe. The Fe K-edge FT-EXAFS curves displayed in **Figure**

26b for d-Ni$_3$FeN/Ni$_3$Fe and Ni$_3$FeN show that the Fe-N atomic interaction in d-Ni$_3$FeN/Ni$_3$Fe is about 1.6 Å, whereas the Fe-Ni/Fe-Fe atomic interaction spans between 2.2 and 2.5 Å. These values indicate that the atomic interactions in d-Ni$_3$FeN/Ni$_3$Fe are weaker compared to those in Ni$_3$FeN. This suggests the presence of significant defects and lattice alterations in d-Ni$_3$FeN/Ni$_3$Fe.

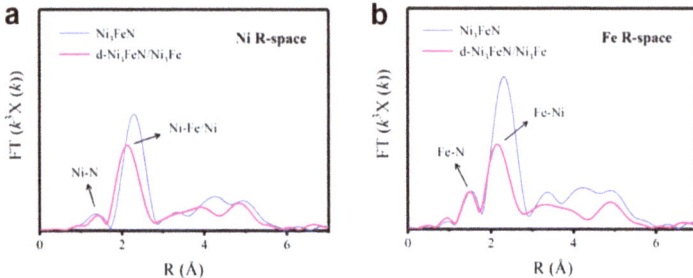

Figure 26. EXAFS spectra for (a) Ni R-space and (b) Fe R-space of d-Ni$_3$FeN/Ni$_3$Fe and Ni$_3$FeN. Reproduced with permission from ref [66]. Copyright 2021, the Royal Society of Chemistry.

The existence of defects and lattice distortions in d-Ni$_3$FeN/Ni$_3$Fe markedly influences its electronic structure, leading to the modulation of reactive site activity and improvement of electrocatalytic properties. Additionally, these defects and distortions contribute to the enhancements in structural resilience by reducing surface energy. Lattice defects, such as dislocations and boundaries, create a large number of unsaturated coordinated atoms. This subsequently impinges on the electronic structure and augments the catalytic efficacy for the OER and HER. The defect-enriched Ni$_3$FeN/Ni$_3$Fe ultrathin nanosheets demonstrate exceptional bifunctional capabilities for the HER and OER, with low overpotentials of 125 and 250 mV, respectively, to produce the current density of 10 mA cm^{-2}, as illustrated in **Figure 27a-b**.

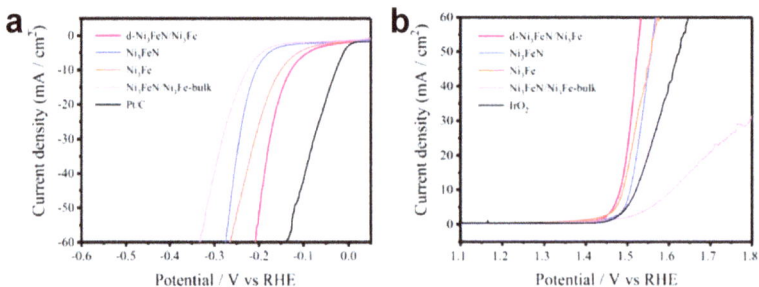

Figure 27. LSV curves for the (a) HER and (b) OER. Reproduced with permission from ref [66]. Copyright 2021, the Royal Society of Chemistry.

Materials Research Forum LLC
https://doi.org/10.21741/9781644903070

Zhang et al. [67] developed a unique method to create biaxially stretched MoS_2 nanoshells. These nanoshells possess a core-shell architecture with a single-crystalline Ni_3S_2 core and a MoS_2 shell via an in situ self-vulcanization process. The Ni_3S_2 nanoparticle core serves as a template to enable the fabrication of full-curved MoS_2 nanoshells through thermodynamics strategies. Post the creation of NiS_2-$MoO_{2.8}$ hetero-nanosheets, NiS_2 nanoparticles are commonly encased by layered $MoO_{2.8}$, setting up the framework for the synthesis of a core-shell construct. Subsequently, exposing the NiS_2-$MoO_{2.8}$ precursors to annealing in a N_2/H_2 (95/5 vol%) atmosphere at 300 °C yields Ni_3S_2@BL MoS_2 nanosheets. Under reducing conditions, S exhibits catalytic behavior and forms a bond with unsaturated Mo by departing O, consequently forming MoS_2. Moreover, NiS_2 transforms Ni_3S_2, as a result of sulfur atom migration.

The MoS_2 nanoshells are anchored to the Ni_3S_2 surface through S atom bridges, delineating the Ni-S-Mo interface, as a consequence of thermodynamic impetus. The newly synthesized MoS_2 nanoshells exhibit a curved morphology, resulting in biaxial strain. As inferred from the HR-TEM image, the lattice fringe of the MoS_2 nanoshell measures 0.64 nm, slightly wider than the lattice fringe of bulk MoS_2 (0.62 nm), as depicted in **Figure 28a**, due to the manifestation of biaxial tensile strain. The lattice spacing of 0.182 nm corresponds to the (211) crystallographic plane of Ni_3S_2. The presence of a continuous curved surface, formed by a layered MoS_2 material enveloping a Ni_3S_2 core, is easily observed. This observation is confirmed by the fast Fourier transform (FFT) image acquired across the interface, as depicted in **Figure 28b**. The heterostructures house MoS_2 nanoshells showing considerable biaxial strain coupled with sulfur vacancies. Drawing from prior research, the Nørskov group has postulated that both strain and S vacancies affect the catalytic potency of Mo sites situated beneath the S vacancies [68]. These exposed Mo sites serve as the genuine active sites for the HER.

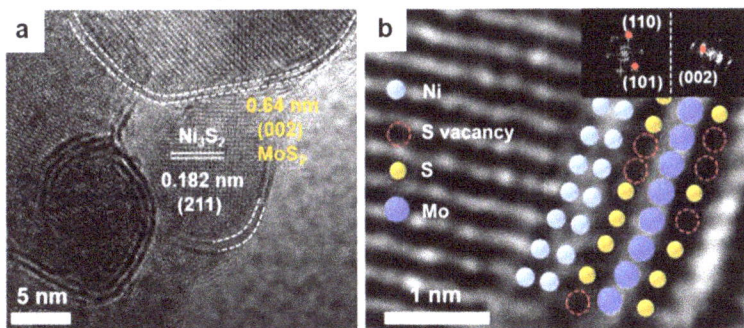

Figure 28. (a) HRTEM images of Ni_3S_2@BL MoS_2. The dashed lines indicate bilayer MoS_2. (b) HRTEM image and model for epitaxial growth of layered MoS_2 on Ni_3S_2. The inset shows the FFT images of Ni_3S_2 (left) and MoS_2 (right). Reproduced with permission from ref [67]. Copyright 2022, Wiley-VCH.

Han and his colleagues [69] observed that the strain induced by curvature accelerates the formation of vacancies in tungsten dichalcogenides, as highlighted in **Figure 29a**. With the density of S vacancies increasing, the coordinate structure of the exposed Mo site would change accordingly, as illustrated in **Figure 29b**. The findings for the ΔG_H calculations are depicted in **Figure 29c**. The ΔG_H of S sites, without the induction of strain, is calculated to be 2.028 eV. This indicates that the adsorption of H onto the pristine MoS_2 structure is weak. Both uniaxial strain and biaxial strain can augment the interaction between H intermediates and adsorption sites. However, the biaxial strain features a considerably larger impact on amplifying the affinity of S sites for H intermediates relative to the same strain level of uniaxial strain. More precisely, S sites experiencing less than 5% biaxial strain exhibit an augmented catalytic efficiency of 1.155 eV.

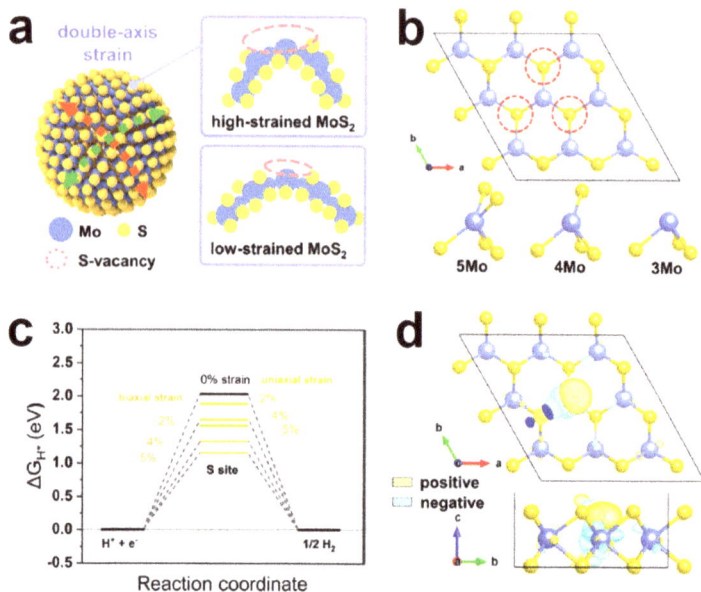

Figure 29. (a) Schematic of the S-vacancies in MoS_2 model induced by different strain conditions. (b) Illustration of Mo adsorption sites with different coordination structures. (c) Calculated free-energy diagram of HER for pure MoS_2 model under 2%, 4%, and 5% uniaxial/biaxial strain conditions. (d) The charge density difference of the H adsorbed 4Mo structure; the yellow and blue regions represent electron accumulation and depletion, respectively. Reproduced with permission from ref [67]. Copyright 2022, Wiley-VCH.

In addition, the Bader analysis and charge density disparity analysis are implemented to scrutinize the exact process of charge transfer between the H intermediate and Mo site, as shown in **Figure 29d**. According to the Bader analysis, the H adsorbate uptakes 0.35 electrons from the 4Mo structure under a 5% strain condition. This value is lower than that of the 3Mo structure under identical conditions (0.43 electrons), indicating a relatively weaker contact between the H adsorbate and Mo site in the 4Mo structure compared to the 3Mo structure, as corroborated by the ΔG_H data. The Ni_3S_2 core particle fabricated within its original position serves as a framework for the development of intensely curved MoS_2 nanoshells throughout the synthesis process. Additionally, it facilitates the transport of electrons to MoS_2 due to the disparity in work function. The electrode coated with bilayer MoS_2 nanoshells possesses a substantial HER performance, achieving an overpotential of merely 78.2 mV at a current density of 10 mA cm^{-2}.

5.   Conclusion and Prospects

This chapter delves into the crucial role played by catalyst surface structures in enhancing the efficiency and performance of the electrocatalytic hydrogen production process. It underscores the significance of employing surface engineering techniques to tailor catalyst properties and achieve optimized electrocatalytic activity. The discussion revolves around various strategies for surface structure modulation, including geometrical morphology, chemical state, and electronic structure. The chapter also highlights the value of theoretical simulations and computational methods in predicting and optimizing the effects of surface structure modulation, providing essential guidance and theoretical support for experimental studies. By utilizing these approaches, it becomes possible to finely tune the catalyst's surface composition, electronic structure, and surface energy, thereby improving the HER capability. This includes accelerating reaction kinetics, reducing overpotentials, and enhancing the catalyst's stability.

While significant progress has been made in developing superior hydrogen evolution catalysts through surface structure engineering, there remains a gap between fundamental research and industrial applications. Future research endeavors should focus on exploring morphology, chemical state, and electronic structure to bridge this gap. Ongoing studies are expected to concentrate on investigating novel materials, employing advanced characterization techniques, and employing computational modelling to further optimize catalyst performance. Furthermore, the integration of surface engineering with other strategies, such as catalyst support modification and electrolyte engineering, holds tremendous potential for achieving groundbreaking advancements in electrocatalytic hydrogen production. By continuing to explore these avenues, researchers can contribute to the development of highly efficient, cost-effective, and sustainable catalysts, facilitating the widespread adoption of renewable energy technologies and propelling the transition towards a cleaner and greener energy future.

References

[1]   J. Zhang, A.R. Woldu, X. Zhao, X. Peng, Y. Song, H. Xia, F. Lu, Plasmon-enhanced hydrogen evolution on Pt-anchored titanium nitride nanowire arrays, Appl. Surf. Sci., 598 (2022) 153745. https://doi.org/10.1016/j.apsusc.2022.153745

[2]   X. Peng, X. Jin, N. Liu, P. Wang, Z. Liu, B. Gao, L. Hu, P.K. Chu, A high-performance electrocatalyst composed of nickel clusters encapsulated with a carbon network on TiN nanaowire arrays for the oxygen evolution reaction, Appl. Surf. Sci., 567 (2021) 150779. https://doi.org/10.1016/j.apsusc.2021.150779

[3]   M. Qiang, X. Zhang, H. Song, C. Pi, X. Wang, B. Gao, Y. Zheng, X. Peng, P.K. Chu, K.

Huo, General synthesis of nanostructured Mo_2C electrocatalysts using a carbon template for electrocatalytic applications, Carbon, 197 (2022) 238-245. https://doi.org/10.1016/j.carbon.2022.06.016

[4]   X. Peng, S. Xie, S. Xiong, R. Li, P. Wang, X. Zhang, Z. Liu, L. Hu, Ultralow-voltage hydrogen production and simultaneous Rhodamine B beneficiation in neutral wastewater, J. Energy Chem., 81 (2023) 574-582. https://doi.org/10.1016/j.jechem.2023.03.022

[5]   X. Wang, X. Zhang, Y. Xu, H. Song, X. Min, Z. Tang, C. Pi, Heterojunction Mo-based binary and ternary nitride catalysts with Pt-like activity for the hydrogen evolution reaction, Chem. Eng. J., 473 (2023) 144370. https://doi.org/10.1016/j.cej.2023.144370

[6]   C. Huang, C. Pi, X. Zhang, K. Ding, P. Qin, J. Fu, X. Peng, In Situ Synthesis of MoP Nanoflakes Intercalated N-Doped Graphene Nanobelts from MoO_3-Amine Hybrid for High-Efficient Hydrogen Evolution Reaction, Small, 14 (2018) 1800667. https://doi.org/10.1002/smll.201800667

[7]   Z. Huang, A.R. Woldu, X. Peng, P. Chu, Remarkably boosted water oxidation activity and dynamic stability at large-current–density of $Ni(OH)_2$ nanosheet arrays by Fe ion association and underlying mechanism, Chem. Eng. J., 477 (2023) 147155. https://doi.org/10.1016/j.cej.2023.147155

[8]   X. Peng, Y. Yan, S. Xiong, Y. Miao, J. Wen, Z. Liu, B. Gao, L. Hu, P.K. Chu, Se-$NiSe_2$ hybrid nanosheet arrays with self-regulated elemental Se for efficient alkaline water splitting, J. Mater. Sci. Technol., 118 (2022) 136-143. https://doi.org/10.1016/j.jmst.2021.12.022

[9]   Y. Li, L. Hu, W. Zheng, X. Peng, Liu, Ni/Co-based nanosheet arrays for efficient oxygen evolution reaction, Nano Energy, 52 (2018) 360-368. https://doi.org/10.1016/j.nanoen.2018.08.010

[10]   X. Peng, A.M. Qasim, W. Jin, L. Wang, L. Hu, Y. Miao, Ni-doped amorphous iron phosphide nanoparticles on TiN nanowire arrays: an advanced alkaline hydrogen evolution electrocatalyst, Nano Energy, 53 (2018) 66-73. https://doi.org/10.1016/j.nanoen.2018.08.028

[11]   S. Xie, Y. Yan, S. Lai, J. He, Z. Liu, B. Gao, Javanbakht, Ni^{3+}-enriched nickel-based electrocatalysts for superior electrocatalytic water oxidation, Appl. Surf. Sci., 605 (2022) 154743. https://doi.org/10.1016/j.apsusc.2022.154743

[12]   L. Xiong, Y. Qiu, X. Peng, Z. Liu, P.K. Chu, Electronic structural engineering of transition metal-based electrocatalysts for the hydrogen evolution reaction, Nano Energy, 104 (2022) 107882. https://doi.org/10.1016/j.nanoen.2022.107882

[13]   Z. Liu, B. Li, Y. Feng, D. Jia, C. Li, Q. Sun, Y. Zhou, Strong Electron Coupling of Ru and Vacancy-Rich Carbon Dots for Synergistically Enhanced Hydrogen Evolution Reaction, Small, 17 (2021) 2102496. https://doi.org/10.1002/smll.202102496

[14]   X. Peng, S. Xie, X. Wang, C. Pi, Z. Liu, B. Gao, L. Hu, W. Xiao, Energy-saving

hydrogen production by the methanol oxidation reaction coupled with the hydrogen evolution reaction co-catalyzed by a phase separation induced heterostructure, J. Mater. Chem., 10 (2022) 20761-20769. https://doi.org/10.1039/D2TA02955C

[15] Y. Yang, H. Meng, C. Kong, W. Ma, H. Zhu, F. Ma, C. Wang, Z. Hu, Template-free synthesis of 1D hollow Fe doped CoP nanoneedles as highly activity electrocatalysts for overall water splitting, Int. J. Hydrogen Energ., 46 (2021) 28053-28063. https://doi.org/10.1016/j.ijhydene.2021.06.047

[16] R. Li, S. Xie, Y. Zeng, Q. Zhao, M. Mao, Z. Liu, P.K. Chu, X. Peng, Synergistic dual-regulating the electronic structure of NiMo selenides composite for highly efficient hydrogen evolution reaction, Fuel, 358 (2024) 130203. https://doi.org/10.1016/j.fuel.2023.130203

[17] L. Yang, T. Yang, E. Wang, X. Yu, K. Wang, Z. Du, S. Cao, K.-C. Chou, X. Hou, Bifunctional hierarchical NiCoP@FeNi LDH nanosheet array electrocatalyst for industrial-scale high-current-density water splitting, J. Mater. Sci. Technol., 159 (2023) 33-40. https://doi.org/10.1016/j.jmst.2023.02.050

[18] S. Xie, Y. Yan, S. Lai, J. He, Z. Liu, B. Gao, M. Javanbakht, X. Peng, P.K. Chu, Ni^{3+}-enriched nickel-based electrocatalysts for superior electrocatalytic water oxidation, Appl. Surf. Sci., 605 (2022) 154743. https://doi.org/10.1016/j.apsusc.2022.154743

[19] L. Zhao, M. Wen, Y. Tian, Q. Wu, Y. Fu, A novel structure of quasi-monolayered NiCo-bimetal-phosphide for superior electrochemical performance, J. Energy Chem., 74 (2022) 203-211. https://doi.org/10.1016/j.jechem.2022.07.017

[20] J. Hou, Y. Wu, S. Cao, Y. Sun, L. Sun, Active sites intercalated ultrathin carbon sheath on nanowire arrays as integrated core–shell architecture: highly efficient and durable electrocatalysts for overall water splitting, Small, 13 (2017) 1702018. https://doi.org/10.1002/smll.201702018

[21] W. Liu, P. Geng, S. Li, W. Liu, D. Fan, H. Lu, Z. Lu, Y. Liu, Tuning electronic configuration of WP_2 nanosheet arrays via nickel doping for high-efficiency hydrogen evolution reaction, J. Energy Chem., 55 (2021) 17-24. https://doi.org/10.1016/j.jechem.2020.06.068

[22] B. Zhang, Y.H. Lui, H. Ni, S. Hu, Bimetallic $(Fe_xNi_{1-x})_2P$ nanoarrays as exceptionally efficient electrocatalysts for oxygen evolution in alkaline and neutral media, Nano Energy, 38 (2017) 553-560. https://doi.org/10.1016/j.nanoen.2017.06.032

[23] Y. Zhou, J. Zhang, H. Ren, Y. Pan, Y. Yan, F. Sun, X. Wang, S. Wang, J. Zhang, Mo doping induced metallic CoSe for enhanced electrocatalytic hydrogen evolution, Appl. Catal. B-Environ., 268 (2020) 118467. https://doi.org/10.1016/j.apcatb.2019.118467

[24] L. Chen, P. Wu, C. Zhu, S. Yang, K. Qian, N. Ullah, W. Wei, C. Sun, Y. Xu, J. Xie, Fabrication of carbon nanotubes encapsulated cobalt phosphide on graphene: cobalt promoted hydrogen evolution reaction performance, Electrochim. Acta, 330 (2020) 135213. https://doi.org/10.1016/j.electacta.2019.135213

[25]   X. Lv, S. Yin, CoP-embedded nitrogen and phosphorus co-doped mesoporous carbon nanotube for efficient hydrogen evolution, Appl. Surf. Sci., 537 (2021) 147834. https://doi.org/10.1016/j.apsusc.2020.147834

[26]   Z. Tang, S. Wei, Y. Wang, L. Dai, Three-dimensional reduced graphene oxide decorated with cobalt metaphosphate as high cost-efficiency electrocatalysts for the hydrogen evolution reaction, RSC Adv., 12 (2022) 10522-10533. https://doi.org/10.1039/D2RA01271E

[27]   F. Meng, Y. Yu, D. Sun, L. Li, S. Lin, L. Huang, W. Chu, S. Ma, B. Xu, Three-dimensional flower-like WP_2 nanowire arrays grown on Ni foam for full water splitting, Appl. Surf. Sci., 546 (2021) 148926. https://doi.org/10.1016/j.apsusc.2021.148926

[28]   S.J. Patil, N.R. Chodankar, S.-K. Hwang, P.A. Shinde, G.S.R. Raju, K.S. Ranjith, Y.S. Huh, Y.-K. Han, Co-metal–organic framework derived $CoSe_2@MoSe_2$ core–shell structure on carbon cloth as an efficient bifunctional catalyst for overall water splitting, Chem. Eng. J., 429 (2022) 132379. https://doi.org/10.1016/j.cej.2021.132379

[29]   L. Yang, L. Huang, Y. Yao, L. Jiao, In-situ construction of lattice-matching $NiP_2/NiSe_2$ heterointerfaces with electron redistribution for boosting overall water splitting, Appl. Catal. B-Environ., 282 (2021) 119584. https://doi.org/10.1016/j.apcatb.2020.119584

[30]   S.-K. Park, Y.C. Kang, MOF-templated N-doped carbon-coated $CoSe_2$ nanorods supported on porous CNT microspheres with excellent sodium-ion storage and electrocatalytic properties, ACS Appl. Mater. Interfaces, 10 (2018) 17203-17213. https://doi.org/10.1021/acsami.8b03607

[31]   Z. Chen, W. Wang, S. Huang, P. Ning, Y. Wu, C. Gao, T.-T. Le, J. Zai, Y. Jiang, Z. Hu, Well-defined $CoSe_2@MoSe_2$ hollow heterostructured nanocubes with enhanced dissociation kinetics for overall water splitting, Nanoscale, 12 (2020) 326-335. https://doi.org/10.1039/C9NR08751F

[32]   Z. Huang, S. Yuan, T. Zhang, B. Cai, B. Xu, X. Lu, L. Fan, F. Dai, D. Sun, Selective selenization of mixed-linker Ni-MOFs: $NiSe_2@$ NC core-shell nano-octahedrons with tunable interfacial electronic structure for hydrogen evolution reaction, Appl. Catal. B-Environ., 272 (2020) 118976. https://doi.org/10.1016/j.apcatb.2020.118976

[33]   Z. Fang, L. Peng, H. Lv, Y. Zhu, C. Yan, S. Wang, P. Kalyani, X. Wu, G. Yu, Metallic transition metal selenide holey nanosheets for efficient oxygen evolution electrocatalysis, ACS Nano, 11 (2017) 9550-9557. https://doi.org/10.1021/acsnano.7b05481

[34]   C. Yuan, H.B. Wu, Y. Xie, X.W. Lou, Mixed transition-metal oxides: design, synthesis, and energy-related applications, Angew. Chem. Int. Ed., 53 (2014) 1488-1504. https://doi.org/10.1002/anie.201303971

[35]   M.I. Godinho, M.A. Catarino, M. da Silva Pereira, M. Mendonça, F. Costa, Effect of the partial replacement of Fe by Ni and/or Mn on the electrocatalytic activity for oxygen evolution of the $CoFe_2O_4$ spinel oxide electrode, Electrochim. Acta, 47 (2002) 4307-4314. https://doi.org/10.1016/S0013-4686(02)00434-6

[36]   X. Zhao, Y. Fu, J. Wang, Y. Xu, J.-H. Tian, R. Yang, Ni-doped $CoFe_2O_4$ hollow nanospheres as efficient bi-functional catalysts, Electrochim. Acta, 201 (2016) 172-178. https://doi.org/10.1016/j.electacta.2016.04.001

[37]   R. Singh, N. Singh, J. Singh, G. Balaji, N. Gajbhiye, Effect of partial substitution of Cr on electrocatalytic properties of $CoFe_2O_4$ towards O_2-evolution in alkaline medium, Int. J. Hydrogen Energ., 31 (2006) 701-707. https://doi.org/10.1016/j.ijhydene.2005.07.003

[38]   Z. Wang, P. Guo, S. Cao, H. Chen, S. Zhou, H. Liu, H. Wang, J. Zhang, S. Liu, S. Wei, D. Sun, X. Lu, Contemporaneous inverse manipulation of the valence configuration to preferred Co^{2+} and Ni^{3+} for enhanced overall water electrocatalysis, Appl. Catal. B-Environ., 284 (2021) 119725. https://doi.org/10.1016/j.apcatb.2020.119725

[39]   J.K. Nørskov, T. Bligaard, A. Logadottir, J.R. Kitchin, J.G. Chen, S. Pandelov, U. Stimming, Trends in the Exchange Current for Hydrogen Evolution, J. Electrochem. Soc., 152 (2005) J23. https://doi.org/10.1149/1.1856988

[40]   J. Greeley, T.F. Jaramillo, J. Bonde, I. Chorkendorff, J.K. Nørskov, Computational high-throughput screening of electrocatalytic materials for hydrogen evolution, Nat. Mater., 5 (2006) 909-913. https://doi.org/10.1038/nmat1752

[41]   F. Calle-Vallejo, J.I. Martínez, J. Rossmeisl, Density functional studies of functionalized graphitic materials with late transition metals for oxygen reduction reactions, Phys. Chem. Chem. Phys., 13 (2011) 15639-15643. https://doi.org/10.1039/c1cp21228a

[42]   T. Kou, T. Smart, B. Yao, I. Chen, D. Thota, Y. Ping, Y. Li, Theoretical and Experimental Insight into the Effect of Nitrogen Doping on Hydrogen Evolution Activity of Ni_3S_2 in Alkaline Medium, Adv. Energy Mater., 8 (2018) 1703538. https://doi.org/10.1002/aenm.201703538

[43]   R. Hang, Y. Liu, L. Zhao, A. Gao, L. Bai, X. Huang, X. Zhang, B. Tang, P.K. Chu, Fabrication of Ni-Ti-O nanotube arrays by anodization of NiTi alloy and their potential applications, Sci. Rep., 4 (2014) 7547. https://doi.org/10.1038/srep07547

[44]   G. Fazio, L. Ferrighi, D. Perilli, C. Di Valentin, Computational electrochemistry of doped graphene as electrocatalytic material in fuel cells, Int. J. Quantum Chem., 116 (2016) 1623-1640. https://doi.org/10.1002/qua.25203

[45]   X. Peng, A.M. Qasim, W. Jin, L. Wang, L. Hu, Y. Miao, W. Li, Y. Li, Z. Liu, K. Huo, K.-y. Wong, P.K. Chu, Ni-doped amorphous iron phosphide nanoparticles on TiN nanowire arrays: An advanced alkaline hydrogen evolution electrocatalyst, Nano Energy, 53 (2018) 66-73. https://doi.org/10.1016/j.nanoen.2018.08.028

[46]   P. Zhou, L. Wang, J. Lv, R. Li, F. Gao, X. Huang, Y. Lu, G. Wang, Tuning the electronic structure of Co@N–C hybrids via metal-doping for efficient electrocatalytic hydrogen evolution reaction, J. Mater. Chem. A, 10 (2022) 4981-4991. https://doi.org/10.1039/D1TA08226D

[47]   L.B. Huang, L. Zhao, Y. Zhang, Y.Y. Chen, Q.H. Zhang, H. Luo, X. Zhang, T. Tang, L.

Gu, J.S. Hu, Self-Limited on-Site Conversion of MoO_3 Nanodots into Vertically Aligned Ultrasmall Monolayer MoS_2 for Efficient Hydrogen Evolution, Adv. Energy Mater., 8 (2018) 1800734. https://doi.org/10.1002/aenm.201800734

[48] M. Kuang, P. Han, Q. Wang, J. Li, G. Zheng, CuCo Hybrid Oxides as Bifunctiona Electrocatalyst for Efficient Water Splitting, Adv. Funct. Mater., 26 (2016) 8555-8561. https://doi.org/10.1002/adfm.201604804

[49] W.L. Kwong, E. Gracia-Espino, C.C. Lee, R. Sandström, T. Wågberg, J. Messinger, Cationic Vacancy Defects in Iron Phosphide: A Promising Route toward Efficient and Stable Hydrogen Evolution by Electrochemical Water Splitting, ChemSusChem, 10 (2017) 4544-4551. https://doi.org/10.1002/cssc.201701565

[50] D. Lancet, I. Pecht, Spectroscopic and immunochemical studies with nitrobenzoxadiazolealanine, a fluorescent dinitrophenyl analog, Biochemistry, 16 (1977) 5150-5157. https://doi.org/10.1021/bi00642a031

[51] Z. Chen, T.T. Fan, X. Yu, Q.L. Wu, Q.H. Zhu, L.Z. Zhang, J.H. Li, W.P. Fang, X.D. Yi, Gradual carbon doping of graphitic carbon nitride towards metal-free visible light photocatalytic hydrogen evolution, J. Mater. Chem. A, 6 (2018) 15310-15319. https://doi.org/10.1039/C8TA03303J

[52] T. Zhang, M.-Y. Wu, D.-Y. Yan, J. Mao, H. Liu, W.-B. Hu, X.-W. Du, T. Ling, S.-Z. Qiao, Engineering oxygen vacancy on NiO nanorod arrays for alkaline hydrogen evolution, Nano Energy, 43 (2018) 103-109. https://doi.org/10.1016/j.nanoen.2017.11.015

[53] T. Ling, D.Y. Yan, H. Wang, Y. Jiao, Z. Hu, Y. Zheng, L. Zheng, J. Mao, H. Liu, X.W. Du, M. Jaroniec, S.Z. Qiao, Activating cobalt(II) oxide nanorods for efficient electrocatalysis by strain engineering, Nat. Commun., 8 (2017) 1509. https://doi.org/10.1038/s41467-017-01872-y

[54] A. Karmakar, K. Karthick, S.S. Sankar, S. Kumaravel, M. Ragunath, S. Kundu, Oxygen vacancy enriched NiMoO4 nanorods via microwave heating: a promising highly stable electrocatalyst for total water splitting, J. Mater. Chem. A, 9 (2021) 11691-11704. https://doi.org/10.1039/D1TA02165F

[55] Y. Liu, H.T.D. Bui, A.R. Jadhav, T. Yang, S. Saqlain, Y. Luo, J. Yu, A. Kumar, H. Wang, L. Wang, V.Q. Bui, M.G. Kim, Y.D. Kim, H. Lee, Revealing the Synergy of Cation and Anion Vacancies on Improving Overall Water Splitting Kinetics, Adv. Funct. Mater., 31 (2021) 2010718. https://doi.org/10.1002/adfm.202010718

[56] G. Zhou, Y. Shan, L. Wang, Y. Hu, J. Guo, F. Hu, J. Shen, Y. Gu, J. Cui, L. Liu, X. Wu, Photoinduced semiconductor-metal transition in ultrathin troilite FeS nanosheets to trigger efficient hydrogen evolution, Nat. Commun., 10 (2019) 399. https://doi.org/10.1038/s41467-019-08358-z

[57] Z. Dong, F. Lin, Y. Yao, L. Jiao, Crystalline $Ni(OH)_2$/Amorphous $NiMoO_x$ Mixed-Catalyst with Pt-Like Performance for Hydrogen Production, Adv. Energy Mater., 9

(2019) 1902703. https://doi.org/10.1002/aenm.201902703

[58]	J. Zhang, T. Wang, P. Liu, Y. Liu, J. Ma, D. Gao, Enhanced Catalytic Activities of Metal-Phase-Assisted 1T@2H-MoSe$_2$ Nanosheets for Hydrogen Evolution, Electrochim. Acta, 217 (2016) 181-186. https://doi.org/10.1016/j.electacta.2016.09.076

[59]	W. Xiao, P. Liu, J. Zhang, W. Song, Y.P. Feng, D. Gao, J. Ding, Dual-Functional N Dopants in Edges and Basal Plane of MoS$_2$ Nanosheets Toward Efficient and Durable Hydrogen Evolution, Adv. Energy Mater., 7 (2017) 1602086. https://doi.org/10.1002/aenm.201602086

[60]	S. Deng, M. Luo, C. Ai, Y. Zhang, B. Liu, L. Huang, Z. Jiang, Q. Zhang, L. Gu, S. Lin, X. Wang, L. Yu, J. Wen, J. Wang, G. Pan, X. Xia, J. Tu, Synergistic Doping and Intercalation: Realizing Deep Phase Modulation on MoS$_2$ Arrays for High-Efficiency Hydrogen Evolution Reaction, Angew. Chem. Int. Ed., 58 (2019) 16289-16296. https://doi.org/10.1002/anie.201909698

[61]	B. Gao, Y. Zhao, X. Du, Y. Chen, B. Guan, Y. Li, Y. Li, S. Ding, H. Zhao, C. Xiao, Z. Song, Facile phase transition engineering of MoS$_2$ for electrochemical hydrogen evolution, J. Mater. Chem. A, 9 (2021) 8394-8400. https://doi.org/10.1039/D0TA12076F

[62]	P. Strasser, S. Koh, T. Anniyev, J. Greeley, K. More, C. Yu, Z. Liu, S. Kaya, D. Nordlund, H. Ogasawara, M.F. Toney, A. Nilsson, Lattice-strain control of the activity in dealloyed core–shell fuel cell catalysts, Nat. Chem., 2 (2010) 454-460. https://doi.org/10.1038/nchem.623

[63]	D. Voiry, H. Yamaguchi, J. Li, R. Silva, D.C.B. Alves, T. Fujita, M. Chen, T. Asefa, V.B. Shenoy, G. Eda, M. Chhowalla, Enhanced catalytic activity in strained chemically exfoliated WS$_2$ nanosheets for hydrogen evolution, Nat. Mater., 12 (2013) 850-855. https://doi.org/10.1038/nmat3700

[64]	J. Wang, M. Yan, K. Zhao, X. Liao, P. Wang, X. Pan, W. Yang, L. Mai, Field Effect Enhanced Hydrogen Evolution Reaction of MoS$_2$ Nanosheets, Adv. Mater., 29 (2017) 1604464. https://doi.org/10.1002/adma.201604464

[65]	H. Wang, S. Xu, C. Tsai, Y. Li, C. Liu, J. Zhao, Y. Liu, H. Yuan, F. Abild-Pedersen, F.B. Prinz, J.K. Nørskov, Y. Cui, Direct and continuous strain control of catalysts with tunable battery electrode materials, Science, 354 (2016) 1031-1036. https://doi.org/10.1126/science.aaf7680

[66]	Z. Li, H. Jang, D. Qin, X. Jiang, X. Ji, M.G. Kim, L. Zhang, X. Liu, J. Cho, Alloy-strain-output induced lattice dislocation in Ni$_3$FeN/Ni$_3$Fe ultrathin nanosheets for highly efficient overall water splitting, J. Mater. Chem. A, 9 (2021) 4036-4043. https://doi.org/10.1039/D0TA11618A

[67]	T. Zhang, Y. Liu, J. Yu, Q. Ye, L. Yang, Y. Li, H.J. Fan, Biaxially Strained MoS$_2$ Nanoshells with Controllable Layers Boost Alkaline Hydrogen Evolution, Adv. Mater., 34 (2022) 2202195. https://doi.org/10.1002/adma.202202195

[68]   H. Li, C. Tsai, A.L. Koh, L. Cai, A.W. Contryman, A.H. Fragapane, J. Zhao, H.S. Han, H.C. Manoharan, F. Abild-Pedersen, J.K. Nørskov, X. Zheng, Activating and optimizing MoS_2 basal planes for hydrogen evolution through the formation of strained sulphur vacancies, Nat. Mater., 15 (2016) 48-53. https://doi.org/10.1038/nmat4465

[69]   W. Han, Z. Liu, Y. Pan, G. Guo, J. Zou, Y. Xia, Z. Peng, W. Li, A. Dong, Designing Champion Nanostructures of Tungsten Dichalcogenides for Electrocatalytic Hydrogen Evolution, Adv. Mater., 32 (2020) 2002584. https://doi.org/10.1002/adma.202002584

Electrocatalytic Hydrogen Production: Catalysts and Applications   Materials Research Forum LLC
Materials Research Foundations **165** (2024)   https://doi.org/10.21741/9781644903070

CHAPTER 8

Heterostructure Catalysts for Electrocatalytic Hydrogen Production

Xiang Peng*

Hubei Key Laboratory of Plasma Chemistry and Advanced Materials, Engineering Research Center of Phosphorus Resources Development and Utilization of Ministry of Education, School of Materials Science and Engineering, Wuhan Institute of Technology, Wuhan 430205, China

xpeng@wit.edu.cn

Abstract

Interface engineering plays a vital role in improving hydrogen evolution reaction (HER) performance. Various concepts have been proposed to describe heterostructures, such as metal/metal oxides, metal/semiconductor, Mott-Schottky heterostructures, and other heterostructures, have been proposed. The influence of heterostructure catalysts on HER catalytic performance has been discussed, providing a detailed insight into the pivotal role played by heterostructures in enhancing HER performance.

Keywords

Heterostructure, Hydrogen Evolution Reaction, Metal/Metal Oxide Heterostructure, Metal/Semiconductor Heterostructure, Mott-Schottky Heterostructure

1. Introduction

The widespread application of high-performance catalysts based on precious metals is hindered by their scarcity and high cost. To overcome this challenge, researchers have dedicated significant efforts to explore hydrogen evolution catalysts that are not only efficient but also abundant and economically viable [1-3]. In this pursuit, it has become increasingly evident that the hydrogen evolution reaction (HER) primarily takes place at the surface interface of the catalyst, highlighting the critical role of the electronic structure and catalytic properties of this interface in determining the catalyst's activity. Adjusting the electronic structure of catalysts is considered an effective strategy to enhance their electrocatalytic activity, as schematically illustrated in **Figure 1** [4]. When two components come into direct contact, the mismatch of Fermi levels leads to charge redistribution at the surface interface, and the resulting built-in electric field plays a crucial role in charge separation [5].

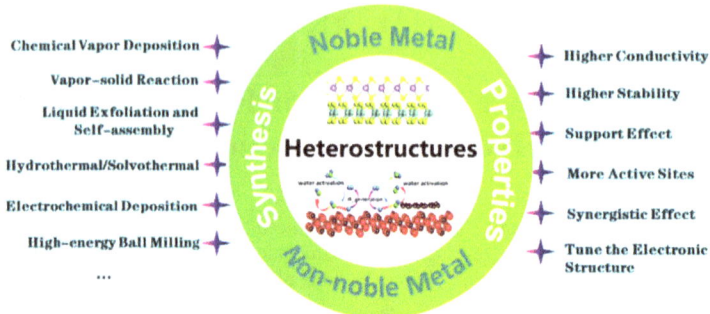

Figure 1. *Synthesis methods and catalytic properties of heterostructure catalysts for HER. Reproduced with permission from ref [4]. Copyright 2020, the Royal Society of Chemistry.*

Regarding the geometric structure, lattice constant defects induced by strain, whether through compression or stretching, cause a displacement of the d-band center position, thereby triggering defect formation. Furthermore, differences in electron affinity can induce charge transfer through chemical bonding at the surface interface, effectively modulating the catalyst's electronic structure and enhancing its intrinsic activity [6, 7]. Therefore, interface engineering plays a vital role in improving HER performance. Various concepts have been proposed to describe heterostructures, such as metal/metal oxides [8-12], metal/semiconductor [13-15], Mott-Schottky heterostructures [16-18], and other heterostructures [19-22], have been proposed. In this chapter, we delve into the influence of heterostructure catalysts on HER catalytic performance, providing a detailed insight into the pivotal role played by heterostructures in enhancing HER performance.

2.   Heterostructure Catalysts for Electrocatalytic Hydrogen Production

2.1.   Metal/Metal Oxide Heterostructure Catalysts

Over the past decade, the utilization of metal oxides in electrolytic hydrogen evolution has witnessed a gradual increase. However, this progress has been accompanied by certain challenges, including high electrocatalytic activation energy, catalyst deactivation, and poor corrosion resistance [23, 24]. To overcome these obstacles, researchers have proposed the introduction of metal particles to form heterostructure catalysts consisting of metal/metal oxide combinations [25]. Metal particles offer advantages such as high conductivity, enhanced catalytic activity, and a large surface area [26, 27]. Introducing metal particles into the catalyst system can facilitate electron transfer, increase active sites, modulate surface electron structures, suppress oxygen evolution reactions, and enhance catalyst stability. The formation of heterostructure catalysts with metal particles and metal oxides significantly enhances hydrogen evolution performance by optimizing interface effects and achieving synergistic effects between the metals and metal oxides [28, 29]. This design concept presents a promising approach to address the limitations of metal oxides in electrocatalysis, thus advancing the development of sustainable and efficient water electrolysis for hydrogen production.

For example, Chen et al. [8] developed a cost-effective, highly active, and efficient heterogeneous hydrogen evolution catalyst based on interactions between metals, metal oxides, and carbon support. The process involved spin-coating an ordered mesoporous carbon (CMK-3) precursor onto nickel foam, followed by heat treatment to obtain a stable carbon-based substrate on the nickel foam. Subsequently, $NiMoO_x$ nanoparticles (NPs) were deposited on the surface of CMK-3, resulting in the formation of the $NiMoO_x@CMK$-3 catalyst. Notably, the high conductivity of CMK-3 facilitated the uniform dispersion of $NiMoO_x$ NPs during the electrodeposition process, enabling the exposure of active sites. This design ensured efficient electron transfer and rapid adsorption/desorption of H_2 during the hydrogen evolution process. High-resolution transmission electron microscopy (HR-TEM) analysis revealed characteristic lattice fringes corresponding to the (310) and (220) planes of Ni_4Mo and the (023) plane of $NiMoO_4$, indicating the formation of a $Ni_4Mo/NiMoO_4$ heterostructure on CMK-3, as depicted in **Figure 2a-b**.

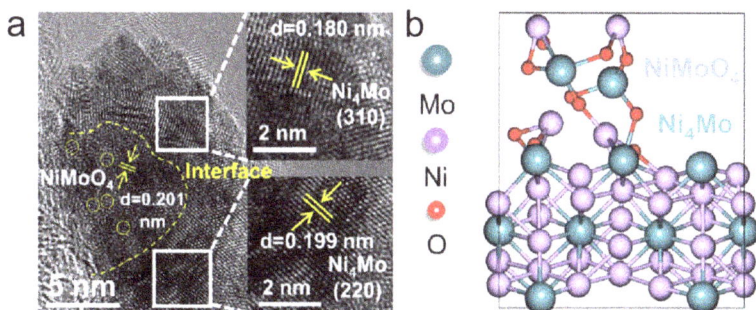

Figure 2. *(a) HR-TEM image of $NiMoO_x@CMK$-3. (b) The theoretical model for the designed heterogeneous $NiMoO_x$. Reproduced with permission from ref [8]. Copyright 2023, Elsevier.*

To gain further insights into the surface electronic configuration and surface chemistry of the $NiMoO_x@CMK$-3 hierarchical composite catalyst, X-ray photoelectron spectroscopy (XPS) was employed. The high-resolution Ni-$2p$ spectrum depicted in **Figure 3a** reveals two peaks at 852.6 and 869.6 eV, corresponding to Ni^0-$2p_{3/2}$ and $2p_{1/2}$, respectively. Additionally, the peaks appear at 856.3 and 874.1 eV, accompanied by two broad satellite peaks at 861.5 and 879.8 eV, which can be attributed to Ni^{2+}-$2p_{3/2}$ and $2p_{1/2}$ [30, 31]. Similarly, the high-resolution Mo $3d$ spectrum can be deconvoluted into three sets of peaks, as shown in **Figure 3b**. The binding energies of Mo^0-$3d_{5/2}$ and $3d_{3/2}$ are measured at 227.98 and 230.9 eV, while those of Mo^{4+}-$3d_{5/2}$ and $3d_{3/2}$ are found at 229.6 and 233.9 eV, and Mo^{6+}-$3d_{5/2}$ and $3d_{3/2}$ at 232.3 and 235.5 eV, respectively. The higher oxidation states of Mo^{6+} and Ni^{2+} correspond to the formation of $NiMoO_4$, whereas the lower oxidation state of Mo^{4+} is attributed to the surface adsorption of hydroxyl groups [32]. Moreover, the prominent signals of Mo^0 and Ni^0 strongly confirm the formation of Ni_4Mo in the hybrid $NiMoO_x@CMK$-3.

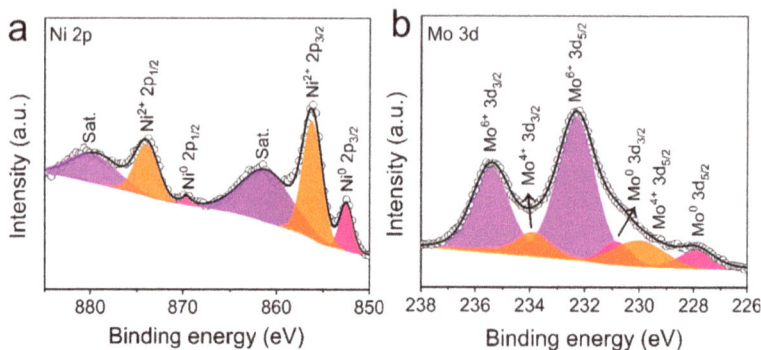

Figure 3. *High-resolution XPS signals of (a) Ni-2p and (b) Mo-3d for the NiMoO$_x$@CMK-3 composite catalyst. Reproduced with permission from ref [8]. Copyright 2023, Elsevier.*

Through X-ray absorption near-edge structure (XANES) and extended X-ray absorption fine structure (EXAFS) spectroscopy, a detailed investigation was conducted to examine the local atomic coordination and electronic environment of the NiMoO$_x$@CMK-3 catalyst. The Ni K-edge spectrum displayed in **Figure 4a** reveals that the intensity of the magenta line peak falls between that of Ni foil and NiO reference material, indicating a positive charge on Ni in NiMoO$_x$@CMK-3 [33]. Simultaneously, the XANES spectrum of the Mo K-edge shown in **Figure 4b** indicates that the magenta line peak intensity shifts lower compared to Mo foil, suggesting a reduction in the oxidation state of Mo. Due to the higher electronegativity of Mo in the Ni$_4$Mo alloy [34], the Mo atoms in NiMoO$_x$@CMK-3 absorb electrons from Ni atoms.

Further investigation of the Fourier-transformed (FT) k^3-weighted EXAFS spectra at the Ni and Mo K-edges provided insights into the radial structure functions and interactions between Ni and Mo, as shown in **Figure 4c-d**. In the FT-EXAFS spectrum at the Ni K-edge shown in **Figure 4c**, a minor peak at approximately 1.7 Å is attributed to the Ni-O bond, followed by a peak at 2.48 ± 0.01 Å, possibly originating from Ni-Ni/Mo, which is consistent with previous studies [35, 36]. Correspondingly, at the Mo K-edge, the peak at around 1.8 Å corresponds to the Mo-O scattering, and the peak at 2.72 ± 0.01 Å corresponds to the Mo-Mo pathway. In conclusion, these results indicate that the heterogeneous NiMoO$_x$ catalyst consists of Ni$_4$Mo and NiMoO$_4$ components.

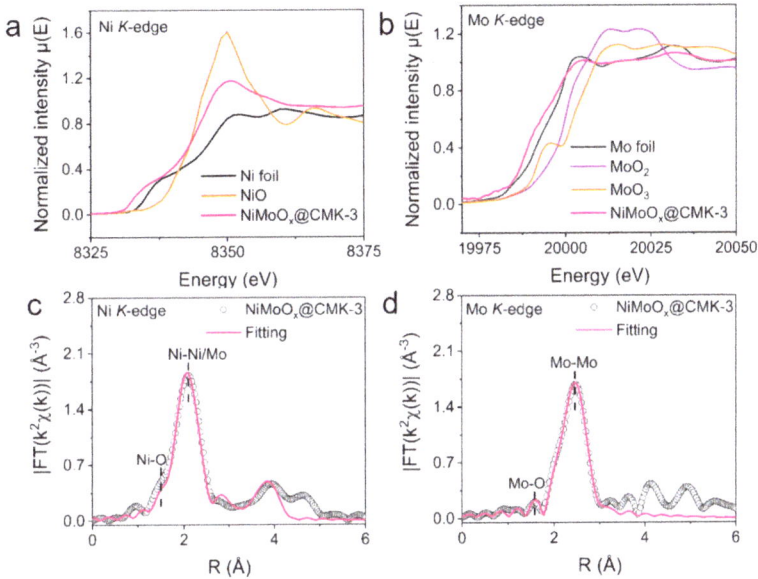

Figure 4. *XANES spectra of (a) Ni K-edge and (b) Mo K-edge for the Ni foil, NiO, and NiMoO_x@CMK-3, respectively. FT-EXAFS spectra for (c) Ni K-edge and (d) Mo K-edge, respectively. Reproduced with permission from ref [8]. Copyright 2023, Elsevier.*

To explore the electrocatalytic benefits of the hierarchical heterogeneous structure $NiMoO_x$@CMK-3 in the HER, a conventional three-electrode system was employed. Linear sweep voltammetry (LSV) at a scan rate of 5 mV s^{-1} was performed to evaluate the electrocatalytic activity of $NiMoO_x$@CMK-3 and other comparative catalysts, including bare nickel foam (NF), CMK-3, $NiMoO_x$, and commercial Pt/C catalysts, in a 1.0 M KOH electrolyte. As shown in **Figure 5a**, the $NiMoO_x$@CMK-3 catalyst exhibited the highest HER catalytic activity among all the tested samples. It required extremely low overpotentials (η) of only 7, 46, and 81 mV to achieve the current densities of 10, 50, and 100 mA cm^{-2}, respectively, as indicated in **Figure 5b**. The Tafel slope of the $NiMoO_x$@CMK-3 catalyst was measured to be 27.7 mV dec^{-1}, indicating that the Tafel process is the rate-determining step for the HER kinetics of the prepared $NiMoO_x$@CMK-3 catalyst. This confirms the high intrinsic activity of $NiMoO_x$ and the rapid charge transfer at the unique interface between CMK-3 and $NiMoO_x$.

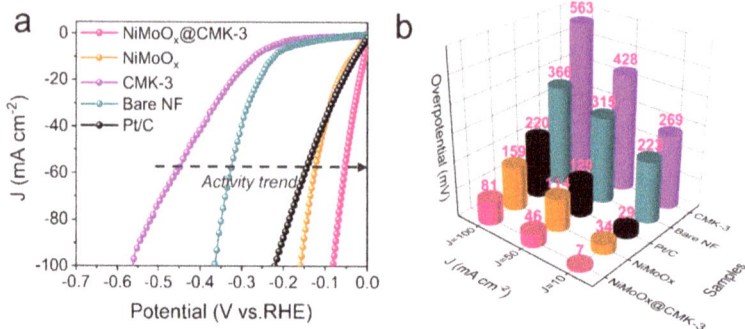

Figure 5. (a) HER polarization curves (iR-corrected). (b) Comparison of the overpotentials required to achieve the current densities of 10, 50, and 100 mA cm^{-2} for the NiMoO$_x$@CMK-3, NiMoO$_x$, CMK-3, bare NF, and Pt/C catalysts in 1.0 M KOH. Reproduced with permission from ref [8]. Copyright 2023, Elsevier.

Determining the electrochemically active surface area (ECSA) through double-layer capacitance (C_{dl}), a significant increase in the C_{dl} value was observed from 1.86 mF cm^{-2} for NiMoO$_x$ to 151.1 mF cm^{-2} for NiMoO$_x$@CMK-3, as shown in **Figure 6a**. This indicates that the deposition of NiMoO$_x$ NPs on CMK-3 exposes more active sites. Interestingly, while the C_{dl} value of CMK-3 is similar to that of NiMoO$_x$@CMK-3, its HER performance is significantly poorer. Therefore, it can be reasonably inferred that NiMoO$_x$ NPs are the intrinsic active material, and CMK-3, as a conductive substrate, facilitates rapid charge transfer and provides a larger surface area to increase the number of active sites.

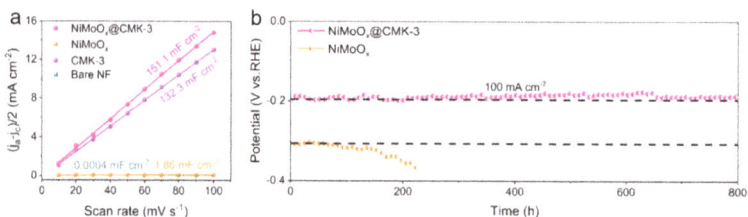

Figure 6. (a) C_{dl} of the different catalysts. (b) Chronopotentiometry stability tests of the NiMoO$_x$@CMK-3 and NiMoO$_x$ catalysts at the high current density of 100 mA cm^{-2} (without iR correction). Reproduced with permission from ref [8]. Copyright 2023, Elsevier.

The turnover frequency (TOF) was calculated to study the intrinsic catalytic capability of the catalysts. NiMoO$_x$@CMK-3 exhibited a TOF of 11.05 s^{-1}/Mo active site at the overpotential of 100 mV, which is three times higher than that of Pt/C catalyst (3.14 s^{-1}). Furthermore, the long-term HER performance of the NiMoO$_x$@CMK-3 catalyst was investigated using chronopotentiometry. At a large current density of 100 mA cm^{-2}, the NiMoO$_x$@CMK-3 catalyst demonstrated stable hydrogen generation for 800 hours without significant degradation, while the NiMoO$_x$ catalyst exhibited a noticeable current drop after 200 hours, as indicated in **Figure 6b**. Thus, the strong synergistic effect between heterogeneous NiMoO$_x$ and CMK-3 significantly accelerates mass transport, enhances interface electron density, and contributes to improved HER activity.

In another study conducted by Zhou et al. [10], they investigated the use of a Pt single-atom immobilized NiO/Ni heterostructure on Ag nanowires (Pt$_{SA}$-NiO/Ni@Ag NWs) as a catalyst for alkaline hydrogen evolution. The catalyst preparation process is illustrated in **Figure 7**. Firstly, Ag nanowires (Ag NWs) were synthesized on a flexible cloth substrate through a hydrothermal reaction, forming a conductive network. Subsequently, Ni/NiO composite material is electrodeposited onto the Ag network to create uniformly distributed nanosheets. Finally, the sequential cyclic voltammetry method is employed to achieve NiO/Ni with individually anchored Pt atoms (Pt$_{SA}$-NiO/Ni). During the electroreduction of single Pt atoms, hydrogen bubbles were generated and released due to the higher cathodic potential between 0 and −0.50 V *vs.* the reversible hydrogen electrode (RHE) under alkaline conditions [37].

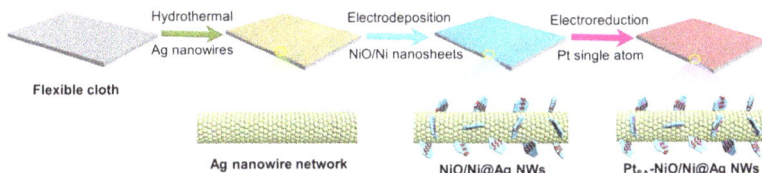

Figure 7. The synthesis process of Pt single atom anchored NiO/Ni heterostructure nanosheets on Ag nanowires network. Reproduced with permission from ref [10]. Copyright 2021, Springer Nature.

The morphology of Pt$_{SA}$-NiO/Ni nanosheets on Ag NWs remained unchanged compared to the original NiO/Ni, indicating excellent structural stability for HER applications. Additionally, the exposed NiO/Ni nanosheets provided additional Pt atomic anchoring sites, leading to enhanced HER performance. This confirmed the secure fixation of individually dispersed Pt atoms within the NiO/Ni nanosheets and validated the formation of a composition comprising NiO/Ni anchored with single Pt atoms, without compromising the structural integrity of NiO/Ni.

The evolution of the electronic state of single Pt atoms in NiO/Ni, NiO, and Ni supports was explored using XPS, as shown in **Figure 8**. The Pt-4f spectra of Pt$_{SA}$-NiO/Ni, Pt$_{SA}$-NiO, and Pt$_{SA}$-Ni closely resemble Pt^0 but exhibited varying degrees of positive shifts compared to Pt foil. This confirmed the electrochemical reduction of PtCl$_6^{2-}$ and the subsequent charge transfer from Pt sites to the carriers (NiO/Ni, NiO, and Ni) [38, 39], indicating the presence of electronic interactions. Notably, Pt$_{SA}$-NiO displayed the most significant positive shift in the Pt-4f spectrum, suggesting the maximum electron loss in Pt species [40, 41]. Furthermore, the fitting curves of the XPS spectra revealed the presence of Pt(IV) species in the samples, originating from the surface adsorption of PtCl$_6^{2-}$ ions [42, 43].

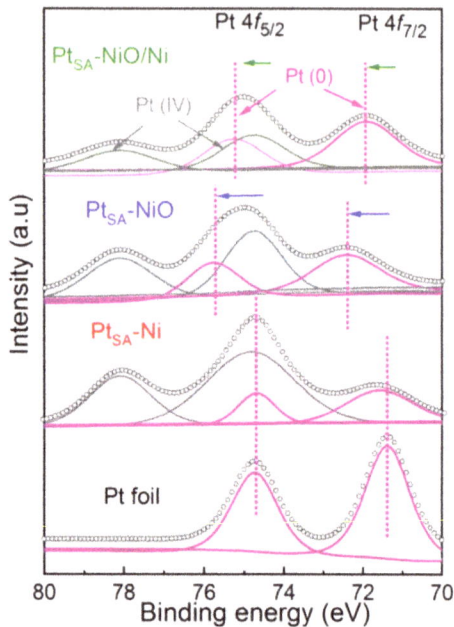

Figure 8. XPS spectra of Pt-4f. Reproduced with permission from ref [10]. Copyright 2021, Springer Nature.

Based on the above structural characterization, the catalyst prepared with the coupled NiO/Ni heterostructure and Pt single-atom clusters (Pt SAC) exhibited superior HER activity in an alkaline medium. The electrocatalytic performance of Pt_{SA}-NiO/Ni for alkaline HER was evaluated in a 1.0 M KOH solution. As shown in **Figure 9a**, the Pt_{SA}-NiO/Ni catalyst demonstrated the highest HER performance among all catalysts, requiring significantly lower overpotentials of 26 mV and 85 mV to achieve current densities of 10 and 100 mA cm^{-2}, respectively, as shown in **Figure 9b**. This indicated that the coupling of single Pt atoms with the NiO/Ni heterostructure maximized the alkaline HER activity of Pt-based catalysts. Additionally, the Tafel slope of Pt_{SA}-NiO/Ni was measured to be 27.07 mV dec^{-1}, suggesting a typical Volmer-Tafel mechanism for alkaline HER. It indicated that the rate-determining step for the Pt_{SA}-NiO/Ni catalyst was H_2 desorption (Tafel step) rather than H_2O dissociation (Volmer step) [24, 44]. Meanwhile, the Pt_{SA}-NiO/Ni catalyst exhibited high durability in alkaline electrolytes, with negligible performance loss after 5000 cycles or 30 hours of HER testing. These results collectively demonstrated the excellent coupling of the NiO/Ni heterostructure with Pt single atoms, promoting rapid alkaline HER.

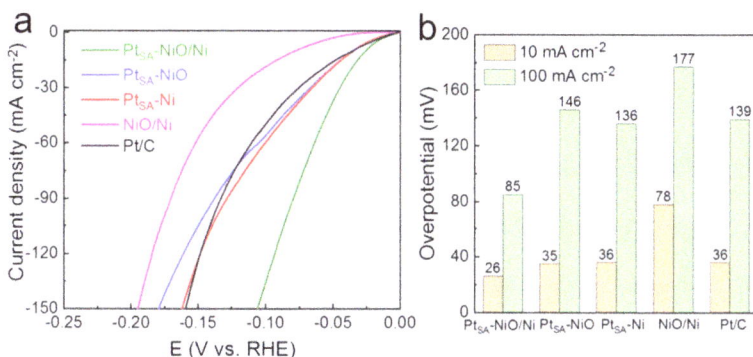

Figure 9. (a) HER polarization curves of the Pt_{SA}-NiO/Ni, Pt_{SA}-NiO, Pt_{SA}-Ni, NiO/Ni, and Pt/C catalysts. (b) The comparison of overpotentials required to achieve the current densities of 10 and 100 mA cm^{-2} for various catalysts. Reproduced with permission from ref [10]. Copyright 2021, Springer Nature.

Peng et al. [11] introduced the concept of the "chimney effect" to explain the synergistic effects at the interface of metal oxide/metal composite catalysts, contributing to a deeper understanding of their catalytic behavior. **Figure 10a-b** illustrates the differences between the metal oxide/metal interface and the non-interface regions. In these composite catalysts, two distinct pathways for the HER can occur: one near the interface and the other away from the interface. Initially, H_2O^* adsorbed on the metal or metal oxide surface undergoes dissociation, resulting in the formation of OH* and H*. The generated H* can then adsorb either near or far from the interface, leading to different mechanisms that influence the overall catalytic behavior and activity of the catalyst.

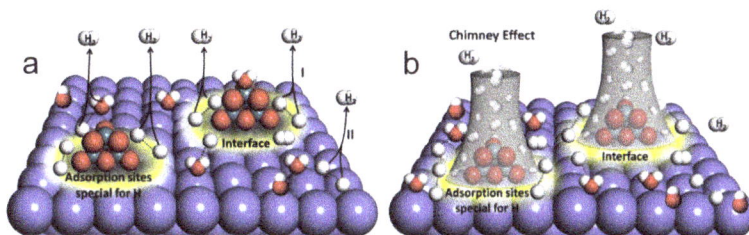

Figure 10. Schematic diagrams of the proposed HER mechanism on the RuO_2/Ni composite catalysts. (a) Two possible HER pathways. (b) "Chimney Effect" at the interface. Reproduced with permission from ref [11]. Copyright 2019, Elsevier.

The NiO/Ni heterostructures were prepared through a hydrothermal and annealing process. Initially, the Ni^{2+} precursor was synthesized via a hydrothermal reaction, followed by heat treatment to obtain NiO nanosheets. NiO/Ni nanosheet catalysts with varying NiO content were obtained by adjusting the reduction temperature under a hydrogen atmosphere. X-ray diffraction

(XRD) patterns, as shown in **Figure 11**, revealed that as the reduction temperature increased from 240 °C to 400 °C, the intensity of the Ni peak increased while that of the NiO peak decreased. This indicated a gradual reduction in the NiO content in the NiO/Ni catalysts with higher reduction temperatures. At 400 °C, NiO was completely transformed into Ni. The Ni(111) crystal planes of the samples showed a slight downward shift with increasing NiO content, as illustrated in the inset of **Figure 11**, suggesting an altered crystal structure of Ni influenced by the NiO content.

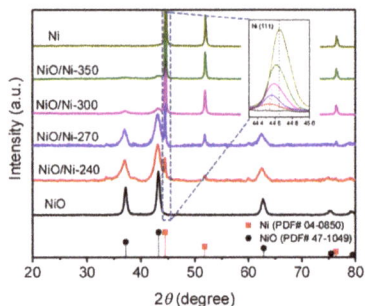

Figure 11. *XRD patterns of Ni, NiO, and NiO/Ni catalysts with different NiO contents varying with reduction temperature. Reproduced with permission from ref [11]. Copyright 2019, Elsevier.*

To investigate the catalytic activity of the NiO/Ni catalysts, HER performance of was tested in 1.0 M NaOH solution. The linear sweep voltammetry (LSV) curves shown in **Figure 12a** demonstrated that all NiO/Ni catalysts exhibited higher HER activity compared to pure NiO and Ni catalysts. Among them, the NiO/Ni-300 catalyst exhibited the best catalytic activity, with an overpotential of 194 mV to afford a current density of 10 mA cm^{-2} in a 1.0 M NaOH solution. **Figure 12b** shows a consistent trend between the HER activity of NiO/Ni samples and the variation in Ni interface atomic content, indicating a strong correlation between HER activity and the quantity of NiO/Ni interfaces.

Increasing the reduction temperature resulted in the gradual transformation of NiO into Ni, leading to the forming of NiO/Ni interfaces and enhancing the catalytic activity of NiO/Ni. At the optimal reduction temperature of 300 °C, the NiO/Ni interface content reached its maximum value (28.06%), and NiO/Ni exhibited the highest catalytic activity among the different samples. Further increasing the reduction temperature caused excessive reduction of NiO content, reducing the NiO/Ni interface content and ultimately resulting in a decline in NiO/Ni catalytic activity. The presence of a higher content of metal oxide-metal interfaces created more "chimneys" along the interfaces, increasing the efficiency of the metal oxide/metal system in HER. Thus, the experimental results confirmed that the "chimney effect" is the underlying mechanism behind the synergistic effects induced by interfaces in metal oxide/metal composite catalysts for catalyzing HER.

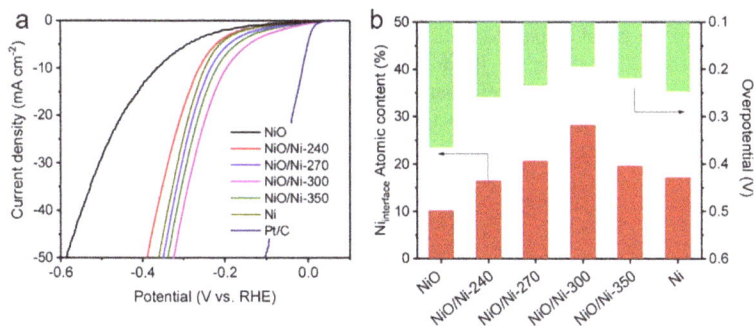

Figure 12. *(a) Polarization curves of the Ni catalysts carrying different amounts of NiO in a 1.0 M NaOH electrolyte. (b) Comparisons of the content of Ni$_{interface}$ species on the surfaces of the prepared samples and the overpotentials in HER. Reproduced with permission from ref [11]. Copyright 2019, Elsevier.*

2.2.   Metal/Semiconductor Heterostructure Catalysts

In recent years, semiconductor materials have shown great potential in catalysis [45-47], electronics [48, 49], and sensor applications [50-52] due to their unique electronic structure [53, 54]. Metal sulfides [55, 56], spinels [57, 58], and perovskites [59, 60] are among the semiconductor catalysts that have gained significant attention in research. However, these semiconductor catalysts face challenges such as poor stability in aqueous solutions, low electrical conductivity, and relatively low catalytic activity, which limit their long-term applications in electrocatalysis [61]. To overcome these limitations, the integration of metals to form metal-semiconductor heterojunctions has been employed [62].

By carefully designing metal-semiconductor heterojunctions, the limitations of semiconductor catalysts can be overcome, leading to more efficient and stable catalytic performance. The introduction of metals not only enhances catalytic activity but also improves the stability and resistance to catalyst poisoning [63, 64]. This integrated strategy provides a promising pathway for developing high-performance and sustainable electrocatalytic materials, with the potential to drive the application of semiconductor materials in energy conversion and environmental fields.

For instance, Shan et al. [14] designed and developed a Pt-MoS$_2$ heterostructure, which exhibited high activity and stability for the HER. The researchers synthesized MoS$_2$ nanosheets using a room temperature chemical solution method and then mixed them with H$_2$PtCl$_6$ aqueous solution and polyvinylpyrrolidone (PVP). The resulting mixture was stirred, followed by the addition of NaBH$_4$ powder. After 30 minutes, the product was collected through centrifugation.

The crystal phases of the Pt-MoS$_2$ sample were analyzed by XRD. The XRD patterns in **Figure 13a** showed the dominant diffraction peak corresponding to the (002) plane of 2H-MoS$_2$ (JCPDS card No. 37-1492), indicating a preferred orientation along the (001) direction. The presence of Pt NPs on the Pt-MoS$_2$ sample was not pronounced in the XRD pattern due to their low loading, small size, and good dispersion. However, an observed weak peak at 39.7° corresponded to the (111) plane of the metal Pt phase (JCPDS card No. 04-0802), as shown in **Figure 13b**. The lower angle shift of peaks in the Pt-MoS$_2$ sample, as illustrated in **Figure 13c**, suggested a certain degree of lattice strain or deformation in MoS$_2$, which can induce micro-strains, increase surface energy, and strengthen the interaction between the deposited metal and the support.

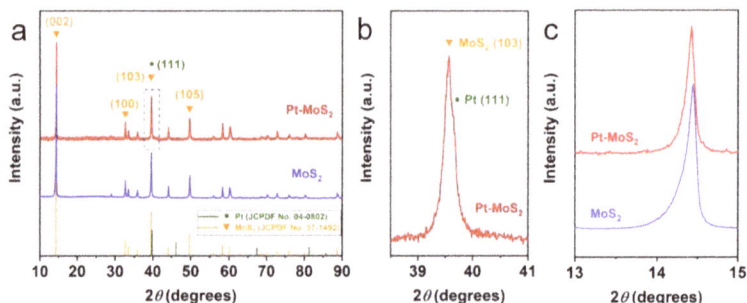

Figure 13. *(a) XRD of the Pt-MoS$_2$ (blue) and MoS$_2$ (red) catalysts. (b) Enlarged XRD with the value of 2θ goes from 38° to 41°. (c) Enlarged XRD spectra with the value of 2θ go from 13° to 15°. Reproduced with permission from ref [14]. Copyright 2022, Elsevier.*

XPS was utilized to investigate the composition and valence states of the Pt-MoS$_2$ sample. The surface Pt content was confirmed to be 3.8 wt% by XPS analysis. The binding energies of Mo^{4+}-3$d_{5/2}$ and Mo^{4+}-3$d_{3/2}$ electron peaks in the Pt-MoS$_2$ sample are shown in **Figure 14a**, observed at 229.6 eV and 232.7 eV, respectively. The Mo^{4+}-3d peaks in the Pt-MoS$_2$ sample were shifted downward by 0.1 eV compared to unloaded MoS$_2$, indicating electron transfer from Pt to Mo atoms. The binding energy at 226.9 eV shown in **Figure 14a** can be assigned to S-2s. The Pt-4f spectrum in **Figure 14b** showed that both Pt-4$f_{7/2}$ and Pt-4$f_{5/2}$ can be divided into two peaks. The peaks at 72.4 eV and 75.8 eV belong to Pt$^{δ+}$ species, indicating Pt atoms connected to S atoms. The downward shift of the characteristic peaks of S-2p confirmed the formation of Pt-S bonds in Pt-MoS$_2$ [65], indicating strong electronic metal-support interaction (EMSI) between Pt and MoS$_2$, which facilitated electron transfer.

The electrocatalytic activity and stability of Pt-MoS$_2$, Pt NPs, and pure MoS$_2$ catalysts were compared. The Pt-MoS$_2$ heterostructure exhibited a significantly lower overpotential of 67.4 mV to produce a current density of 10 mA cm^{-2} compared to Pt NPs (100.0 mV), as shown in **Figure 15a**. As shown in **Figure 15b**, the Tafel slope of Pt-MoS$_2$ was measured to be 76.2 mV-dec^{-1}, which was significantly lower than other catalysts, indicating improved catalytic efficiency. Pt-MoS$_2$ also demonstrated good stability, showing almost no decay in current density after a 24-hour constant-pressure electrolysis test at −0.3 V *vs*. RHE, as depicted in **Figure 15c**.

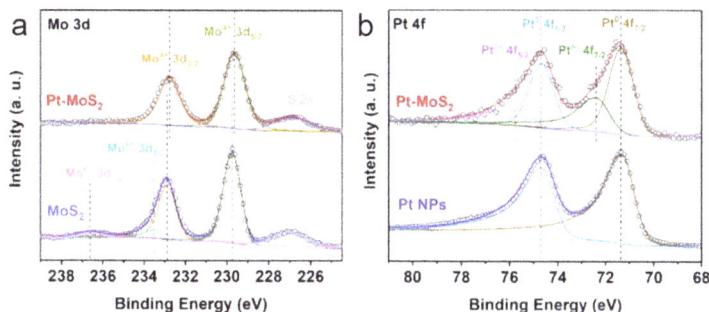

Figure 14. (a) XPS spectra of Mo-3d for Pt-MoS$_2$ and pure MoS$_2$. (b) Pt-4f core level for Pt-MoS$_2$ and Pt nanoparticles. Reproduced with permission from ref [14]. Copyright 2022, Elsevier.

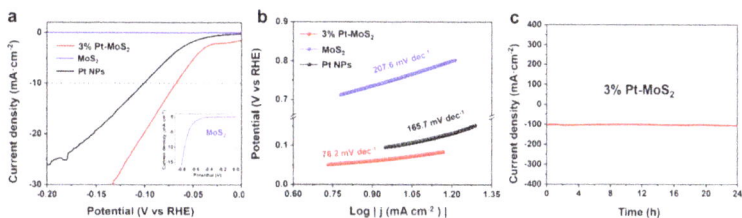

Figure 15. (a) Polarization curves and (b) Tafel plots for the Pt-MoS$_2$ heterostructures, MoS$_2$ nanosheet, and Pt NPs catalysts. (c) Stability test of the Pt-MoS$_2$ heterostructure catalyst. Reproduced with permission from ref [14]. Copyright 2022, Elsevier.

Liu et al. [13] reported the synthesis of a Pt/GaN heterostructure catalyst using electrodeposition technology for alkaline HER. The synthesis process of Pt/GaN is illustrated in **Figure 16**. The synthesis process involved mixing gallium nitrate and melamine, followed by calcination to obtain GaN. Pt NPs were then electrodeposited onto the GaN substrate. XRD analysis confirmed that the crystal structure of GaN was preserved after the electrodeposition of Pt nanoparticles.

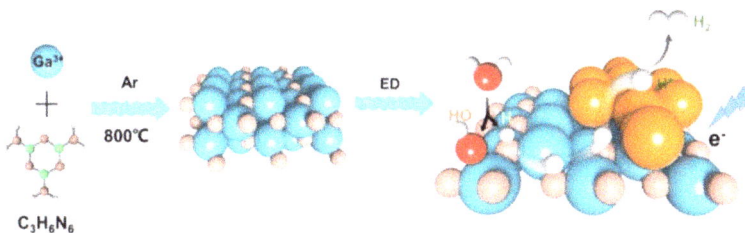

Figure 16. Schematic illustration of synthesis and HER procession on the surface of Pt/GaN catalyst. Reproduced with permission from ref [13]. Copyright 2023, Wiley-VCH.

The XPS characterization was performed on the samples before and after electrodeposition. The results exhibited characteristic signals for Ga, N, O, and Pt elements. As shown in **Figure 17a**, compared to GaN, the Ga-$2p_{1/2}$ and Ga-$2p_{2/3}$ peaks from Pt/GaN shifted to lower binding energies. Additionally, after loading Pt nanoparticles, the O-$1s$ peak also shifted towards lower binding energies, as indicated in **Figure 17b**, suggesting the introduction of oxygen during the electrodeposition process. The Ga-$2p$ peaks and N-$1s$ peaks in the XPS spectra shifted to lower binding energies after Pt deposition, indicating the introduction of oxygen during the electrodeposition process. The two characteristic peaks of Pt-$4f_{7/2}$ and $4f_{5/2}$ were located at 71.58, 74.9 eV, and 72.9, 76.2 eV, respectively, corresponding to Pt^0 and Pt^{2+}, as shown in **Figure 17c**. The Pt-$4f$ peaks indicated the presence of metallic Pt and suggested strong chemical interactions between Pt NPs and the GaN substrate, forming a Pt/GaN heterostructure with pronounced heterostructure interactions [66].

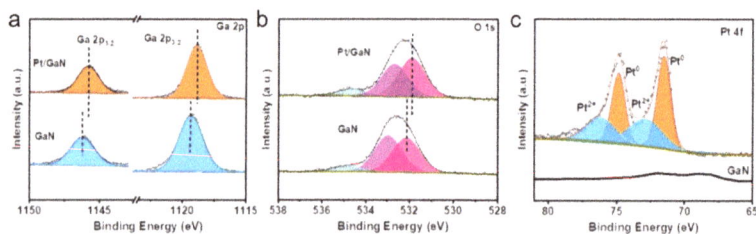

Figure 17. High-solution XPS of (a) Ga-2p, (b) O-1s, and (c) Pt-4f. Reproduced with permission from ref [13]. Copyright 2023, Wiley-VCH.

In order to understand the catalytic performance of the Pt/GaN catalyst, extensive testing was conducted on the obtained samples in a 1.0 M KOH solution. The Pt/GaN catalyst exhibited remarkable HER performance in HER, exhibiting significantly reduced overpotentials. Specifically, the overpotentials required to achieve current densities of 10 and 1000 mA cm^{-2} were lowered by 24 and 441 mV, respectively, compared to other catalysts. This highlights the competitive HER performance of the synthesized Pt/GaN catalyst. Notably, the Tafel slope of Pt/C decreased from 120 to 48.9 mV dec^{-1} after loading Pt NPs on GaN. This observation emphasizes the substantial enhancement of the Volmer step facilitated by the GaN substrate during the HER process in alkaline media. Moreover, the Pt/GaN catalyst exhibited higher mass activity, with a value of 7.34 mg$_{Pt}$$^{-1}$ at an overpotential of 100 mV, surpassing both commercial Pt/C and synthesized Pt/CP catalysts. These findings clearly demonstrate the pronounced improvement in the intrinsic activity of the Pt/GaN catalyst during the HER process, which can be attributed to the synergistic interplay between the GaN support and Pt nanoparticles.

In addition to its outstanding performance, the Pt/GaN catalyst exhibited robust stability in alkaline conditions. This was confirmed through meticulous evaluation involving 1000 cycles of cyclic voltammetry (CV) testing. The catalyst demonstrated consistent and reliable performance throughout the testing period, further affirming its efficacy. These impressive results highlight the potential of the Pt/GaN catalyst for sustained and high-performance applications in alkaline HER.

Materials Research Forum LLC
https://doi.org/10.21741/9781644903070

To gain insights into the superior alkaline HER performance of the Pt/GaN catalyst, first-principles calculations using density functional theory (DFT) were conducted. The theoretical analysis involved a Pt cluster consisting of 13 Pt atoms. Two crucial processes govern the efficiency of alkaline HER in the catalyst system: the effective dissociation of water molecules (Volmer step) and the facile coupling of H-H (Heyrovsky or Tafel step). The adsorption and dissociation of water molecules on the Pt_{13}@GaN surface were examined. The results revealed that water molecules preferentially bind to the Ga sites of Pt_{13}@GaN, with an adsorption energy of 1.02 eV. **Figure 18a** illustrates that the adsorbed H_2O can readily undergo dissociation at Ga and N sites through heterolytic cleavage, with a low dissociation barrier of 0.07 eV and a substantial energy gain of 1.32 eV. These findings confirm the high catalytic activity of Pt_{13}@GaN for H_2O dissociation on the GaN substrate. Furthermore, in the alkaline HER process, the timely removal of adsorbed hydroxyl (OH) at the electrocatalytic active sites is crucial. **Figure 18b** depicts the desorption pathway for OH, where a proton is provided by a water molecule, leading to the formation of adsorbed H_2O and the release of OH^- ions [67]. This proton exchange pathway significantly reduces the desorption barrier for OH to only 1.10 eV, facilitating the efficient removal of OH species.

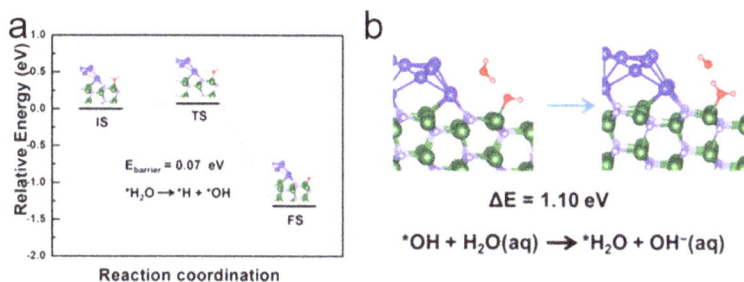

Figure 18. *(a) Dissociation of the H_2O molecule on Pt_{13}@GaN. (b) Desorption of OH from N sites via the water-assisted mechanism. Reproduced with permission from ref [13]. Copyright 2023, Wiley-VCH.*

Additionally, the possibility of hydrogen spillover from the GaN substrate into the Pt cluster was investigated. The calculations indicated that the binding energy of H around the N sites on the Pt cluster is more negative compared to the pure GaN(100) surface. This suggests that the generated H intermediates have a propensity to migrate towards the Pt cluster. **Figure 19a** illustrates a potential pathway for hydrogen spillover across the GaN/Pt_{13} interface, indicating that a moderate energy barrier of only 0.53 eV is required to facilitate H overflow from N sites to Pt sites. This demonstrates the kinetic feasibility of H spillover, further enhancing the catalytic activity of the Pt/GaN system.

Finally, the electrical conductivity of GaN(100), O-doped GaN (GaNO), and Pt_{13}@GaN were analyzed by calculating their density of states (DOS), as shown in **Figure 19b**. The original GaN(100) surface is a wide bandgap semiconductor. However, O doping introduces gap states near the Fermi level, effectively enhancing the conductivity. After loading Pt_{13} clusters, the Pt metal states significantly fill the gap in GaN, leading to improved electrical conductivity of the $Pt_{13}@GaN$ catalyst and further accelerating the HER process.

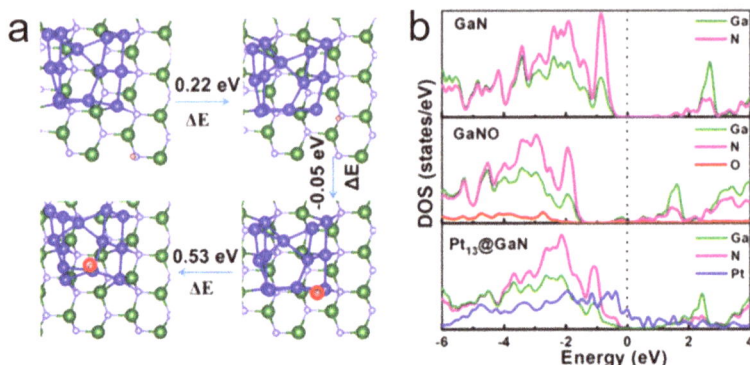

Figure 19. *(a) Possible hydrogen spillover route from the N sites to the Pt sites on Pt13@GaN. (b) The DOS for GaN (100), O-doped GaN (100) and $Pt_{13}@GaN$ systems. Reproduced with permission from ref [13]. Copyright 2023, Wiley-VCH.*

2.3.   Mott-Schottky Heterostructure Catalysts

The Mott-Schottky heterostructure is an exceptional class of electrocatalytic materials known for their ability to modulate their electronic structure and surface properties by manipulating charge transfer phenomena through an external field [68-70]. This design concept draws inspiration from the Mott-Schottky effect, where the introduction of an external electric field at the material's surface effectively modifies its band structure, thereby fine-tuning its electrocatalytic performance [71]. Mott-Schottky heterostructures have demonstrated remarkable performance in various electrocatalytic processes, including oxygen reduction reactions [72, 73] and HER [74, 75]. This exceptional performance positions them as promising candidates for applications in fields such as fuel cells [76] and hydrogen production through water electrolysis. Consequently, obtaining a comprehensive understanding of Mott-Schottky heterostructures and optimizing their electronic properties are crucial steps toward advancing the development of efficient and controllable electrocatalytic materials.

For example, Peng et al. [17] explored a straightforward approach to fabricating an efficient HER catalyst by incorporating ruthenium species onto tungsten oxide, thereby creating a Mott-Schottky heterojunction. Ru-$WO_{2.72}$ (WR) nanoflowers were prepared through a combination of hydrothermal and solution methods, as depicted in **Figure 20a**. The HR-TEM image in **Figure 20b** confirmed the presence of Ru, firmly anchored on the surface of $WO_{2.72}$ nanowires. This observation confirmed the successful reduction and modification of ruthenium on the $WO_{2.72}$ nanowires through a redox reaction, without the need for additional reducing agents. To investigate the formation of the WR heterojunction, the researchers employed Raman spectroscopy, as depicted in **Figure 20c**. In addition to displaying typical O-WO and W-O vibrational modes observed in $WO_{2.72}$ and WR, the WR sample exhibited a distinctive Ru-O-W vibration at around 400 cm^{-1} [77]. This characteristic vibration provided evidence of the formation of a Mott-Schottky heterojunction between ruthenium and tungsten oxide.

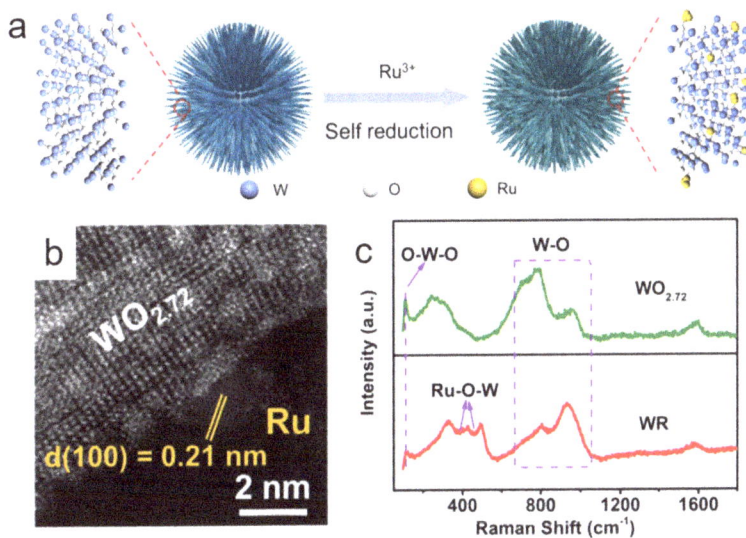

Figure 20. (a) Schematic synthesis of WR heterojunction. (b) HR-TEM images of WR composite. (c) Raman spectra of WO$_{2.72}$ and WR. Reproduced with permission from ref [17]. Copyright 2022, Elsevier.

XPS analysis of WO$_{2.72}$ and WR samples provides insights into their electronic structures. In **Figure 21a**, the W-4$f_{5/2}$ and W-4$f_{7/2}$ peaks are observed. The W-4$f_{5/2}$ peaks of W(VI) exhibit two main peaks at 36.3 eV and 38.2 eV, while W(V) shows two main peaks at 34.7 eV [78]. The presence of both W(VI) and W(V) species can be attributed to the angular and edge-sharing W sites within the [WO$_6$] octahedra [79]. Comparing the WO$_{2.72}$ sample to the WR sample, the proportion of W(V) decreases significantly in the WR sample, indicating the crucial role of W as an electron donor in the Mott-Schottky heterostructure. Additionally, the Ru-3d XPS signals display two distinct peaks at 280.6 and 282.05 eV, corresponding to Ru-3$d_{5/2}$, with an additional peak at 284.8 eV for Ru-3$d_{3/2}$ (**Figure 21b**) [80]. The WR sample predominantly exhibits metallic Ru^0, and the proportion of Ru^0 is highest in the WR sample. This suggests that after loading Ru, there are no significant changes in the electronic structure of oxygen (O), indicating the transfer of electrons from the WO$_{2.72}$ substrate to Ru.

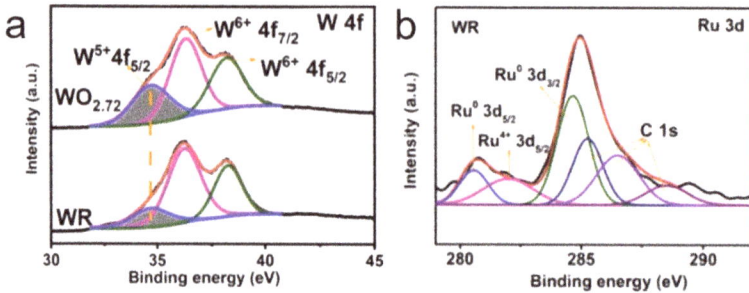

Figure 21. *(a) W-4f, and (b) Ru-3d XPS spectra of WR and WO$_{2.27}$. Reproduced with permission from ref [17]. Copyright 2022, Elsevier.*

In this study, an acidic electrolyte and a typical three-electrode system were employed to evaluate the electrocatalytic activity of the samples using the LSV technique. The WR composite, with a Ru loading of 2.73 wt%, demonstrated significantly improved electrocatalytic performance compared to the WO$_{2.72}$ sample. It exhibited an ultra-low onset potential of 10 mV and an overpotential of 40 mV to achieve a current density of 10 mA cm^{-2} (**Figure 22a**), outperforming commercial Ru/C catalysts. The Tafel slope of WR was remarkably low at 50 mV dec^{-1}, indicating improved reaction kinetics following the classic Volmer-Heyrovsky pathway (**Figure 22b**) [81, 82]. The double-layer capacitance of WR was determined to be 91.63 mF cm^{-2}, suggesting a high electrochemically active surface area. At 100 mV, the TOF was calculated to be 5.29 s^{-1}, further confirming the high intrinsic activity of WR. Additionally, electrochemical impedance spectroscopy (EIS) analysis revealed a significant reduction in the charge transfer resistance (R_{ct}) of the WR catalyst, indicating an enhanced electron transfer rate at the reaction surface/interface with the electrolyte [83, 84].

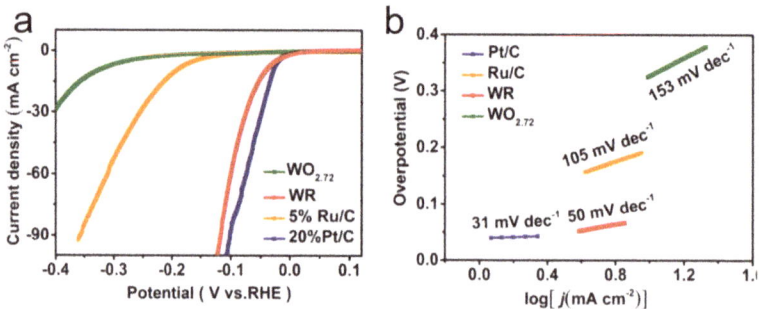

Figure 22. *(a) LSV curves and (b) Tafel slopes of WO$_{2.72}$, WR (2.73 wt% Ru), Ru/C (5 wt% Ru) and Pt/C (20 wt% Pt). Reproduced with permission from ref [17]. Copyright 2022, Elsevier.*

The electrochemical stability of WR was found to be excellent, as demonstrated by its ability to maintain a current density of over 90% for more than 10 hours, as indicated in the inset of

Figure 23a. After 24 hours of electrochemical testing, WR underwent SEM, TEM, and XPS characterization. The nanowires of WR were no longer visible, and the morphology of the sea urchin-like nanospheres changed, as illustrated in **Figure 23b-c**. These observations suggested a decrease in exposed active sites, which consequently led to a decline in performance. Moreover, the oxidation state of tungsten exhibited noticeable changes, as shown in **Figure 23d**, with the majority of W(V) being converted to W(VI). Concurrently, there was a positive shift in the proportion of Ru^{4+}, indicating a certain level of oxidation of WR during the durability test, contributing to a partial decline in HER performance.

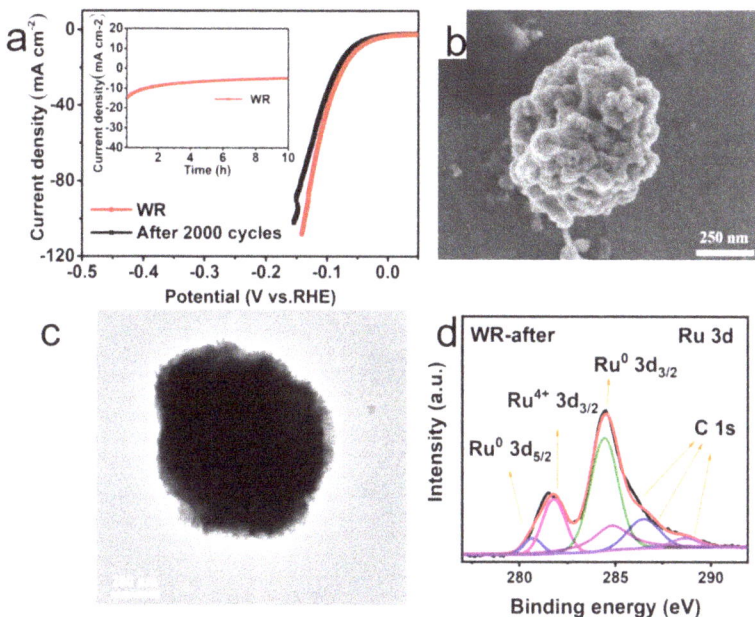

Figure 23. (a) LSV curves of WR before and after 2000 cycles (inset: stability test of WR at 50 mV). (b) SEM, (c) TEM images, and (d) Ru-3d XPS spectra of WR after chronopotentiometric test. Reproduced with permission from ref [17]. Copyright 2022, Elsevier.

The Mott-Schottky heterojunction interface at the metal-metal oxide interface induces an electric field effect that contributes to the promotion of asymmetric charge distribution, resulting in enhanced reactivity and selectivity of the catalyst, thereby expediting the HER process [7, 69, 85, 86]. A recent study by Chen et al. [18] presented a theoretically designed Mott-Schottky heterojunction catalyst, Ru/CeO_2, with a strong built-in electric field (BEF) effect based on a cerium oxide substrate.

Firstly, the impact of BEF on the electronic structure was investigated, and DFT calculations were performed using CeO_2 as the substrate to determine the work functions of three different models: $CeO_2(111)$, $Ru(101)$, and Ru/CeO_2. The work function (Φ) was defined as the energy change between the Fermi level and the electrostatic potential [87, 88]. Upon deposition of Ru onto the CeO_2 surface, electrons were found to transfer from Ru to CeO_2 [89, 90]. Consequently, the constructed BEF resulted in Ru nanoclusters pointing towards CeO_2 [91], visually representing the surface charge characteristics through the electron density difference, which determined the direction of electron transfer [92].

Figure 24a displays the contour plot of $\Delta\rho$, illustrating Ru/CeO_2, with the average values of $\Delta\rho$ plotted along the z-axis. The Ru interface exhibited a positive charge, while the CeO_2 surface displayed a negative charge, indicating a clear electron transfer from Ru to CeO_2. This observation aligns with the conclusion drawn from the work function analysis, confirming that BEF has a significant influence on the redistribution of charges, thereby optimizing the adsorption energy of hydrogen intermediates at active sites. To illustrate the local charge redistribution around the Mott-Schottky heterojunction interface driven by the BEF, **Figure 24b** presents a schematic representation, showing the flow of electrons from Ru nanoclusters to CeO_2.

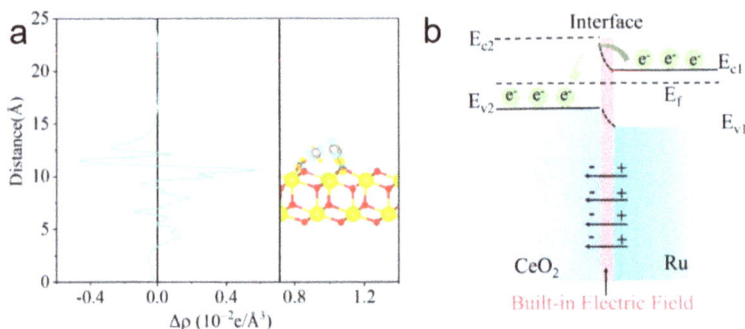

Figure 24. *(a) The planar-averaged electron density difference $\Delta\rho$ and corresponding side view of the electron density difference over Ru/CeO_2. (b) Schematic diagram of the Mott-Schottky heterojunction. Reproduced with permission from ref [18]. Copyright 2023, Elsevier.*

To comprehensively assess the impact of the Mott-Schottky heterojunction on the HER performance, the electronic density of states (DOS) was calculated for both CeO_2 and Ru/CeO_2 configurations. The DOS analysis of Ru/CeO_2 indicated the formation of an interface-induced BEF between Ru nanoclusters and CeO_2. This resulted in a left shift of the conduction band near the Fermi level, exhibiting metallic characteristics. The elimination of the bandgap suggests that Ru/CeO_2 exhibits improved electrical conductivity, facilitating electron transfer within the catalyst [93, 94]. Additionally, the Gibbs free energy (ΔG_H) for the adsorption of hydrogen intermediates on Ru/CeO_2, $CeO_2(111)$, and $Ru(101)$ were computed. **Figure 25** displays the ΔG_H values, which are -0.28 eV for Ru/CeO_2, 1.5 eV for $CeO_2(111)$, and -0.34 eV for $Ru(101)$. The ΔG_H value for Ru/CeO_2 is closer to 0, indicating a stronger HER capability. These findings

Materials Research Forum LLC
https://doi.org/10.21741/9781644903070

suggest that inducing electron transfer can optimize the adsorption of hydrogen intermediates, leading to the formation of a relatively stable BEF.

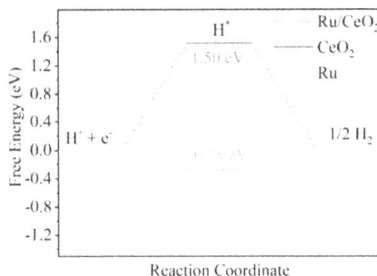

Figure 25. *Gibbs free energy diagram of the CeO_2 (111) plane, Ru (101) plane, and Ru/CeO₂. Reproduced with permission from ref [18]. Copyright 2023, Elsevier.*

Experimental verification of the Ru/CeO_2 Mott-Schottky heterojunction was conducted to validate the previous theoretical calculations. Catalysts with different Ru contents were prepared using hydrothermal and solution methods. XRD analysis confirmed the crystal phase composition of the catalysts, as shown in **Figure 26a**. Notably, all samples exhibited distinct diffraction peaks corresponding to the (111), (200), (220), and (311) lattice planes of CeO_2 (JCPDS card No. 43-1002) at 28.5°, 33.1°, 47.5°, and 56.3°, respectively. No significant diffraction peaks related to ruthenium compounds were observed, suggesting that Ru may exist in smaller particle sizes, possibly dispersed on the surface of CeO_2.

XPS spectra confirmed that Ru/CeO_2 is composed of Ce, O, and Ru elements. In **Figure 26b**, Ce-$3d$ XPS spectra of pure CeO_2 and Ru/CeO_2 displayed contributions from eight peaks corresponding to Ce^{4+} peaks, Ce^{3+} peaks, and satellite peaks. It is worth noting that the Ce^{4+} peak in Ru/CeO_2 exhibited a positive shift (approximately 0.1 eV) compared to pristine CeO_2. This shift indicates strong charge trapping and further confirms the formation of the interface BEF between Ru clusters and CeO_2. The Ru-$3p$ XPS fitting curve showed four peaks corresponding to Ru^{4+} and Ru^0, suggesting that the oxidation state of Ru varies between 0 and 4 due to charge transfer from the metal to the support, as indicated in **Figure 26c** [95]. The strong metal-support interaction (SMSI) between Ru and CeO_2 optimized the adsorption strength of H* and H_2O, possibly resulting in electron-deficient Ru atoms and ultimately enhancing the HER activity. In summary, all characterizations confirmed the successful preparation of BEF at the Ru/CeO_2 interface, leading to the formation of electron-deficient Ru clusters and electron-rich CeO_2. These experimental findings are consistent with the theoretical calculations, validating the existence of the Mott-Schottky heterojunction at the Ru/CeO_2 interface.

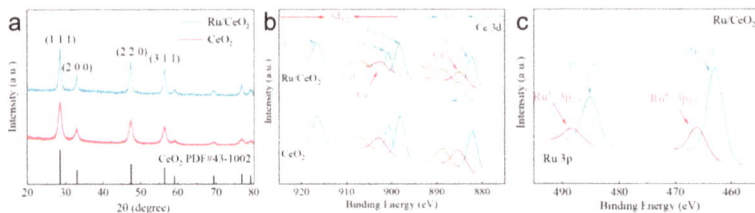

Figure 26. *(a) XRD patterns of CeO₂ and Ru/CeO₂. High-resolution XPS spectra of (b) Ce-3d and (c) Ru-3p. Reproduced with permission from ref [18]. Copyright 2023, Elsevier.*

In a nitrogen-saturated 1.0 M KOH solution, the catalytic performance of catalysts with different Ru loadings in HER was evaluated using LSV, as shown in **Figure 27a**. The catalysts were tested using a conventional three-electrode system, with a graphite rod as the counter electrode, the working electrode consisting of the sample loaded on a glassy carbon electrode with a loading of 0.26 mg cm^{-2}, and a saturated calomel electrode as the reference electrode. To achieve a current density of 10 mA cm^{-2}, the Ru/CeO₂ catalyst required an overpotential of only 55 mV, outperforming the commercial Pt/C catalyst (61 mV) and pristine CeO₂ (667 mV). The Tafel slope of Ru/CeO₂ was 43.3 mV dec^{-1}, which was lower than other electrode materials, as indicated in **Figure 27b**. The lower Tafel slope suggests significantly accelerated HER kinetics for Ru/CeO₂, following the classical Volmer-Heyrovsky pathway [96, 97].

The ECSA of the catalyst was calculated using double-layer capacitance, and Ru/CeO₂ exhibited a high C_{dl} value of 10.21 mF cm^{-2}, which was significantly higher than other catalysts. This indicates that Ru/CeO₂ has the highest density of active sites [98]. Furthermore, the stability of the Ru/CeO₂ catalyst was evaluated using the chronoamperometry method. As shown in **Figure 27c**, there was no significant increase in potential during continuous operation at 20 mA cm^{-2} for 20 hours. The inset in **Figure 27c** represents the LSV curves after 5000 cycles, showing no apparent changes. These results collectively demonstrate that Ru/CeO₂ is a highly active and stable catalyst for the HER.

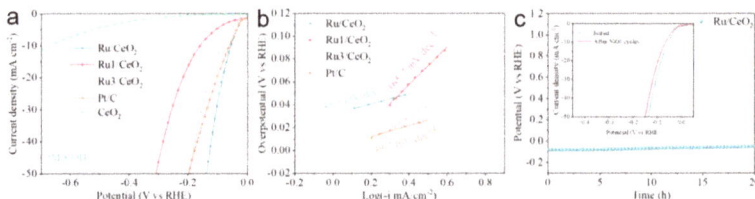

Figure 27. *The HER test of different catalysts in 1.0 M KOH. (a) LSVs for different catalysts. (b) Tafel plots. (c) LSVs before and after 5000 CV cycles and chronoamperometric curves of Ru/CeO₂. Reproduced with permission from ref [18]. Copyright 2023, Elsevier.*

2.4.   Other Heterostructure Catalysts

Transition metal compounds, including nitrides [99, 100], carbides [101], sulfides [102], and oxides [103], have gained significant attention in heterogeneous catalyst research due to their high activity, low cost, and excellent stability. Researchers have been able to achieve more complex and ordered heterostructures through clever design and synthesis.   These heterostructures play a crucial role in regulating the electronic structure and catalytic performance, offering solutions for enhancing the performance of transition metal-based catalysts in various reactions, such as hydrogen evolution.   The optimization of these heterostructures holds the promise of developing more efficient and sustainable electrocatalytic materials in the future.

For instance, Shen et al. [104] demonstrated a controlled phosphorization treatment on ultrathin $Co_{0.85}Se$ nanosheets, resulting in the formation of an orthogonal crystal phase $CoSe_2$/amorphous CoP heterojunction ($CoSe_2$/a-CoP). The amorphous structure of CoP facilitated the creation of a rich interface, exposing a higher density of active sites. The synthesis of the $CoSe_2$/a-CoP sample involved a hydrothermal and thermal treatment process.   Initially, 1 mmol $Co(NO_3)_2 \cdot 6H_2O$ and 1 mmol Na_2SeO_3 were dissolved in 38 mL of ethylene glycol. The resulting mixture underwent a hydrothermal reaction at 180 °C for 24 hours, followed by cooling, washing, and drying to obtain $Co_{0.85}Se$ nanosheets. Subsequently, 40 mg $Co_{0.85}Se$ and 800 mg NaH_2PO_2 were dispersed on a quartz boat, with $Co_{0.85}Se$ placed downstream and NaH_2PO_2 upstream in a tube furnace. The furnace was heated to 350 °C at a rate of 2 °C min^{-1} and maintained for 2 hours under a high-purity argon gas flow. After cooling to room temperature, the $CoSe_2$/a-CoP sample was obtained. Further adjustment of the heat treatment temperature to 550 °C resulted in $CoSe_2$/c-CoP.

SEM characterization was performed to analyze the morphology of the samples. As shown in **Figure 28a**, $Co_{0.85}Se$ was composed of ultrathin nanosheets, while $CoSe_2$/a-CoP exhibited a microsphere morphology comprising rough and dense nanosheets (**Figure 28b**), with a material surface roughness of 0.546 nm. This implies that $CoSe_2$/a-CoP possesses a larger specific surface area, which is conducive to exposing a larger number of electrochemical reaction active sites.

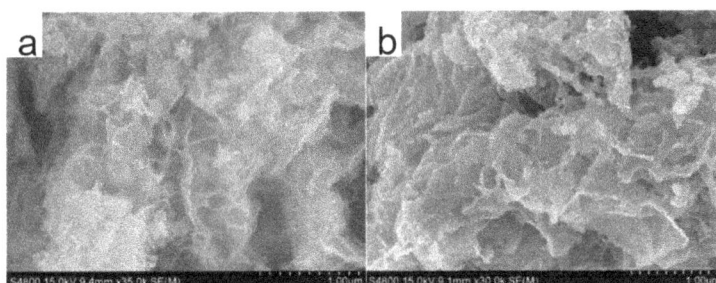

Figure 28. *SEM images of (a) $Co_{0.85}Se$ and (b) $CoSe_2$/a-CoP. Reproduced with permission from ref [104]. Copyright 2022, Wiley-VCH.*

X-ray absorption spectroscopy (XAS) characterization played a crucial role in providing information about the electronic and local structure of the samples. **Figure 29a-b** depicts the Co K-edge XANES and FT-EXAFS spectra of $CoSe_2$, $CoSe_2$/a-CoP, and Co foil. The Co K-edge XANES spectra reveal that the absorption edge of $CoSe_2$/a-CoP has shifted to higher energy compared to $CoSe_2$, indicating an increase in the oxidation state of Co after the formation of the heterojunction. In addition, the characteristic Co-Se bond peak at 2.09 Å is observed in both $CoSe_2$ and $CoSe_2$/a-CoP spectra, confirming the presence of Co-Se bonds [105]. In the case of $CoSe_2$/a-CoP, an additional peak corresponding to the Co-P bond (1.70 Å) is observed, indicating the presence of amorphous CoP in the heterojunction [106]. This peak confirms the formation of the $CoSe_2$/amorphous CoP heterojunction and provides evidence for the incorporation of the Co-P bond in the structure. Furthermore, the Se K-edge XANES spectra in **Figure 29c-d** show a difference between $CoSe_2$ and $CoSe_2$/a-CoP. The peak at approximately 2.09 Å corresponds to the Se-Co bond. Notably, the Se-Co bond length in $CoSe_2$/a-CoP is shorter than that in $CoSe_2$, indicating a strong electron coupling between $CoSe_2$ and amorphous CoP.

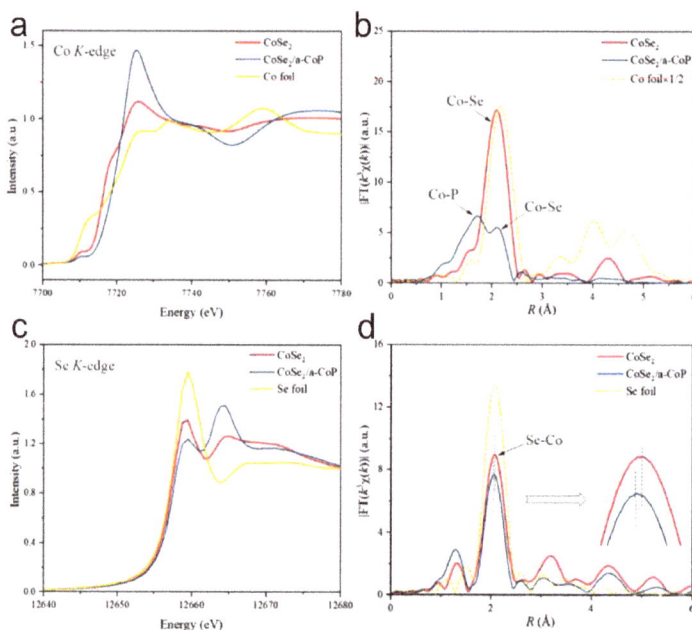

Figure 29. (a) The normalized XANES at the Co K-edge of CoSe₂, CoSe₂/a-CoP, and Co foil. (b) FT-EXAFS spectra. (c) The normalized XANES at the Se K-edge of CoSe₂, CoSe₂/a-CoP, and Se foil. (d) FT-EXAFS spectra. Reproduced with permission from ref [104]. Copyright 2022, Wiley-VCH.

XPS was employed to investigate the changes in oxidation states within the samples. The survey spectra of CoSe$_2$/a-CoP, as depicted in **Figure 30a**, reveal the presence of Co, Se, and additional P elements. In the case of CoSe$_2$, the peaks at 778.7 and 793.7 eV are attributed to Co-2$p_{3/2}$ and Co-2$p_{1/2}$ [107], respectively, with satellite peaks observed at 784.6 and 802.1 eV, as illustrated in **Figure 30b**. The remaining peaks in the spectrum originate from surface cobalt oxide.

In the CoSe$_2$/a-CoP spectrum, the Co-2$p_{3/2}$ and 2$p_{1/2}$ peak shift to 778.4 and 793.4 eV, respectively. This shift indicates a higher oxidation state of Co in CoSe$_2$/a-CoP compared to CoSe$_2$, consistent with the conclusions drawn from XANES analysis. Furthermore, the binding energies of Se-3$d_{5/2}$ and 3$d_{3/2}$, initially at 55.0 eV and 55.9 eV of CoSe$_2$ [107], shift to 54.6 eV and 55.4 eV for CoSe$_2$/a-CoP, as shown in **Figure 30c**. This shift implies a reduction in the oxidation state of Se after the formation of the heterojunction, confirming the occurrence of charge transfer between CoSe$_2$ and CoP. The increase in the oxidation state of Co and the decrease in the oxidation state of Se enhance the columbic force between them, leading to a reduction in the Se-Co bond length, consistent with the conclusions drawn from EXAFS analysis. Regarding the high-resolution XPS spectrum of phosphorus displayed in **Figure 30d**, the peak at 129.6 eV arises from the P-2p in CoP [108]. In comparison to elemental phosphorus, it shifts towards lower binding energy, indicating a negative oxidation state. Additionally, the peaks at 134.2 and 134.7 eV correspond to phosphorus oxide and MLL Auger electrons of Se, respectively [109].

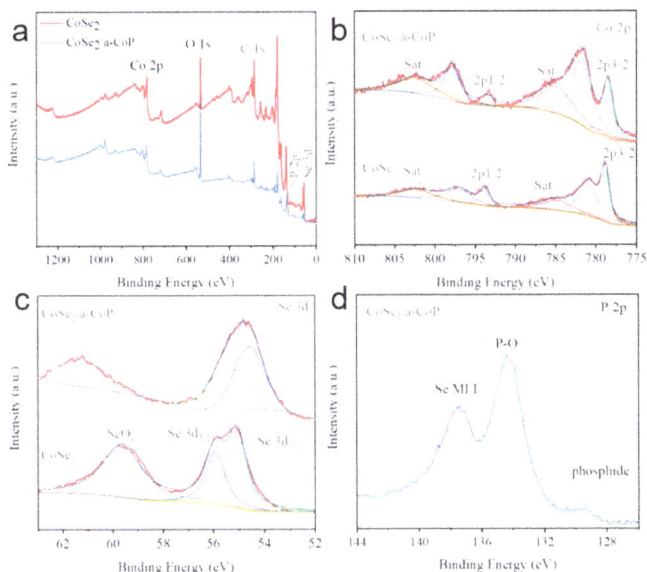

Figure 30. *XPS spectra of survey (a), Co-2p (b), Se-3d (c), and P-2p (d) for CoSe$_2$ and CoSe$_2$/a-CoP. Reproduced with permission from ref [104]. Copyright 2022, Wiley-VCH.*

The electrochemical measurements were performed to assess the HER activity of the samples. In 0.5 M H_2SO_4, $CoSe_2$/a-CoP exhibited a near Pt/C overpotential, with an overpotential of only 65 mV to generate a current density of 10 mA cm^{-2}. The overpotential was significantly lower than the corresponding values for $CoSe_2$ (173 mV) and $CoSe_2$/c-CoP (127 mV). The Tafel slope of $CoSe_2$/a-CoP was measured to be 54 mV dec^{-1}, indicating that its HER process follows the Volmer-Heyrovsky mechanism [110]. The fitted charge transfer resistance (R_{ct}) value for $CoSe_2$/a-CoP was 17 Ω, suggesting that the constructed heterojunction has the minimum R_{ct} value, which favors the electrocatalytic reaction.

To comprehensively evaluate the HER performance of the samples in electrolytes with different pH values, in 1.0 M KOH electrolyte, $CoSe_2$/a-CoP exhibited an overpotential (η_{10}) value of 151 mV with a Tafel slope of 80 mV dec^{-1}, demonstrating outstanding HER performance and long-term durability. In 1 M phosphate-buffered solution (PBS) electrolyte, $CoSe_2$/a-CoP exhibited an η_{10} value of 185 mV with a Tafel slope of 100 mV dec^{-1}. These results indicate that $CoSe_2$/a-CoP performs optimally in neutral electrolytes. Overall, $CoSe_2$/a-CoP proves to be an excellent HER catalyst with superior performance compared to $CoSe_2$ and $CoSe_2$/c-CoP.

To understand the enhanced electrocatalytic HER activity, first-principles calculations were conducted. The calculations revealed that for isolated $CoSe_2$ and CoP, Co typically loses partial electron density, exhibiting a positive charge state, while Se and P accumulate electron density, displaying a negative charge state, as illustrated in **Figure 31a-d**. After the formation of the heterojunction, the charge of CoP decreases, and the charge of $CoSe_2$ increases for both $CoSe_2$/c-CoP and $CoSe_2$/a-CoP, indicating charge transfer from CoP to $CoSe_2$. Notably, in $CoSe_2$/a-CoP, the charge transfer significantly increases compared to $CoSe_2$/c-CoP, suggesting that the amorphous structure of CoP contributes to strengthening the electronic coupling at the interface.

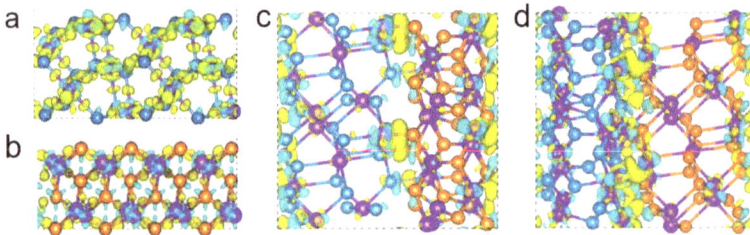

Figure 31. (a-d) Charge density difference of $CoSe_2$, CoP, $CoSe_2$/c-CoP, and $CoSe_2$/a-CoP. The purple, grey-green, and brown spheres denote Co, Se, and P atoms, respectively. The yellow and cyan contours represent change accumulation and dilution, respectively. Reproduced with permission from ref [104]. Copyright 2022, Wiley-VCH.

The adsorption/desorption free energy of H on a catalyst is a crucial indicator for analyzing its catalytic activity. A ΔG_H value closer to zero signifies higher activity, indicating that the catalyst can efficiently adsorb and desorb H. In the case of $CoSe_2$/a-CoP, the ΔG_H at Co sites is closest to zero, indicating that Co serves as the catalytically active center, as shown in **Figure 32a**. Furthermore, compared to the Co sites in CoP ($\Delta G_H = -0.215$ eV), the Se sites in $CoSe_2$ ($\Delta G_H =$

0.307 eV), and the Co sites in $CoSe_2/c$-CoP ($\Delta G_H = -0.122$ eV), the ΔG_H of $CoSe_2/a$-CoP (-0.107 eV) suggests its superior activity, consistent with the results from electrochemical experiments.

Additionally, the negative ΔG_H for CoP and $CoSe_2/c$-CoP implies overly strong H adsorption, which can hamper the desorption process. On the other hand, the intermediate value for $CoSe_2/a$-CoP indicates improved H absorption and desorption characteristics. To gain a deeper understanding of the electronic structure, DOS calculations were performed. The projected DOS graphs reveal a small band gap in $CoSe_2$, indicating poor metal conductivity. The DOS distribution of CoP indicates strong spin polarization, with the DOS near the Fermi level primarily contributed by the spin-down electrons of Co, while the spin-up portion of Co and the electron density of P are nearly zero. This suggests that Co is the main catalytically active center in CoP, aligning with the results of H adsorption-free energy. For $CoSe_2/a$-CoP, the DOS at the Fermi level is still predominantly contributed by Co, as shown in **Figure 32b**, indicating that Co remains the catalytically active center. This further supports the notion that the enhanced catalytic activity of $CoSe_2/a$-CoP is attributed to the active Co sites.

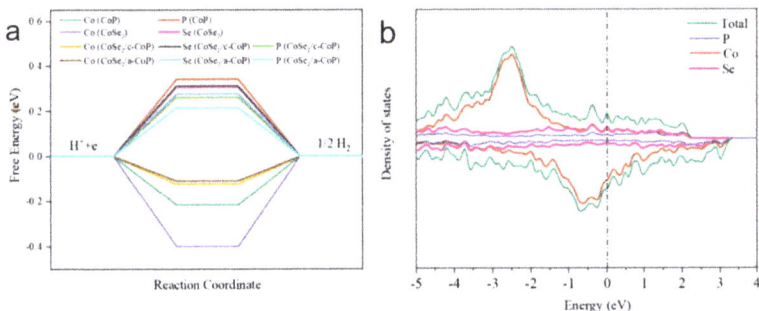

Figure 32. *(a) ΔG_H on different sites of $CoSe_2$, CoP, $CoSe_2/c$-CoP, and $CoSe_2/a$-CoP. (b) Projected DOS of $CoSe_2/a$-CoP. Reproduced with permission from ref [104]. Copyright 2022, Wiley-VCH.*

3. Conclusion and Prospects

In this chapter, the focus is on exploring the impact of heterogeneous structure catalysis on the HER and improving its catalytic performance. Different types of heterogeneous structures are designed and constructed to enhance the electrocatalytic activity of the HER. Firstly, the heterogeneous structure of metal/metal oxide leverages the synergistic effects of two materials to improve catalytic activity. Secondly, the metal/semiconductor heterogeneous structure effectively promotes the HER process by controlling the electronic structure and electron transfer effects. Thirdly, the introduction of the Mott-Schottky heterostructure structure is particularly important in optimizing intermediate adsorption and desorption steps through interface effects, thereby improving catalytic activity. Through systematic studies of these heterogeneous structures, valuable insights are gained into their mechanisms in the HER, providing theoretical guidance and experimental foundations for the design of more efficient catalysts.

However, heterogeneous structure catalysts also face challenges that need to be addressed in future research. Some areas of focus include:

(i) Material design and synthesis: Exploring more material combinations and synthesis methods to achieve complex and efficient heterogeneous structure catalysts suitable for practical applications.

(ii) Stability and activity under different environmental conditions: Investigating the stability and activity of heterogeneous structure catalysts under various operating conditions to ensure their sustainability and long-term durability in real-world applications.

(iii) In-situ characterization techniques and real-time monitoring: Continuously exploring new material combinations and synthesis strategies, coupled with in-situ characterization techniques, to provide real-time monitoring of the evolution process of heterogeneous structures during catalytic reactions. This approach aims to offer more authentic and reliable data for catalyst design, enabling further innovation in heterogeneous structure catalysts and making a significant contribution to the sustainable development of renewable energy.

References

[1]   P. Xiao, W. Chen, X. Wang, A review of phosphide-based materials for electrocatalytic hydrogen evolution, Adv. Energy Mater., 5 (2015) 1500985. https://doi.org/10.1002/aenm.201500985

[2]   Y. Zheng, Y. Jiao, L.H. Li, T. Xing, Y. Chen, M. Jaroniec, S.Z. Qiao, Toward design of synergistically active carbon-based catalysts for electrocatalytic hydrogen evolution, ACS Nano, 8 (2014) 5290–5296. https://doi.org/10.1021/nn501434a

[3]   S. Wang, J. Wang, M. Zhu, X. Bao, B. Xiao, D. Su, H. Li, Y. Wang, Molybdenum-carbide-modified nitrogen-doped carbon vesicle encapsulating nickel nanoparticles: a highly efficient, low-cost catalyst for hydrogen evolution reaction, J. Am. Chem. Soc., 137 (2015) 15753-15759. https://doi.org/10.1021/jacs.5b07924

[4]   H. Wang, W. Fu, X. Yang, Z. Huang, J. Li, H. Zhang, Y. Wang, Recent advancements in heterostructured interface engineering for hydrogen evolution reaction electrocatalysis, J. Mater. Chem. A, 8 (2020) 6926-6956. https://doi.org/10.1039/C9TA11646J

[5]   M. Luo, W. Sun, B.B. Xu, H. Pan, Y. Jiang, Interface engineering of air electrocatalysts for rechargeable zinc–air batteries, Adv. Energy Mater., 11 (2020) 2002762. https://doi.org/10.1002/aenm.202002762

[6]   X. Du, J. Huang, J. Zhang, Y. Yan, C. Wu, Y. Hu, C. Yan, T. Lei, W. Chen, C. Fan, J. Xiong, Modulating electronic structures of inorganic nanomaterials for efficient electrocatalytic water splitting, Angew. Chem. Int. Ed., 58 (2019) 4484-4502. https://doi.org/10.1002/anie.201810104

[7]   L. Zhai, X. She, L. Zhuang, Y. Li, R. Ding, X. Guo, Y. Zhang, Y. Zhu, K. Xu, H.J. Fan, S.P. Lau, Modulating built-in electric field via variable oxygen affinity for robust hydrogen evolution reaction in neutral media, Angew. Chem. Int. Ed., 61 (2022)

e202116057. https://doi.org/10.1002/anie.202116057

[8] Y. Chen, K. Yue, J.-W. Zhao, Z. Cai, X. Wang, Y. Yan, Effective modulating of the Mo dissolution and polymerization in Ni$_4$Mo/NiMoO$_4$ heterostructure via metal-metal oxide-support interaction for boosting H$_2$ production, Chem. Eng. J., 466 (2023) 143097. https://doi.org/10.1016/j.cej.2023.143097

[9] M.J. Kenney, J.E. Huang, Y. Zhu, Y. Meng, M. Xu, G. Zhu, W.-H. Hung, Y. Kuang, M. Lin, X. Sun, W. Zhou, H. Dai, An electrodeposition approach to metal/metal oxide heterostructures for active hydrogen evolution catalysts in near-neutral electrolytes, Nano Res., 12 (2019) 1431-1435. https://doi.org/10.1007/s12274-019-2379-7

[10] K.L. Zhou, Z. Wang, C.B. Han, X. Ke, C. Wang, Y. Jin, Q. Zhang, J. Liu, H. Wang, H. Yan, Platinum single-atom catalyst coupled with transition metal/metal oxide heterostructure for accelerating alkaline hydrogen evolution reaction, Nat. Commun., 12 (2021) 3783. https://doi.org/10.1038/s41467-021-24079-8

[11] L. Peng, X. Zheng, L. Li, L. Zhang, N. Yang, K. Xiong, H. Chen, J. Li, Z. Wei, Chimney effect of the interface in metal oxide/metal composite catalysts on the hydrogen evolution reaction, Appl. Catal. B: Environ., 245 (2019) 122-129. https://doi.org/10.1016/j.apcatb.2018.12.035

[12] F. Zhou, G. Tao, Structure sensitivity in hydrogen adsorption on Ni/NiO interfaces for the hydrogen evolution reaction, J. Phys. Chem. C, 127 (2023) 23180-23188. https://doi.org/10.1021/acs.jpcc.3c05321

[13] C. Liu, C. Niu, B. Gao, P. Jiang, Q. Lin, W. Wu, Y. Jia, Z. Wei, Q. Xu, High alkaline electrochemical hydrogen evolution on a Pt/GaN heterostructure, ChemNanoMat, 9 (2023) e202300195. https://doi.org/10.1002/cnma.202300195

[14] A. Shan, X. Teng, Y. Zhang, P. Zhang, Y. Xu, C. Liu, H. Li, H. Ye, R. Wang, Interfacial electronic structure modulation of Pt-MoS$_2$ heterostructure for enhancing electrocatalytic hydrogen evolution reaction, Nano Energy, 94 (2022) 106913. https://doi.org/10.1016/j.nanoen.2021.106913

[15] W. Zhang, Y. Zhao, K. He, J. Luo, G. Li, R. Liu, S. Liu, Z. Cao, P. Jing, Y. Ding, Ultrathin nanoporous metal–semiconductor heterojunction photoanodes for visible light hydrogen evolution, Nano Res., 11 (2018) 2046-2057. https://doi.org/10.1007/s12274-017-1821-y

[16] M. Krishnamachari, S. Lenus, K. Pradeeswari, R. Arun pandian, M. Kumar, J.-H. Chang, S.p. Muthu, R. Perumalsamy, Z. Dai, P. Vijayakumar, Review of Mott–Schottky-based nanoscale catalysts for electrochemical water splitting, ACS Appl. Nano Mater., 6 (2023) 16106-16139. https://doi.org/10.1021/acsanm.3c02677

[17] L. Peng, L. Su, X. Yu, R. Wang, X. Cui, H. Tian, S. Cao, B.Y. Xia, J. Shi, Electron redistribution of ruthenium-tungsten oxides Mott-Schottky heterojunction for enhanced hydrogen evolution, Appl. Catal. B: Environ., 308 (2022) 121229. https://doi.org/10.1016/j.apcatb.2022.121229

[18] X. Chen, D. Shi, M. Bi, J. Song, Y. Qin, S. Du, B. Sun, C. Chen, D. Sun, Constructing built-in electric field via ruthenium/cerium dioxide Mott-Schottky heterojunction for highly efficient electrocatalytic hydrogen production, J. Colloid. Interface. Sci., 652 (2023) 653-662. https://doi.org/10.1016/j.jcis.2023.07.203

[19] M. Wang, W. Ma, Z. Lv, D. Liu, K. Jian, J. Dang, Co-doped Ni_3N nanosheets with electron redistribution as bifunctional electrocatalysts for efficient water splitting, J. Phys. Chem. Lett., 12 (2021) 1581-1587. https://doi.org/10.1021/acs.jpclett.0c03804

[20] L. Gan, J. Lai, Z. Liu, J. Luo, S. Zhang, Q. Zhang, Interfacial engineering of heterojunction copper-cobalt-nickel nitride as binder-free electrode for efficient water splitting, J. Alloys Compd., 905 (2022) 164200. https://doi.org/10.1016/j.jallcom.2022.164200

[21] Y. Yang, X. Xu, X. Wang, Synthesis of Mo-based nanostructures from organic-inorganic hybrid with enhanced electrochemical for water splitting, Sci. China Mater., 58 (2015) 775-784. https://doi.org/10.1007/s40843-015-0088-4

[22] H. Wang, X. Wang, R. Chen, H. Zhang, X. Wang, J. Wang, J. Zhang, L. Mu, K. Wu, F. Fan, X. Zong, C. Li, Promoting photocatalytic H_2 evolution on organic–inorganic hybrid perovskite nanocrystals by simultaneous dual-charge transportation modulation, ACS Energy Lett., 4 (2018) 40-47. https://doi.org/10.1021/acsenergylett.8b01830

[23] J. Verma, S. Goel, Cost-effective electrocatalysts for hydrogen evolution reactions (HER): Challenges and prospects, Int. J. Hydrog. Energy, 47 (2022) 38964-38982. https://doi.org/10.1016/j.ijhydene.2022.09.075

[24] Y. Shi, B. Zhang, Recent advances in transition metal phosphide nanomaterials: synthesis and applications in hydrogen evolution reaction, Chem. Soc. Rev., 45 (2016) 1529-1541. https://doi.org/10.1039/C5CS00434A

[25] Y. Xu, K. Fan, Y. Zou, H. Fu, M. Dong, Y. Dou, Y. Wang, S. Chen, H. Yin, M. Al-Mamun, P. Liu, H. Zhao, Rational design of metal oxide catalysts for electrocatalytic water splitting, Nanoscale, 13 (2021) 20324-20353. https://doi.org/10.1039/D1NR06285A

[26] Q.-L. Zhu, Q. Xu, Immobilization of Ultrafine metal nanoparticles to high-surface-area materials and their catalytic applications, Chem, 1 (2016) 220-245. https://doi.org/10.1016/j.chempr.2016.07.005

[27] C. Gao, F. Lyu, Y. Yin, Encapsulated Metal Nanoparticles for Catalysis, Chem. Rev., 121 (2021) 834-881. https://doi.org/10.1021/acs.chemrev.0c00237

[28] W. Xu, B. Wang, X. Ni, H. Liu, W. Wang, L. Zhang, H. Zhang, Z. Peng, Z. Liu, Heterogeneous synergetic effect of metal-oxide interfaces for efficient hydrogen evolution in alkaline solutions, ACS Appl. Mater. Interfaces, 13 (2021) 13838-13847. https://doi.org/10.1021/acsami.1c00945

[29] M. Gong, W. Zhou, M.C. Tsai, J. Zhou, M. Guan, M.C. Lin, B. Zhang, Y. Hu, D.Y.

Wang, J. Yang, S.J. Pennycook, B.J. Hwang, H. Dai, Nanoscale nickel oxide/nickel heterostructures for active hydrogen evolution electrocatalysis, Nat. Commun., 5 (2014) 4695. https://doi.org/10.1038/ncomms5695

[30]   Y.Y. Chen, Y. Zhang, X. Zhang, T. Tang, H. Luo, S. Niu, Z.H. Dai, L.J. Wan, J.S. Hu, Self-templated fabrication of $MoNi_4/MoO_{3-x}$ nanorod arrays with dual active components for highly efficient hydrogen evolution, Adv. Mater., 29 (2017) 1703311. https://doi.org/10.1002/adma.201703311

[31]   W. Du, Y. Shi, W. Zhou, Y. Yu, B. Zhang, Unveiling the in situ dissolution and polymerization of Mo in Ni_4Mo alloy for promoting the hydrogen evolution reaction, Angew. Chem. Int. Ed., 60 (2021) 7051-7055. https://doi.org/10.1002/anie.202015723

[32]   H. Shi, Y.T. Zhou, R.Q. Yao, W.B. Wan, X. Ge, W. Zhang, Z. Wen, X.Y. Lang, W.T. Zheng, Q. Jiang, Spontaneously separated intermetallic Co_3Mo from nanoporous copper as versatile electrocatalysts for highly efficient water splitting, Nat. Commun., 11 (2020) 2940. https://doi.org/10.1038/s41467-020-16769-6

[33]   B. Xiong, W. Zhao, H. Tian, W. Huang, L. Chen, J. Shi, Nickel-tungsten nano-alloying for high-performance hydrogen electro-catalytic oxidation, Chem. Eng. J., 432 (2022) 134189. https://doi.org/10.1016/j.cej.2021.134189

[34]   J. Mao, C.T. He, J. Pei, W. Chen, D. He, Y. He, Z. Zhuang, C. Chen, Q. Peng, D. Wang, Y. Li, Accelerating water dissociation kinetics by isolating cobalt atoms into ruthenium lattice, Nat. Commun., 9 (2018) 4958. https://doi.org/10.1038/s41467-018-07288-6

[35]   Y. Zhang, G. Fan, I.. Yang, L. Zheng, F. Li, Cooperative effects between Ni-Mo alloy sites and defective structures over hierarchical Ni-Mo bimetallic catalysts enable the enhanced hydrodeoxygenation activity, ACS Sustain. Chem. Eng., 9 (2021) 11604-11615. https://doi.org/10.1021/acssuschemeng.1c04762

[36]   M. Wang, H. Yang, J. Shi, Y. Chen, Y. Zhou, L. Wang, S. Di, X. Zhao, J. Zhong, T. Cheng, W. Zhou, Y. Li, Alloying nickel with molybdenum significantly accelerates alkaline hydrogen electrocatalysis, Angew. Chem. Int. Ed., 60 (2021) 5771-5777. https://doi.org/10.1002/anie.202013047

[37]   P.-C. Hsu, S.-K. Seol, T.-N. Lo, C.-J. Liu, C.-L. Wang, C.-S. Lin, Y. Hwu, Hydrogen bubbles and the growth morphology of ramified zinc by electrodeposition, J. Electrochem. Soc., 155 (2008) D400-D407. https://doi.org/10.1149/1.2894189

[38]   S. Ye, F. Luo, Q. Zhang, P. Zhang, T. Xu, Q. Wang, D. He, L. Guo, Y. Zhang, C. He, X. Ouyang, M. Gu, J. Liu, X. Sun, Highly stable single Pt atomic sites anchored on aniline-stacked graphene for hydrogen evolution reaction, Energy Environ. Sci., 12 (2019) 1000-1007. https://doi.org/10.1039/C8EE02888E

[39]   C. Yan, W. An, T. Shen, L. Ma, M. Zhang, F. Gao, T. Yang, C. Wang, G. Huang, S. Xu, Monodispersed PtCo alloy nanoparticles with a modulated d-band center exhibiting highly efficient hydrogen evolution, J. Mater. Chem. A, 11 (2023) 26812-26820. https://doi.org/10.1039/D3TA05750J

[40] S.T. Hunt, M. Milina, Z. Wang, Y. Román-Leshkov, Activating earth-abundant electrocatalysts for efficient, low-cost hydrogen evolution/oxidation: sub-monolayer platinum coatings on titanium tungsten carbide nanoparticles, Energy Environ. Sci., 9 (2016) 3290-3301. https://doi.org/10.1039/C6EE01929C

[41] X. Huang, Z. Zhao, L. Cao, Y. Chen, E. Zhu, Z. Lin, M. Li, A. Yan, A. Zettl, Y.M. Wang, X. Duan, T. Mueller, Y. Huang, High-performance transition metal–doped Pt$_3$Ni octahedra for oxygen reduction reaction, Science, 348 (2015) 1230-1234. https://doi.org/10.1126/science.aaa8765

[42] K.L. Zhou, C. Wang, Z. Wang, C.B. Han, Q. Zhang, X. Ke, J. Liu, H. Wang, Seamlessly conductive Co(OH)$_2$ tailored atomically dispersed Pt electrocatalyst with a hierarchical nanostructure for an efficient hydrogen evolution reaction, Energy Environ. Sci., 13 (2020) 3082-3092. https://doi.org/10.1039/D0EE01347A

[43] A. Romanchenko, M. Likhatski, Y. Mikhlin, X-ray photoelectron spectroscopy (XPS) study of the products formed on sulfide minerals upon the interaction with aqueous platinum (IV) chloride complexes, Minerals, 8 (2018) 578. https://doi.org/10.3390/min8120578

[44] F. Li, G.F. Han, H.J. Noh, J.P. Jeon, I. Ahmad, S. Chen, C. Yang, Y. Bu, Z. Fu, Y. Lu, J.B. Baek, Balancing hydrogen adsorption/desorption by orbital modulation for efficient hydrogen evolution catalysis, Nat. Commun., 10 (2019) 4060. https://doi.org/10.1038/s41467-019-12012-z

[45] R.M. Navarro Yerga, M.C. Alvarez Galvan, F. del Valle, J.A. Villoria de la Mano, J.L. Fierro, Water splitting on semiconductor catalysts under visible-light irradiation, ChemSusChem, 2 (2009) 471-485. https://doi.org/10.1002/cssc.200900018

[46] P. Chen, W.J. Ong, Z. Shi, X. Zhao, N. Li, Pb-based halide perovskites: Recent advances in photo(electro)catalytic applications and looking beyond, Adv. Funct. Mater., 30 (2020) 1909667. https://doi.org/10.1002/adfm.201909667

[47] K. Chang, X. Hai, J. Ye, Transition metal disulfides as noble-metal-alternative Co-catalysts for solar hydrogen production, Adv. Energy Mater., 6 (2016) 1502555. https://doi.org/10.1002/aenm.201502555

[48] S.R. Forrest, M.E. Thompson, Introduction: Organic electronics and optoelectronics, Chem. Rev., 107 (2007) 923-925. https://doi.org/10.1021/cr0501590

[49] F. Roccaforte, F. Giannazzo, F. Iucolano, J. Eriksson, M.H. Weng, V. Raineri, Surface and interface issues in wide band gap semiconductor electronics, Appl. Surf. Sci., 256 (2010) 5727-5735. https://doi.org/10.1016/j.apsusc.2010.03.097

[50] M.V. Nikolic, V. Milovanovic, Z.Z. Vasiljevic, Z. Stamenkovic, Semiconductor gas sensors: Materials, technology, design, and application, sensors, 20 (2020) 6694. https://doi.org/10.3390/s20226694

[51] G. Korotcenkov, B.K. Cho, Porous Semiconductors: Advanced material for gas sensor

applications, Crit. Rev. Solid State Mater. Sci., 35 (2010) 1-37.
https://doi.org/10.1080/10408430903245369

[52]   Y. Tang, Y. Zhao, H. Liu, Room-temperature semiconductor gas sensors: Challenges and
       opportunities, ACS Sens., 7 (2022) 3582-3597.
       https://doi.org/10.1021/acssensors.2c01142

[53]   S. Chandrasekaran, D. Ma, Y. Ge, L. Deng, C. Bowen, J. Roscow, Y. Zhang, Z. Lin,
       R.D.K. Misra, J. Li, P. Zhang, H. Zhang, Electronic structure engineering on two-
       dimensional (2D) electrocatalytic materials for oxygen reduction, oxygen evolution, and
       hydrogen evolution reactions, Nano Energy, 77 (2020) 105080.
       https://doi.org/10.1016/j.nanoen.2020.105080

[54]   X.Z. Song, W.Y. Zhu, J.C. Ni, Y.H. Zhao, T. Zhang, Z. Tan, L.Z. Liu, X.F. Wang,
       Boosting Hydrogen evolution electrocatalysis via regulating the electronic structure in a
       crystalline-amorphous CoP/CeO_x p-n heterojunction, ACS Appl. Mater. Interfaces, 14
       (2022) 33151−33160. https://doi.org/10.1021/acsami.2c06439

[55]   A. Mondal, A. Vomiero, 2D transition metal dichalcogenides-based electrocatalysts for
       hydrogen evolution reaction, Adv. Funct. Mater., 32 (2022) 2208994.
       https://doi.org/10.1002/adfm.202208994

[56]   A. Gautam, S. Sk, U. Pal, Recent advances in solution assisted synthesis of transition
       metal chalcogenides for photo-electrocatalytic hydrogen evolution, Phys. Chem. Chem.
       Phys., 24 (2022) 20638-20673. https://doi.org/10.1039/D2CP02089K

[57]   M.J. Sadiq Mohamed, S. Caliskan, M.A. Gondal, M.A. Almessiere, A. Baykal, Y.
       Slimani, K.A. Elsayed, M. Hassan, I.A. Auwal, A.Z. Khan, A.A. Tahir, A. Roy, Se-
       Doped magnetic Co–Ni spinel ferrite nanoparticles as electrochemical catalysts for
       hydrogen evolution, ACS Appl. Nano Mater., 6 (2023) 7330-7341.
       https://doi.org/10.1021/acsanm.3c00464

[58]   M. Khalil, M. Lesa, A.G. Juandito, A.R. Sanjaya, T.A. Ivandini, G.T.M. Kadja, M.H.
       Mahyuddin, M. Sookhakian, Y. Alias, A SBA-15-templated mesoporous
       $NiFe_2O_4$/MXene nanocomposite for the alkaline hydrogen evolution reaction, Mater.
       Adv., 4 (2023) 3853-3862. https://doi.org/10.1039/D3MA00289F

[59]   M.S. Alom, C.C.W. Kananke-Gamage, F. Ramezanipour, Perovskite oxides as
       electrocatalysts for hydrogen evolution reaction, ACS Omega, 7 (2022) 7444-7451.
       https://doi.org/10.1021/acsomega.1c07203

[60]   S.A. Ali, T. Ahmad, Treasure trove for efficient hydrogen evolution through water
       splitting using diverse perovskite photocatalysts, Mater. Today Chem., 29 (2023) 101387.
       https://doi.org/10.1016/j.mtchem.2023.101387

[61]   S.M. Thalluri, L. Bai, C. Lv, Z. Huang, X. Hu, L. Liu, Strategies for
       semiconductor/electrocatalyst coupling toward solar-driven water splitting, Adv. Sci., 7
       (2020) 1902102. https://doi.org/10.1002/advs.201902102

[62]   X. Ren, Q. Li, F. Ling, Q. Hu, L. Pang, Construction of MoO_2/MoS_2 heterojunctions on carbon nanotubes as high-efficiency electrocatalysts for H_2 production, CrystEngComm, 25 (2023) 5238-5242. https://doi.org/10.1039/D3CE00674C

[63]   J. Deng, H. Li, J. Xiao, Y. Tu, D. Deng, H. Yang, H. Tian, J. Li, P. Ren, X. Bao, Triggering the electrocatalytic hydrogen evolution activity of the inert two-dimensional MoS_2 surface via single-atom metal doping, Energy Environ. Sci., 8 (2015) 1594-1601. https://doi.org/10.1039/C5EE00751H

[64]   D.A. Kuznetsov, Z. Chen, P.V. Kumar, A. Tsoukalou, A. Kierzkowska, P.M. Abdala, O.V. Safonova, A. Fedorov, C.R. Muller, Single site cobalt substitution in 2D molybdenum carbide (MXene) enhances catalytic activity in the hydrogen evolution reaction, J. Am. Chem. Soc., 141 (2019) 17809-17816. https://doi.org/10.1021/jacs.9b08897

[65]   I.P. Chen, Y.X. Chen, C.W. Wu, C.C. Chiu, Y.C. Hsieh, Large-scale fabrication of a flexible, highly conductive composite paper based on molybdenum disulfide-Pt nanoparticle-single-walled carbon nanotubes for efficient hydrogen production, Chem. Commun., 53 (2016) 380-383. https://doi.org/10.1039/C6CC08050B

[66]   Q. Dang, Y. Sun, X. Wang, W. Zhu, Y. Chen, F. Liao, H. Huang, M. Shao, Carbon dots-Pt modified polyaniline nanosheet grown on carbon cloth as stable and high-efficient electrocatalyst for hydrogen evolution in pH-universal electrolyte, Appl. Catal. B: Environ., 257 (2019) 117905. https://doi.org/10.1016/j.apcatb.2019.117905

[67]   Z.Y. Yu, Y. Duan, X.Y. Feng, X. Yu, M.R. Gao, S.H. Yu, Clean and affordable hydrogen fuel from alkaline water splitting: Past, recent progress, and future prospects, Adv. Mater., 33 (2021) 2007100. https://doi.org/10.1002/adma.202007100

[68]   Q. Xu, J. Zhang, H. Zhang, L. Zhang, L. Chen, Y. Hu, H. Jiang, C. Li, Atomic heterointerface engineering overcomes the activity limitation of electrocatalysts and promises highly-efficient alkaline water splitting, Energy Environ. Sci., 14 (2021) 5228-5259. https://doi.org/10.1039/D1EE02105B

[69]   T. Li, J. Yin, D. Sun, M. Zhang, H. Pang, L. Xu, Y. Zhang, J. Yang, Y. Tang, J. Xue, Manipulation of Mott-Schottky Ni/CeO_2 heterojunctions into N-doped Carbon nanofibers for high-efficiency electrochemical water splitting, Small, 18 (2022) 2106592. https://doi.org/10.1002/smll.202106592

[70]   X. Zhang, H. Xue, J. Sun, N. Guo, T. Song, J. Sun, Y.-R. Hao, Q. Wang, Synergy of phosphorus vacancies and build-in electric field into NiCo/NiCoP Mott-Schottky integrated electrode for enhanced water splitting performance, Chinese Chem. Lett., 35 (2024) 108519. https://doi.org/10.1016/j.cclet.2023.108519

[71]   J. Hou, Y. Sun, Y. Wu, S. Cao, L. Sun, Promoting active sites in core-shell nanowire array as Mott–Schottky electrocatalysts for efficient and stable overall water splitting, Adv. Funct. Mater., 28 (2018) 1704447. https://doi.org/10.1002/adfm.201704447

[72]   Z. Sun, Y. Wang, L. Zhang, H. Wu, Y. Jin, Y. Li, Y. Shi, T. Zhu, H. Mao, J. Liu, C.

Xiao, S. Ding, Simultaneously realizing rapid electron transfer and mass transport in jellyfish-like Mott–Schottky nanoreactors for oxygen reduction reaction, Adv. Funct. Mater., 30 (2020) 1910482. https://doi.org/10.1002/adfm.201910482

[73]   T. Chen, S. Guo, J. Yang, Y. Xu, J. Sun, D. Wei, Z. Chen, B. Zhao, W. Ding, Nitrogen-doped carbon activated in situ by embedded nickel through the Mott-Schottky effect for the oxygen reduction reaction, Chemphyschem, 18 (2017) 3454-3461. https://doi.org/10.1002/cphc.201700834

[74]   J. Chen, J. Zheng, W. He, H. Liang, Y. Li, H. Cui, C. Wang, Self-standing hollow porous Co/a-WOx nanowire with maximum Mott-Schottky effect for boosting alkaline hydrogen evolution reaction, Nano Res., 16 (2022) 4603-4611. https://doi.org/10.1007/s12274-022-5072-1

[75]   Z.H. Xue, H. Su, Q.Y. Yu, B. Zhang, H.H. Wang, X.H. Li, J.S. Chen, Janus Co/CoP nanoparticles as efficient Mott-Schottky electrocatalysts for overall water splitting in wide pH range, Adv. Energy Mater., 7 (2017) 1602355. https://doi.org/10.1002/aenm.201602355

[76]   X. Jiang, K. Elouarzaki, Y. Tang, J. Zhou, G. Fu, J.-M. Lee, Embedded PdFe@N-carbon nanoframes for oxygen reduction in acidic fuel cells, Carbon, 164 (2020) 369-377. https://doi.org/10.1016/j.carbon.2020.04.013

[77]   Y. Tian, S. Cong, W. Su, H. Chen, Q. Li, F. Geng, Z. Zhao, Synergy of $W_{18}O_{49}$ and polyaniline for smart supercapacitor electrode integrated with energy level indicating functionality, Nano Lett., 14 (2014) 2150-2156. https://doi.org/10.1021/nl5004448

[78]   F. Salleh, M.N.A. Tahari, A. Samsuri, T.S.T. Saharuddin, S.S. Sulhadi, M.A. Yarmo, Physical and chemical behaviour of tungsten oxide in the presence of nickel additive under hydrogen and carbon monoxide atmospheres, Int. J. Hydrog. Energy, 46 (2021) 24814-24830. https://doi.org/10.1016/j.ijhydene.2020.08.099

[79]   G. Xi, J. Ye, Q. Ma, N. Su, H. Bai, C. Wang, In situ growth of metal particles on 3D urchin-like WO_3 nanostructures, J. Am. Chem. Soc., 134 (2012) 6508-6511. https://doi.org/10.1021/ja211638e

[80]   G. Chen, S. Desinan, R. Rosei, F. Rosei, D. Ma, Synthesis of Ni-Ru alloy nanoparticles and their high catalytic activity in dehydrogenation of ammonia borane, Chemistry, 18 (2012) 7925-7930. https://doi.org/10.1002/chem.201200292

[81]   Y. Zheng, Y. Jiao, A. Vasileff, S.Z. Qiao, The hydrogen evolution reaction in alkaline solution: From theory, single crystal models, to practical electrocatalysts, Angew. Chem. Int. Ed., 57 (2018) 7568-7579. https://doi.org/10.1002/anie.201710556

[82]   S. Anantharaj, S.R. Ede, K. Karthick, S. Sam Sankar, K. Sangeetha, P.E. Karthik, S. Kundu, Precision and correctness in the evaluation of electrocatalytic water splitting: revisiting activity parameters with a critical assessment, Energy Environ. Sci., 11 (2018) 744-771. https://doi.org/10.1039/C7EE03457A

[83]   K. Xu, Y. Sun, Y. Sun, Y. Zhang, G. Jia, Q. Zhang, L. Gu, S. Li, Y. Li, H.J. Fan, Yin-Yang Harmony: Metal and nonmetal dual-doping boosts electrocatalytic activity for alkaline hydrogen evolution, ACS Energy Lett., 3 (2018) 2750-2756. https://doi.org/10.1021/acsenergylett.8b01893

[84]   C. Lei, Y. Wang, Y. Hou, P. Liu, J. Yang, T. Zhang, X. Zhuang, M. Chen, B. Yang, L. Lei, C. Yuan, M. Qiu, X. Feng, Efficient alkaline hydrogen evolution on atomically dispersed $Ni-N_x$ Species anchored porous carbon with embedded Ni nanoparticles by accelerating water dissociation kinetics, Energy Environ. Sci., 12 (2019) 149-156. https://doi.org/10.1039/C8EE01841C

[85]   J. Yao, W. Huang, W. Fang, M. Kuang, N. Jia, H. Ren, D. Liu, C. Lv, C. Liu, J. Xu, Q. Yan, Promoting electrocatalytic hydrogen evolution reaction and oxygen evolution reaction by fields: Effects of electric field, magnetic field, strain, and light, Small Methods, 4 (2020) 2000494. https://doi.org/10.1002/smtd.202000494

[86]   X. Long, H. Lin, D. Zhou, Y. An, S. Yang, Enhancing full water-splitting performance of transition metal bifunctional electrocatalysts in alkaline solutions by tailoring CeO_2–transition metal oxides–Ni nanointerfaces, ACS Energy Lett., 3 (2018) 290-296. https://doi.org/10.1021/acsenergylett.7b01130

[87]   J. Chen, C. Chen, M. Qin, B. Li, B. Lin, Q. Mao, H. Yang, B. Liu, Y. Wang, Reversible hydrogen spillover in $Ru-WO_{3-x}$ enhances hydrogen evolution activity in neutral pH water splitting, Nat. Commun., 13 (2022) 5382. https://doi.org/10.1038/s41467-022-33007-3

[88]   D. Cahen, A. Kahn, Electron energetics at surfaces and interfaces: Concepts and experiments, Adv. Mater., 15 (2003) 271-277. https://doi.org/10.1002/adma.200390065

[89]   X. Li, S. Song, Y. Gao, L. Ge, W. Song, T. Ma, J. Liu, Identification of the charge transfer channel in cobalt encapsulated hollow nitrogen-doped carbon matrix@CdS heterostructure for photocatalytic hydrogen evolution, Small, 17 (2021) 2101315. https://doi.org/10.1002/smll.202101315

[90]   H. Li, H. Yu, X. Quan, S. Chen, Y. Zhang, Uncovering the key role of the fermi level of the electron mediator in a Z-scheme photocatalyst by detecting the charge transfer process of WO_3-metal-gC_3N_4 (Metal = Cu, Ag, Au), ACS Appl. Mater. Interfaces, 8 (2016) 2111-2119. https://doi.org/10.1021/acsami.5b10613

[91]   W. Mönch, Metal-semiconductor contacts: electronic properties, Surf. Sci., 299 (1994) 928–944. https://doi.org/10.1016/0039-6028(94)90707-2

[92]   P. Xia, S. Cao, B. Zhu, M. Liu, M. Shi, J. Yu, Y. Zhang, Designing a 0D/2D S-scheme heterojunction over polymeric carbon nitride for visible-light photocatalytic inactivation of bacteria, Angew. Chem. Int. Ed., 59 (2020) 5218-5225. https://doi.org/10.1002/anie.201916012

[93]   M. Qu, Y. Jiang, M. Yang, S. Liu, Q. Guo, W. Shen, M. Li, R. He, Regulating electron density of NiFe-P nanosheets electrocatalysts by a trifle of Ru for high-efficient overall

water splitting, Appl. Catal. B: Environ., 263 (2020) 118324.
https://doi.org/10.1016/j.apcatb.2019.118324

[94]   H. Zhang, Y. Lv, C. Chen, C. Lv, X. Wu, J. Guo, D. Jia, Inter-doped ruthenium–nickel oxide heterostructure nanosheets with dual active centers for electrochemical-/solar-driven overall water splitting, Appl. Catal. B: Environ., 298 (2021) 120611.
https://doi.org/10.1016/j.apcatb.2021.120611

[95]   Z. Jiang, S. Song, X. Zheng, X. Liang, Z. Li, H. Gu, Z. Li, Y. Wang, S. Liu, W. Chen, D. Wang, Y. Li, Lattice Strain and Schottky Junction dual regulation boosts ultrafine ruthenium nanoparticles anchored on a N-modified carbon catalyst for H_2 production, J. Am. Chem. Soc., 144 (2022) 19619-19626. https://doi.org/10.1021/jacs.2c09613

[96]   S. Park, J. Park, H. Abroshan, L. Zhang, J.K. Kim, J. Zhang, J. Guo, S. Siahrostami, X. Zheng, Enhancing catalytic activity of MoS_2 basal plane S-vacancy by Co cluster addition, ACS Energy Lett., 3 (2018) 2685-2693.
https://doi.org/10.1021/acsenergylett.8b01567

[97]   F. Zhou, R. Sa, X. Zhang, S. Zhang, Z. Wen, R. Wang, Robust ruthenium diphosphide nanoparticles for pH-universal hydrogen evolution reaction with platinum-like activity, Appl. Catal. B: Environ., 274 (2020) 119092.
https://doi.org/10.1016/j.apcatb.2020.119092

[98]   S. Wang, M. Wang, Z. Liu, S. Liu, Y. Chen, M. Li, H. Zhang, Q. Wu, J. Guo, X. Feng, Z. Chen, Y. Pan, Synergetic function of the single-atom Ru-N_4 Site and Ru nanoparticles for hydrogen production in a wide pH range and seawater electrolysis, ACS Appl. Mater. Interfaces, 14 (2022) 15250-15258. https://doi.org/10.1021/acsami.2c00652

[99]   P. Chen, J. Ye, H. Wang, L. Ouyang, M. Zhu, Recent progress of transition metal carbides/nitrides for electrocatalytic water splitting, J. Alloys Compd., 883 (2021) 160833. https://doi.org/10.1016/j.jallcom.2021.160833

[100]   H. Guo, A. Wu, Y. Xie, H. Yan, D. Wang, L. Wang, C. Tian, 2D porous molybdenum nitride/cobalt nitride heterojunction nanosheets with interfacial electron redistribution for effective electrocatalytic overall water splitting, J. Mater. Chem. A, 9 (2021) 8620-8629.
https://doi.org/10.1039/D0TA11997K

[101]   Z. Kou, T. Wang, Q. Gu, M. Xiong, L. Zheng, X. Li, Z. Pan, H. Chen, F. Verpoort, A.K. Cheetham, S. Mu, J. Wang, Rational design of holey 2D nonlayered transition metal carbide/nitride heterostructure nanosheets for highly efficient water oxidation, Adv. Energy Mater., 9 (2019) 1803768. https://doi.org/10.1002/aenm.201803768

[102]   Y. Guo, T. Park, J.W. Yi, J. Henzie, J. Kim, Z. Wang, B. Jiang, Y. Bando, Y. Sugahara, J. Tang, Y. Yamauchi, Nanoarchitectonics for transition-metal-sulfide-based electrocatalysts for water splitting, Adv. Mater., 31 (2019) 1807134.
https://doi.org/10.1002/adma.201807134

[103]   J.T. Ren, L. Wang, L. Chen, X.L. Song, Q.H. Kong, H.Y. Wang, Z.Y. Yuan, Interface metal oxides regulating electronic state around nickel species for efficient alkaline

hydrogen electrocatalysis, Small, 19 (2023) 2206196.
https://doi.org/10.1002/smll.202206196

[104] S. Shen, Z. Wang, Z. Lin, K. Song, Q. Zhang, F. Meng, L. Gu, W. Zhong, Crystalline-amorphous interfaces coupling of $CoSe_2$/CoP with optimized d-band center and boosted electrocatalytic hydrogen evolution, Adv. Mater., 34 (2022) 2110631. https://doi.org/10.1002/adma.202110631

[105] K. Jiang, B. Liu, M. Luo, S. Ning, M. Peng, Y. Zhao, Y.R. Lu, T.S. Chan, F.M.F. de Groot, Y. Tan, Single platinum atoms embedded in nanoporous cobalt selenide as electrocatalyst for accelerating hydrogen evolution reaction, Nat. Commun., 10 (2019) 1743. https://doi.org/10.1038/s41467-019-09765-y

[106] Y. Lin, K. Sun, S. Liu, X. Chen, Y. Cheng, W.C. Cheong, Z. Chen, L. Zheng, J. Zhang, X. Li, Y. Pan, C. Chen, Construction of CoP/NiCoP nanotadpoles heterojunction interface for wide pH Hydrogen evolution electrocatalysis and supercapacitor, Adv. Energy Mater., 9 (2019) 1901213. https://doi.org/10.1002/aenm.201901213

[107] J.K. Kim, G.D. Park, J.H. Kim, S.K. Park, Y.C. Kang, Rational design and synthesis of extremely efficient macroporous $CoSe_2$-CNT composite microspheres for hydrogen evolution reaction, Small, 13 (2017) 1700068. https://doi.org/10.1002/smll.201700068

[108] J. Tian, Q. Liu, A.M. Asiri, X. Sun, Self-supported nanoporous cobalt phosphide nanowire arrays: an efficient 3D hydrogen-evolving cathode over the wide range of pH 0-14, J. Am. Chem. Soc., 136 (2014) 7587-7590. https://doi.org/10.1021/ja503372r

[109] T. Zhu, J. Ding, Q. Shao, Y. Qian, X. Huang, P, Se-Codoped MoS_2 Nanosheets as accelerated electrocatalysts for hydrogen evolution, ChemCatChem, 11 (2018) 689-692. https://doi.org/10.1002/cctc.201801541

[110] N. Xue, Z. Lin, P. Li, P. Diao, Q. Zhang, Sulfur-doped $CoSe_2$ porous nanosheets as efficient electrocatalysts for the hydrogen evolution reaction, ACS Appl. Mater. Interfaces, 12 (2020) 28288-28297. https://doi.org/10.1021/acsami.0c07088

Materials Research Forum LLC
https://doi.org/10.21741/9781644903070

CHAPTER 9

Electrocatalytic Hydrogen Production Across a Wide pH Range

Xiang Peng*

Hubei Key Laboratory of Plasma Chemistry and Advanced Materials, Engineering Research Center of Phosphorus Resources Development and Utilization of Ministry of Education, School of Materials Science and Engineering, Wuhan Institute of Technology, Wuhan 430205, China

xpeng@wit.edu.cn

Abstract

Emphasizing the significance of researching hydrogen evolution reaction (HER) across a wide pH range is essential for achieving sustainable development. This research direction focuses on the development of HER catalysts that exhibit outstanding activity and stability under acidic, neutral, and alkaline conditions, and even in complex seawater. The adaptability of HER across a wide pH range becomes evident in its ability to effectively handle diverse environmental conditions, offering greater flexibility for practical applications in renewable energy systems.

Keywords

Hydrogen Evolution Reaction, Universal pH Range, Doping, Heterogeneous Structure, Alloying

1.   Introduction

The utilization of hydrogen evolution reaction (HER) in the energy sector not only represents the forefront of scientific research but also serves as a pivotal driver for advancing the transition to clean energy [1-3]. By catalyzing water electrolysis, HER plays a crucial role in supporting hydrogen generation, and injecting vitality into the renewable energy production chain [4, 5]. Extensive research into renewable energy systems reveals the increasingly prominent importance of HER in this field. Beyond its role in water decomposition, HER significantly influences the overall performance of the system [6, 7]. Emphasizing the significance of researching HER across a wide pH range is essential for achieving sustainable development [8-10]. This research direction focuses on the development of HER catalysts that exhibit outstanding activity and stability under acidic, neutral, and alkaline conditions, and even in complex seawater.

The adaptability of HER across a wide pH range becomes evident in its ability to effectively handle diverse environmental conditions, offering greater flexibility for practical applications in renewable energy systems [11-13]. In actual operation, the electrolyte's acidity or alkalinity may fluctuate, particularly at the electrode surface, due to changing environmental conditions [14-16]. Consequently, HER electrocatalysts with broad pH adaptability can ensure efficient hydrogen production, enhancing system robustness and opening up new possibilities for achieving a clean and sustainable energy supply [17, 18]. The study of HER across a wide pH range is not only a profound scientific exploration but also closely aligned with the practical requirements of various applications [19, 20]. In this cutting-edge field of relentless exploration, researchers anticipate breakthroughs in overcoming the performance limitations of HER electrocatalysts [21, 22].

This chapter provides a thorough discussion of the critical aspects involved in constructing efficient HER electrocatalysts for wide pH range applications. It addresses the challenges faced by catalysts and the specific performance requirements under varying pH conditions, including acidic, neutral, and alkaline environments. Comprehensive discussions on catalyst activity, stability, and selectivity at different pH levels aim to clarify the key issues inherent in the successful development of efficient HER electrocatalysts. The exploration in this chapter covers crucial aspects such as material design [23, 24], synthesis methods [25, 26], catalyst activity and stability [27, 28], and the intricate relationship between catalyst structure and performance. These discussions provide profound insights into addressing pertinent challenges in the field. Additionally, the chapter delves into the design of novel materials, enhancements in synthesis processes, and the formulation of engineering strategies for catalysts adaptable to different pH conditions.

By presenting practical solutions and guidance, this chapter aims to propel new achievements in the field of HER electrocatalysts for wide pH range applications. It equips readers with the knowledge and understanding necessary to navigate the complexities of constructing efficient catalysts. Ultimately, this contribution will advance the development of HER electrocatalysts and contribute to the progress of clean energy technologies.

2.   Mechanisms and Challenges of HER Across a Wide pH Range

2.1.   Catalytic Mechanisms of HER in Various Conditions

Under distinct pH conditions, the efficiency of the HER is governed by diverse mechanisms [29-31]. Understanding the intricacies of these mechanisms under acidic, alkaline, and neutral conditions is crucial, and they are discussed as follows.

(i) Acidic conditions (pH < 7)

In acidic media, the HER mechanism typically involves a Volmer-Heyrovsky pathway[32, 33], as schematically illustrated in **Figure 1**. Initially, the Volmer step involves the adsorption of protons (H^+) on the electrode surface, followed by the discharge of hydrogen ions to form adsorbed hydrogen atoms (H^*)[34]. The Heyrovsky step involves the addition of another proton to the adsorbed hydrogen atom, resulting in the formation of molecular hydrogen (H_2)[35]. The reaction can be expressed as follows.

$$2H^+ + 2e^- \rightarrow H_2 \uparrow$$ (Reaction 1)

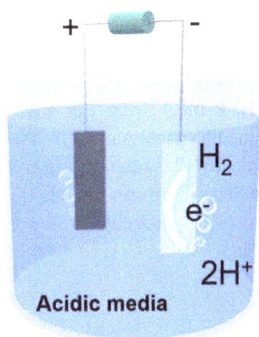

Figure 1. Scheme of HER in acidic media.

This reduction mechanism lays the foundation for efficient hydrogen production in acidic solutions.

(ii) Alkaline Conditions (pH > 7)

In alkaline media, the HER mechanism typically follows the Volmer-Tafel pathway, which is much more difficult compared to that in acidic electrolytes [36, 37], as schematically illustrated in **Figure 2**. The Volmer step involves the adsorption of hydroxide ions (OH⁻) on the electrode surface, followed by the discharge of water molecules to form adsorbed hydroxyl radicals (OH*) [38]. The Tafel step involves the combination of two adsorbed hydroxyl radicals to generate molecular hydrogen (H_2) and water [39, 40]. Water molecules undergo reduction, producing hydrogen gas and hydroxide ions (OH⁻). It can be expressed as follows.

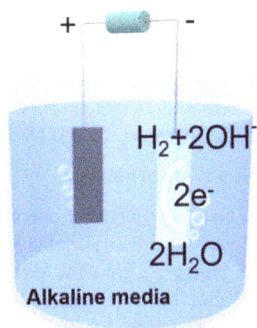

$$2H_2O + 2e^- \rightarrow H_2 \uparrow + 2OH^-$$ (Reaction 2)

Figure 2. Scheme of HER in alkaline media.

This mechanism emphasizes the unique challenges and opportunities presented in alkaline environments.

(iii) Neutral Conditions (pH = 7)

Under neutral pH conditions, the HER mechanism can involve a combination of the pathways observed in acidic and alkaline media, involving both the reduction of protons and water decomposition [18, 41], as schematically illustrated in **Figure 3**. The Volmer step still involves the adsorption of protons, while the Tafel step can include the combination of adsorbed protons or water molecules to generate molecular hydrogen[35]. It can be expressed as follows.

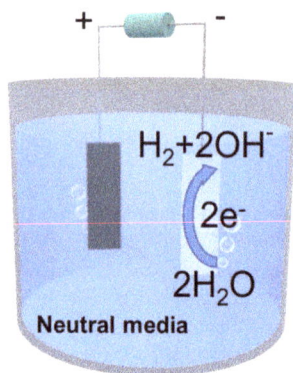

$$2H_2O + 2e^- \rightarrow H_2 \uparrow + 2OH^- \qquad \qquad \text{(Reaction 3)}$$

Figure 3. Scheme of HER in neutral media.

This dual-reduction pathway underscores the nuanced behavior of the HER at neutral pH and the associated complexities in catalyst design.

The understanding of these mechanisms is crucial for designing efficient HER electrocatalysts that can operate effectively under different pH conditions [18, 42]. By tailoring catalyst materials, surface structures, and active sites, researchers can optimize the performance of electrocatalysts to enhance HER efficiency in specific pH environments.

2.2. Challenges of HER Across a Wide pH Range

The construction of electrocatalysts that can effectively operate over a wide pH range presents a critical challenge due to substantial variations in material stability under different pH conditions [43, 44]. These differences arise primarily from the distinct electrochemical properties of materials in varying acidic and alkaline environments. For instance, molybdenum disulfide (MoS_2) is considered one of the most promising catalysts for the HER [45, 46]. However, it is susceptible to corrosion or deactivation in alkaline conditions. The elevated pH in alkaline media can lead to the potential dissolution or alteration of the material structure, consequently weakening its catalytic activity. This corrosion or degradation can limit the long-term stability and durability of MoS_2-based electrocatalysts in alkaline environments [47].

Indeed, destructive mechanisms, such as corrosion, deactivation, acid corrosion, and structural fatigue, directly impact the catalytic activity and durability of electrocatalysts [48, 49]. In alkaline environments, corrosion and deactivation can be particularly detrimental. The corrosive nature of the alkaline medium can result in the dissolution or alteration of active components or catalyst structures. This corrosion and deactivation can lead to a decline in catalytic activity and a decrease in the overall energy conversion efficiency of the electrocatalyst [50, 51]. Similarly, in acidic environments, acid corrosion and structural fatigue pose significant challenges. Acid corrosion can gradually erode the catalyst material, leading to the degradation of its surface and active sites. This corrosion-induced damage can negatively impact the catalytic capability of the electrocatalyst, affecting its performance over time. Additionally, the repetitive cycling of the acidic environment can cause structural fatigue, resulting in the deterioration of the material's properties and a decrease in its long-term lifespan [52-54]. To address these destructive mechanisms and enhance the performance and durability of HER electrocatalysts, it is imperative to have a comprehensive understanding of these challenges under diverse pH conditions.

3. Construction Strategies for HER Catalysts Applied to a Wide pH Range

There is a significant disparity in the activity and stability of catalysts when exposed to different pH conditions, posing challenges in the design and optimization of catalysts [55, 56]. To overcome this issue, scientists have employed a range of innovative strategies, including doping [57, 58], constructing heterogeneous structures [59, 60], and alloying [61, 62]. By cleverly applying these strategies, novel catalysts with high efficiency under various acidic, alkaline, and neutral environments have been successfully developed [63, 64]. These research findings not only provide robust support for the advancement of catalysis but also pave the way for addressing stability issues in real-world catalyst applications. The utilization of innovative catalyst design strategies injects fresh vitality into both fundamental scientific exploration and technological applications in related fields [65, 66].

3.1. Doping

Doping is a widely employed strategy in catalyst design, which involves the introduction of additional elements or compounds to modify the catalytic properties of the catalysts [67-69]. This strategy has proven particularly advantageous in the construction of catalysts for wide pH range applications [30, 70]. By doping, the surface acidity or alkalinity of the catalysts can be modulated, enabling adaptation to different pH conditions while facilitating electron transfer processes [71, 72]. This regulation strategy significantly enhances the activity and stability of the catalysts, especially in addressing variations in activity and stability under diverse pH environments. Carefully designed doping elements optimize the interactions between the doped atoms and the host material, expanding the applicability of the catalysts across a wide pH range and showcasing superior performance [73, 74]. Consequently, doping strategies have emerged as powerful tools for the design of catalysts that span a broad pH range of applications, offering innovative avenues for achieving highly efficient catalytic reactions.

Zhang et al. [75] conducted a study where they successfully incorporated palladium (Pd) into nickel-iron layered double hydroxides (NiFe-LDHs) using a hydrothermal synthesis method. This resulted in the formation of a self-supported material known as Pd/NiFeO$_x$, grown on a nickel foam (NF) substrate. The morphological characteristics of Pd/NiFeO$_x$ were elucidated through scanning electron microscopy (SEM) images presented in **Figure 4a-b.** These images revealed two-dimensional nanoplates with a porous structure. This unique morphology contributes to a significantly enlarged specific surface area and exposure of active sites, thereby positively impacting the electrocatalytic properties. The enlarged surface area facilitates enhanced electrical conductivity, improved diffusion of electrolyte, and facilitated generation and release of bubbles at the catalyst surface. These enhanced electrocatalytic properties of Pd/NiFeO$_x$ are attributed to its distinctive morphological features.

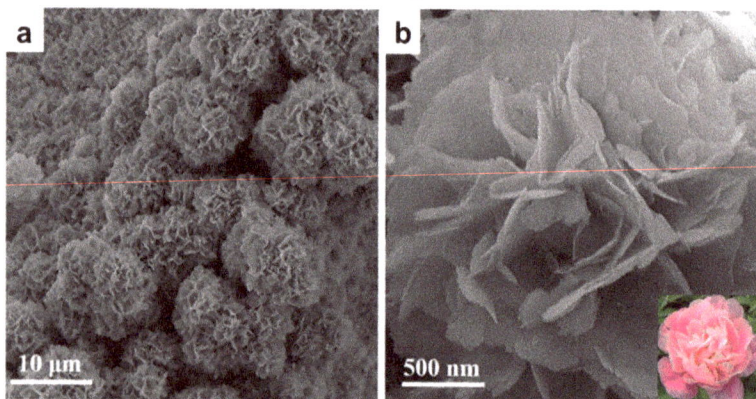

Figure 4. Morphology of Pd/NiFeO$_x$ nanosheets: (a) Low magnification, (b) high magnification. Reproduced with permission from ref [75]. Copyright 2021, Wiley-VCH.

The X-ray diffraction (XRD) pattern shown in **Figure 5** reveals valuable insights into the composition of the catalyst. The pattern indicates the presence of metallic nickel (Ni) species originating from the nickel foam, as well as nickel oxide (NiO) and iron oxide (Fe$_2$O$_3$) species derived from the NiFe-LDHs host material. Moreover, the XRD pattern exhibits distinctive diffraction peaks at specific angles, namely 40.1°, 46.7°, 68.1°, and 82.1°. These peaks correspond to the crystal planes (111), (200), (220), and (311) of palladium (Pd), providing conclusive evidence for the successful incorporation of Pd into the host material.

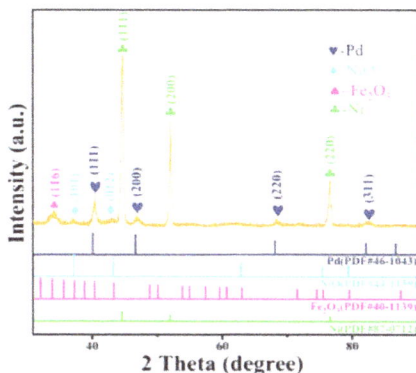

Figure 5. *XRD patterns of Pd/NiFeO$_x$ nanosheets. Reproduced with permission from ref [75]. Copyright 2021, Wiley-VCH.*

Further validation of Pd incorporation is supported by transmission electron microscopy (TEM) and selected area electron diffraction (SAED), as shown in **Figure 6a**. The high-resolution TEM (HR-TEM) analysis demonstrates lattice fringes with spacings of 0.224 and 0.195 nm, which correspond to the (111) and (100) crystal planes of Pd, as illustrated in **Figure 6b-c**. This additional evidence confirms the successful doping of Pd into the NiFe LDHs host, providing further support for the incorporation of Pd in the catalyst.

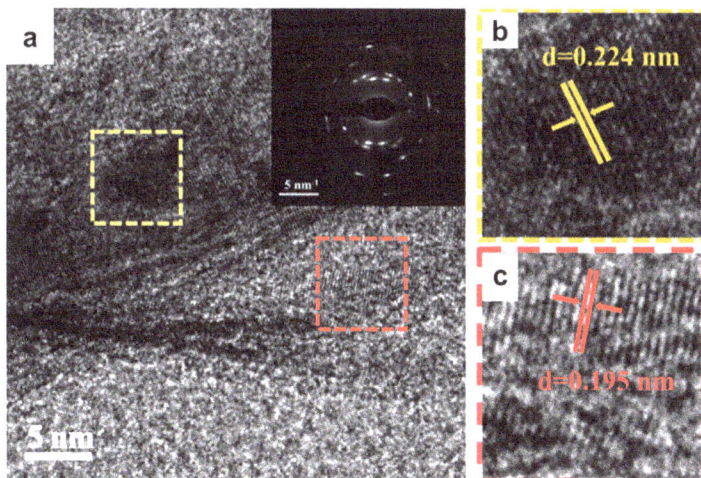

Figure 6. *(a) TEM and (b-c) HR-TEM images of Pd/NiFeO$_x$ nanosheets with the insert in (b) showing the SAED patterns. Reproduced with permission from ref [75]. Copyright 2021, Wiley-VCH.*

To explore the effects of Pd (palladium) doping on the electronic properties of the host material, a thorough characterization of the surface composition and valence states of Pd/NiFeO$_x$ nanosheets was conducted using X-ray photoelectron spectroscopy (XPS). The high-resolution Pd-3d XPS spectrum in **Figure 7a** displays peaks at 336.8 and 342.2 eV, corresponding to the $3d_{3/2}$ and $3d_{1/2}$ levels of Pd, respectively. Due to surface oxidation, additional peaks at 339.1 and 345 eV are observed, attributable to the $3d_{3/2}$ and $3d_{1/2}$ levels of Pd^{2+}. The observed positive shift in binding energy compared to the standard Pd binding energy provides further confirmation of successful Pd doping into the host material and the formation of strong multi-metal interactions. Additionally, the Ni-2$p_{3/2}$ spectrum in **Figure 7b** exhibits two distinct peaks indicating the presence of NiO and NiOOH. In comparison to NiFeO$_x$ nanosheets, a slight negative shift in the binding energy of Ni^{2+} at 853.9 and 871.9 eV in Pd/NiFeO$_x$ nanosheets suggests a robust electron interaction between Pd and NiFeO$_x$. These results collectively demonstrate the presence of strong electronic interactions between Pd and the NiFeO$_x$ host, highlighting the significant impact of Pd doping on the electronic modulation of the catalyst.

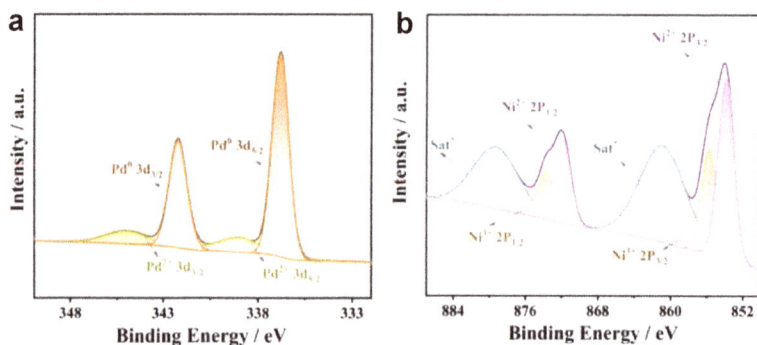

Figure 7. XPS of (a) Pd-3d and (b) Ni-2p for the Pd/NiFeO$_x$ nanosheets. Reproduced with permission from ref [75]. Copyright 2021, Wiley-VCH.

The HER performance of Pd/NiFeO$_x$ was investigated in a three-electrode system using different electrolytes, namely KOH, H$_2$SO$_4$, and PBS. **Figure 8** displays the overpotentials at the current density of 10 mA cm^{-2} and corresponding Tafel slopes under different pH conditions. In 1 M PBS, 0.5 M H$_2$SO$_4$, and 1 M KOH electrolytes, the overpotentials required for achieving a current density of 10 mA cm^{-2} were observed to be 75, 46, and 76 mV, respectively. In addition, the Tafel slopes obtained were 103.01, 38-56, and 78.03 mV dec^{-1} for 1 M PBS, 0.5 M H$_2$SO$_4$, and 1 M KOH electrolytes, respectively. The results highlight the exceptional HER capability of the Pd/NiFeO$_x$ across a wide pH range, indicating multiple beneficial effects.

The outstanding performance of the Pd/NiFeO$_x$ catalyst can be attributed to several factors. Firstly, the incorporation of Pd into the NiFe-LDHs host induces defects, regulates the crystal structure and lattice of the host material, and provides additional active sites, thereby significantly enhancing the intrinsic activity of the catalyst. Secondly, the excellent electronic conductivity of Pd facilitates efficient electronic transport during the catalytic process, further improving the kinetics and energy conversion efficiency of the HER reaction.

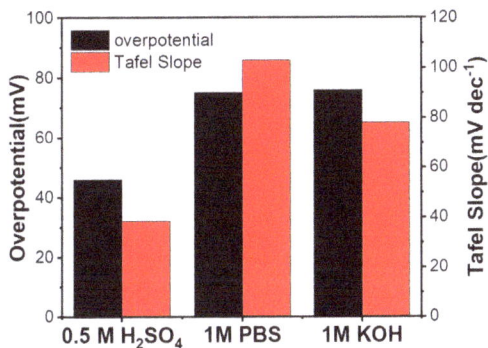

Figure 8. The overpotentials at 10 mA cm^{-2} and Tafel slopes of Pd/NiFeO$_x$ in various electrolytes. Reproduced with permission from ref [75]. Copyright 2021, Wiley-VCH.

3.2.   Constructing Heterogeneous Structures

The heterostructure is composed of multiple components that exhibit synergistic interactions, offering numerous advantages for electrocatalysts used in the HER across a wide pH range [76-78]. Firstly, the presence of multiple components provides the heterostructure with a wealth of active sites [79, 80], enabling it to maintain high activity in diverse acidic and alkaline environments, thus achieving versatile reaction adaptability.   Secondly, the synergistic interactions among components result in the formation of a disordered interface, effectively enhancing the catalytic activity across different pH conditions [81, 82]. This interface promotes efficient charge transfer and facilitates the exchange of species involved in the HER, contributing to improved performance. Additionally, the heterostructure with multiple components optimizes the pathways for electron and ion transport, enhancing the efficiency of these processes. This optimization results in improved overall electrocatalytic performance of the catalyst in a wide pH range [83].   Consequently, the construction of heterostructures offers a flexible and comprehensive solution for the design of HER electrocatalysts tailored for wide pH range applications.

In the realm of hydrogen production through water electrolysis, Mo-based materials have gained significant attention due to their abundant reserves, cost-effectiveness, and versatile tunable electronic structures. Among these materials, MoSe$_2$ has shown high electrocatalytic stability, but it suffers from poor electrical conductivity. On the other hand, molybdenum dioxide (MoO$_2$) possesses excellent conductivity and resistance to acid-base corrosion but lacks efficient HER catalytic capability. Leveraging the unique physical and chemical properties of these materials, Peng et al. [84] successfully synthesized MoO$_3$ nanowires on carbon cloth (CC) using a solution-based method, resulting in MoO$_3$/CC. Subsequently, a phase separation process under a Se atmosphere led to the formation of a core-shell structure known as MoSe$_2$-MoO$_2$/CC

(MSM/CC). The evolution of surface morphology during the phase separation process is illustrated in **Figure 9**. Microscopic studies of MSM/CC were conducted using scanning electron microscopy (SEM), revealing that after selenization through chemical vapor deposition (CVD), the nanowire array structure was preserved, but the surface became rougher and exhibited a flakier morphology. Additionally, there was a noticeable increase in the diameter of the nanowires.

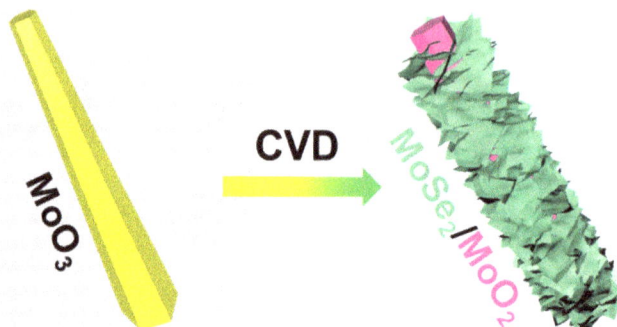

Figure 9. Schematic of the conversion of the MoO₃ nanowire into MSM core-shell structure. Reproduced with permission from ref [84]. Copyright 2023, Elsevier.

The core-shell structure of the final product was confirmed through TEM studies, revealing a core diameter of 200 nm and a shell diameter of 400 nm, as depicted in **Figure 10a**. This core-shell arrangement offers advantages by providing a nanosheet structure that exposes active sites and allows electrolyte access, thereby promoting catalytic properties. The TEM image in **Figure 10b** shows a strong connection between the nanowire core and the nanosheet shell, without sharp boundaries. Furthermore, HR-TEM images provided detailed information about the lattice fringes. Two distinct lattice fringes were observed, with an adjacent distance of 0.24 nm corresponding to the MoO_2 (-211) plane (JCPDS card No. 32-0671) for the nanowire core. Additionally, a lattice spacing of 0.65 nm was observed, corresponding to the $MoSe_2$ (002) plane (JCPDS card No. 77-1715) for the nanosheet shell, as indicated in **Figure 10c**. These HR-TEM observations confirm the presence of different crystalline structures within the core and the shell of the as-prepared catalyst.

XRD analysis was conducted on the MoO_3 sample at 400 °C, as shown in **Figure 11a**. The diffraction pattern of the resulting product (red line) indicates a composite of $MoSe_2$ (JCPDS card No. 77-1715) and MoO_2 (JCPDS card No. 32-0671). The results reveal a phase separation of MoO_3 during the in situ phase separation, leading to the forming of a composite of $MoSe_2$ and MoO_2. When the selenation temperature was increased to 500 °C, a pure phase of $MoSe_2$ (JCPDS card No. 77-1715) was obtained, as depicted in **Figure 11a** (black line). Furthermore, annealing MoO_3/CC at 550 °C for 3 hours without Se powder precursor resulted in a pure phase MoO_2/CC (JCPDS card No. 32-0671), as shown by the blue line in **Figure 11a**.

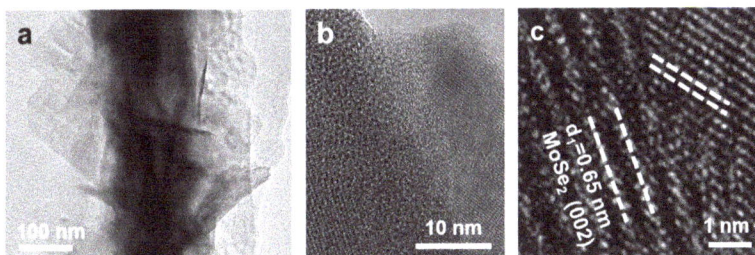

Figure 10. *(a-b) TEM and (c) HR-TEM images of MSM/CC. Reproduced with permission from ref [84]. Copyright 2023, Elsevier.*

Raman scattering studies of the product, as illustrated in **Figure 11b**, further confirmed the composite nature of MSM/CC as MoO_2 and $1T$-$MoSe_2$. In addition, the conductivity test indicated that the pure selenide $MoSe_2$/CC exhibits the lowest conductivity. In contrast, the core-shell structure of MSM/CC exhibited approximately 6 times higher conductivity compared to pure $MoSe_2$/CC and even higher conductivity than the precursor MoO_3/CC. This synergistic effect, combining high exposure of active sites and enhanced conductivity in the MSM/CC composite material, is expected to have a positive impact on the HER performance.

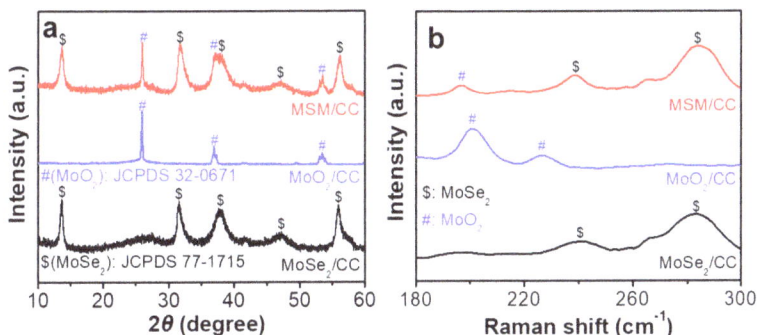

Figure 11. *(a) XRD spectra and (b) Raman scattering spectra of the electrocatalysts. Reproduced with permission from ref [84]. Copyright 2023, Elsevier.*

To investigate the surface chemistry and interaction between the two phases in the MSM/CC heterostructure, XPS analysis was performed. The Mo-$3d$ spectrum reveals that MoO_2/CC is composed of Mo^{4+} (233.2 and 230.0 eV) and Mo^{6+} (235.6 and 232.1 eV), as shown in **Figure 12a**. The presence of Mo^{6+} in MoO_2/CC can be attributed to the surface oxidation that occurs in ambient air, as previously reported. **Figure 12b-c** displays the Mo-$3d$ peaks in MSM/CC and $MoSe_2$/CC, which are composed of Mo^{4+}-$3d_{3/2}$ and $3d_{5/2}$ species. The Mo^{4+} peak in MSM/CC is observed to be 0.56 eV lower than that in MoO_2/CC but 0.20 eV higher than that in $MoSe_2$/CC.

Materials Research Forum LLC
https://doi.org/10.21741/9781644903070

The peak shifts reveal the presence of electronic interactions at the MSM heterojunction interface between MoO_2 and $MoSe_2$.

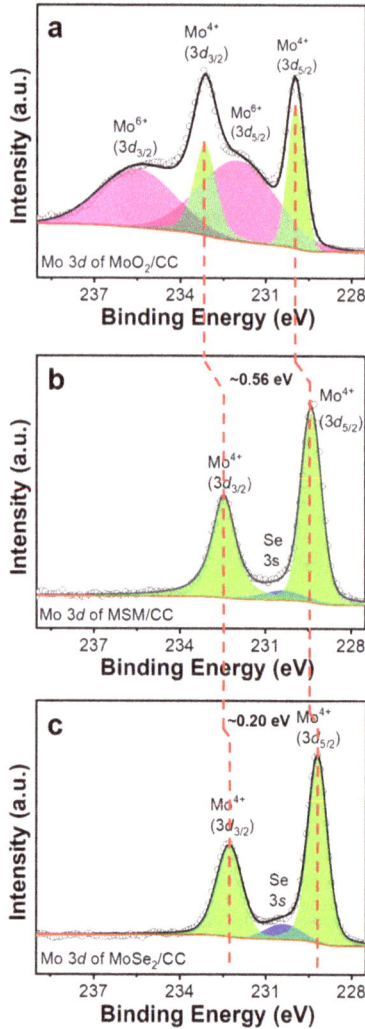

Figure 12. *Mo-3d fine spectra of (a) MoO₂/CC, (b) MSM/CC, (c) MoSe₂/CC. Reproduced with permission from ref [84]. Copyright 2023, Elsevier.*

The HER performance of MSM/CC was systematically investigated across a wide pH range of electrolytes, as shown in **Figure 13a-b.** The overpotentials at a current density of 10 mA cm^{-2} and Tafel slopes are summarized in **Table 1.** In comparison to the HER properties in an alkaline medium (1.0 M KOH), MSM/CC exhibited reduced overpotentials and Tafel slopes in the acidic medium (0.5 M H$_2$SO$_4$). In the neutral electrolyte (0.5 M Na$_2$SO$_4$), the performance of MSM/CC showed distinct characteristics compared to both acidic and alkaline media. It achieved an overpotential of 568 mV at a current density of 10 mA cm^{-2} and a Tafel slope of 174.6 mV dec^{-1}. Moreover, a comparison was made between MSM/CC and the commercial catalyst Pt/C in a neutral electrolyte. As depicted in **Figure 13c**, MSM/CC demonstrated a more rapid increase in current density and lower potential requirements when exceeding a current density of 45 mA cm^{-2} as compared to a commercial Pt/C catalyst. This highlights the promising electrocatalytic properties of MSM/CC, suggesting its potential as an alternative to platinum-based catalysts for the HER.

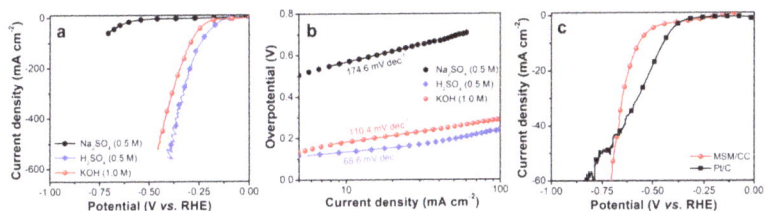

Figure 13. *(a) Polarization curves and (b) Tafel slopes of the MSM/CC electrocatalyst in different electrolytes, (c) Polarization curves of MSM/CC and commercial Pt/C electrocatalysts in 0.5 M Na$_2$SO$_4$. Reproduced with permission from ref [84]. Copyright 2023, Elsevier.*

Table 1. *Required overpotentials and Tafel slopes of the MSM/CC electrocatalyst to generate current densities of 10, 100 and 500 mA cm^{-2} in different electrolytes*

Electrolytes	η_{10} [mV]	η_{100} [mV]	η_{500} [mV]	Tafel slope [mV dec^{-1}]
1.0 M KOH	181	293	449	110.4
0.5 M H$_2$SO$_4$	132	240	382	68-6
0.5 M Na$_2$SO$_4$	568	--	--	174.6

The MSM/CC catalyst exhibits exceptional performance across a wide pH range, attributed to the synergistic interaction between its two phases within the heterogeneous structure. This unique architecture provides several advantages, including an increased number of active sites and the ability to finely tune the surface electronic structure, thereby facilitating catalytic reactions. The remarkable activity of the MoSe$_2$ phase complements the high stability and conductivity provided by MoO$_2$, resulting in effective hydrogen evolution capability under

Materials Research Forum LLC
https://doi.org/10.21741/9781644903070

acidic, alkaline, and neutral conditions. The combination of these two phases within the catalyst leads to enhanced catalytic performance and stability. This discovery not only broadens the application potential of catalysts but also provides valuable insights into the design of electrocatalytic materials with enhanced performance and stability across a wide pH range.

The work of Peng et al. has provided invaluable insights into the strategy of employing heterogeneous structures for the development construction of pH-universal catalysts. Simultaneously, notable advancements in this field have also been accomplished by Kim et al. [85]. They prepared the Co_3O_4-(Am-MOS_x) precursor through a hydrothermal method, followed by subjecting it to thermal treatment in an H_2 atmosphere. By leveraging the in situ generated H_2S gas from MoS_x, they successfully transformed Co_3O_4 into Co_9S_8 nanoparticles, thus creating a Co_9S_8-MoS_2 heterogeneous catalyst capable of operating across a wide pH range for hydrogen evolution.

To gain a deep understanding of the structure and electronic properties of the Co_9S_8-MoS_2 heterostructure, X-ray absorption near edge structure (XANES) and extended X-ray absorption fine structure (EXAFS) analyses were performed. **Figure 14a-b** shows the Co K-edge and Mo K-edge XANES spectra of Co_9S_8-MoS_2. In comparison to pure Co_9S_8, the Co K absorption edge of Co_9S_8-MoS_2 shifts to higher energy, indicating a slight increase in the Co valence state from $Co^{1.78+}$ to $Co^{1.78+\delta}$ due to strong electronic interactions with MoS_2. **Figure 14c** presents the EXAFS spectrum of Co_9S_8-MoS_2 within the 1.5-2.5 nm range, originating from the combined scattering of Co-S and Co-Co bonds, exhibiting enhanced intensity when compared to the pure-phase Co_9S_8. This increase is attributed to the enhanced chelation of cobalt species with the two-dimensional basal plane of MoS_2 in the Co_9S_8-MoS_2 heterostructure. The Mo K-edge EXAFS spectrum of Co_9S_8-MoS_2 in **Figure 14d** displays two peaks representing Mo-S (centered at 1.93 nm) and Mo-Mo (centered at 2.82 nm) bonds. Connecting cobalt nanoparticles to the MoS_2 layer results in a slight reduction in the intensity of the Mo-S peak, while the intensity of the Mo-Mo peak remains unchanged.

HR-TEM studies of Co_9S_8-MoS_2, as depicted in **Figure 15a**, reveal the conversion of Co_3O_4 NPs situated on the basal plane of MoS_2 within the Co_9S_8-MoS_2 heterostructure into Co_9S_8 NPs. The formation of Co_9S_8 nanoparticles on MoS_2, along with their respective (311), (111), (222), and (440) crystal facets, is confirmed by detailed high-resolution scans in **Figure 15b-c**. additionally, the corresponding Fast Fourier Transforms (FFTs) further validate the presence of these crystal facets in the Co_9S_8-MoS_2 structure.

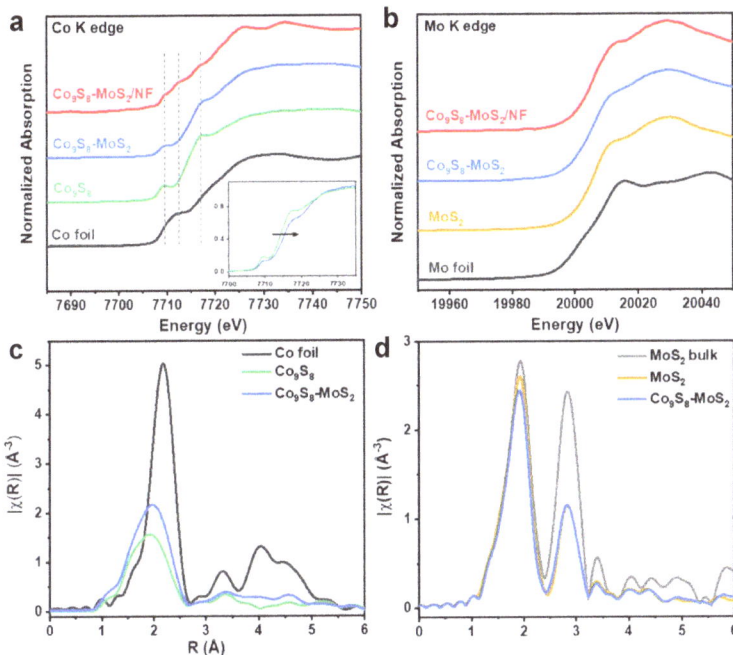

Figure 14. *XANES spectra of (a) Co K-edge and (b) Mo K-edge. Fourier-transform of k^2-weighted EXAFS functions for (c) Co K-edge and (d) Mo K-edge. Reproduced with permission from ref [85]. Copyright 2021, Elsevier.*

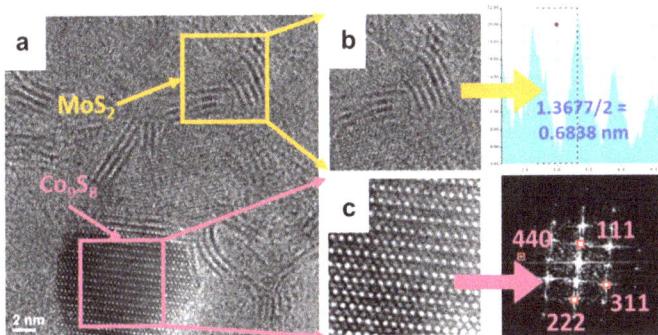

Figure 15. *(a) HR-TEM image, and (b) magnified image and corresponding line profile of MoS_2 in Co_9S_8-MoS_2. (c) Magnified image and FFT analysis for Co_9S_8 in Co_9S_8-MoS_2. Reproduced with permission from ref [85]. Copyright 2021, Elsevier.*

The electrochemical performance of Co_9S_8-MoS_2, Co_3O_4-(am-MoS_x), Co_9S_8, and MoS_2 was evaluated in different electrolytes, namely 1.0 M KOH (pH=14), 0.5 M H_2SO_4 (pH=0), and 0.1 M phosphate-buffered solution (PBS, pH=7), using a standard three-electrode system. The performance of these catalysts was compared to that of a commercial 20 wt% Pt/C catalyst. The polarization curves obtained in 1 M KOH (**Figure 16a**) demonstrate the superior HER performance of the Co_9S_8-MoS_2 catalyst compared to Co_3O_4-(am-MoS_x), pure Co_9S_8 and MoS_2. The overpotential required to generate a current density of 10 mA cm^{-2} is only 167 mV for Co_9S_8-MoS_2, which is significantly lower than that of pure MoS_2 (349 mV), Co_9S_8 (380 mV), and Co_3O_4-(am-MoS_x) (190 mV). This improved performance can be attributed to the activation of the inert MoS_2 substrate by Co_9S_8 and the improvement of electron cloud defects near Mo atoms at the heterogeneous interface. Moreover, the growth of Co_9S_8-MoS_2 on a highly conductive and porous nickel foam (NF) scaffold further promotes HER activity. In **Figure 16b**, the Tafel slopes for Co_9S_8-MoS_2/NF, Co_9S_8-MoS_2, Co_3O_4-(am-MoS_x), MoS_2, and Co_9S_8 are reported as 81.7, 85.5, 86.1, 86.1, and 91.8 mV dec^{-1}, respectively. These Tafel slopes indicate that the HER kinetics of Co_9S_8-MoS_2 and Co_9S_8-MoS_2/NF are faster compared to MoS_2 and Co_9S_8 electrocatalysts. All the catalysts follow the Volmer-Heyrovsky mechanism in an alkaline environment.

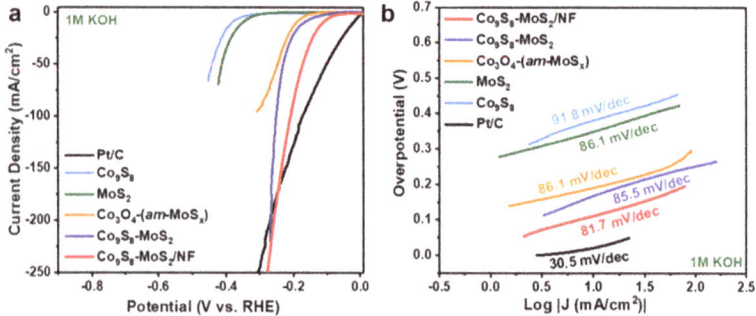

Figure 16. The electrochemical performance of the synthesized catalysts. (a) HER polarization curves and (b) Tafel plots in 1 M KOH solution. Reproduced with permission from ref [85]. Copyright 2021, Elsevier.

The electrochemical performance of all the synthesized electrocatalysts in HER was investigated in a neutral electrolyte (0.1 M PBS). The overpotential values for a current density of 10 mA cm^{-2} (η_{10}) for Co_9S_8-MoS_2/NF, Co_9S_8-MoS_2, Co_3O_4-(am-MoS_x), MoS_2, and Co_9S_8 are reported as 152.1, 161.6, 174.3, 398.4, and 370.7 mV, respectively. Notably, the performance of Co_9S_8-MoS_2/NF stands out, surpassing even that of the commercial Pt/C in the high-current region (>60 mA cm^{-2}). These results indicate that the inherent catalytic activity of MoS_2 can be significantly enhanced by constructing the heterostructure with Co_9S_8. **Figure 17a-b** showcases the linear sweep voltammetry (LSV) curves, as well as the η_{10} and η_{50} values of Co_9S_8-MoS_2 measured in the electrolytes with different pH values. These findings suggest that the Co_9S_8-MoS_2 heterostructure exhibits remarkable HER capability across a wide pH range.

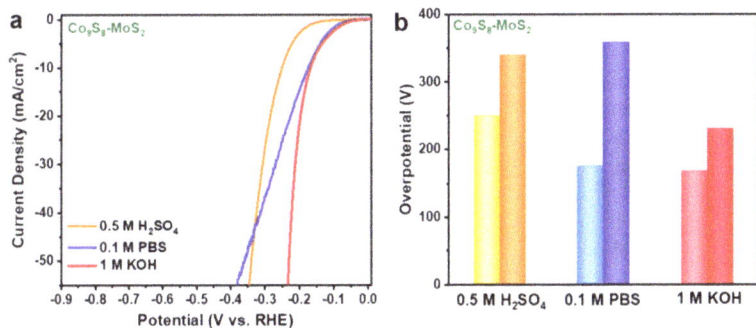

Figure 17. *(a) Polarization curves of Co_9S_8-MoS_2 in the electrolytes with different pH. (b) The η_{10} and η_{50} values in the electrolytes with different pH. Reproduced with permission from ref [85]. Copyright 2021, Elsevier.*

The construction of a catalyst with heterostructures is widely recognized as an effective strategy for achieving exceptional electrocatalytic performance across a wide pH range. The formation of heterostructures brings about several advantages, including optimized electronic structure and enhanced electron transfer efficiency. Furthermore, the synergistic effects between the different components within the heterostructure contribute to improved activity and stability of the catalyst. This design concept not only expands the applicability of electrocatalysts but also provides a viable solution for promoting electrochemical reactions under diverse pH conditions. By harnessing the benefits of heterostructures, researchers can develop catalysts with enhanced performance and broader functionality, enabling various electrochemical processes to be carried out efficiently under different pH environments.

3.3.   Alloying

Alloy materials are created by combining two or more distinct metals or non-metal elements through processes such as melting, dissolution, or mechanical alloying. These materials are designed to integrate the unique characteristics of different elements, resulting in materials with superior performance [86-88]. Alloy materials play a crucial role in the field of electrocatalysis due to their distinctive features, such as the ability to modulate crystal structures and modify electronic energy levels, making them highly relevant in electrochemical applications [89]. The remarkable attributes of alloy materials are particularly significant when it comes to the construction of electrocatalysts for wide pH range applications. The introduction of diverse elements through alloying allows for the modification of crystal structures and surface active sites of the catalyst, leading to enhanced catalytic activity and adaptability to both acidic and alkaline environments [90].   Additionally, alloy materials typically exhibit superior electrocatalytic stability by altering the electronic structure, resulting in a more stable crystal structure that mitigates catalyst deactivation [91, 92]. Therefore, alloy materials provide a powerful approach for the development of electrocatalysts with high activity and stability across diverse pH conditions.

Pt is known for its suitable hydrogen binding energy (HBE), which facilitates the adsorption of hydrogen intermediates and their subsequent interaction to generate H_2, making it an efficient catalyst for acidic HER. However, in alkaline conditions, Pt exhibits significantly lower activity, approximately 2-3 orders of magnitude lower, due to the differences in the HER processes. In alkaline environments, the HER involves slow H_2O dissociation rather than H^+ reduction. To overcome this limitation, Pt-M (where M represents other metal elements) alloys have been developed. These alloys enable individual reaction steps to occur on adjacent sites. The M sites within the alloy exhibit a strong binding affinity with *OH intermediates, facilitating efficient H_2O dissociation. Meanwhile, the neighboring Pt sites possess a suitable hydrogen binding affinity. This arrangement of dual-active sites in ordered intermetallic alloys leads to remarkable performance in the HER. The synergistic effect between the M and Pt sites accelerates both H_2O dissociation and H^+ reduction, resulting in outstanding catalytic activity across various pH conditions. By utilizing Pt-M alloys, researchers have been able to design catalysts with exceptional performance in the HER, demonstrating the potential of these materials to enable efficient hydrogen production at different pH values.

Kuang et al. [88] employed a highly porous carbon material with an ultra-high specific surface area, designated as NMCS-A, as a template to effectively adsorb the $[Fe(bpy)_3][PtCl_6]$ complex through electrostatic interactions. Subsequent in-situ decomposition of the NMCS-A/$[Fe(bpy)_3][PtCl_6]$ composite at high temperature resulted in the formation of Pt_3Fe alloy nanoparticles with a uniformly dispersed distribution. The structural characteristics of Pt_3Fe alloy nanoparticles were extensively investigated using XRD and TEM techniques. **Figure 18a** illustrates the XRD results, showing that the diffraction peak of Pt_3Fe at the (111) plane shifted positively to 40.5° compared to the standard Pt peak at 39.7°. This shift indicates the successful integration of Fe into the Pt lattice, confirming the presence of Fe with a smaller atomic radius. High-angle annular dark-field scanning transmission electron microscopy (HAADF-STEM) images in **Figure 18b** reveal representative Pt_3Fe alloy nanoparticles observed along the [001] direction. Each unit cell of the nanoparticles displays a periodic square array of Fe columns, surrounded by Pt columns at the edges and corners, consistent with the atomic arrangement of Pt_3M alloys. These findings provide clear confirmation of the successful synthesis of ordered Pt_3Fe alloy nanoparticles with a well-defined structure.

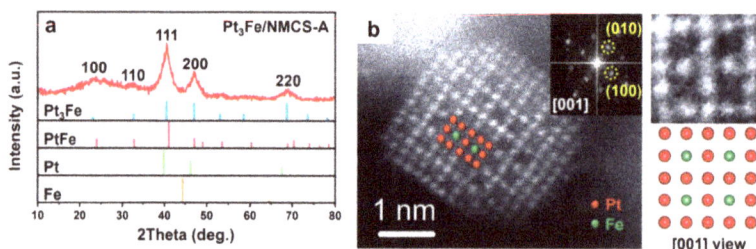

Figure 18. (a) XRD patterns of the Pt_3Fe/NMCS-A. (b) Atomic-resolution HAADF-STEM image and simulated atomic model of the Pt_3Fe alloy nanoparticle (inset is the SAED pattern). Reproduced with permission from ref [88]. Copyright 2023, Wiley-VCH.

The surface chemistry of the Pt$_3$Fe/NMCS-A catalyst was thoroughly investigated using XPS. Notably, the XPS spectra of Pt-4f and Fe-2p provided valuable insights into the electronic interaction between Pt and Fe in the Pt$_3$Fe alloy, as illustrated in **Figure 19a-b**. In the XPS analysis, it was observed that the binding energy of Pt-4f in Pt$_3$Fe/NMCS-A exhibited a shift towards lower values compared to that of Pt/NMCS-A. Simultaneously, the binding energy of Fe-2p in Pt$_3$Fe/NMCS-A showed a shift towards higher values in comparison to Fe/NMCS-A. This significant shift in the XPS spectra indicated a pronounced electronic interaction between Pt and Fe within the Pt$_3$Fe alloy. The negative shift in the Pt-4f spectrum indicated a decrease in the d-value of the Pt-5d orbital, signifying the migration of electrons from Fe to Pt atoms. The robust electronic interaction between Pt-5d and Fe-3d orbitals induced electron redistribution, playing a crucial role in enhancing the HER activity of Pt$_3$Fe/NMCS-A in pH-universal electrolytes.

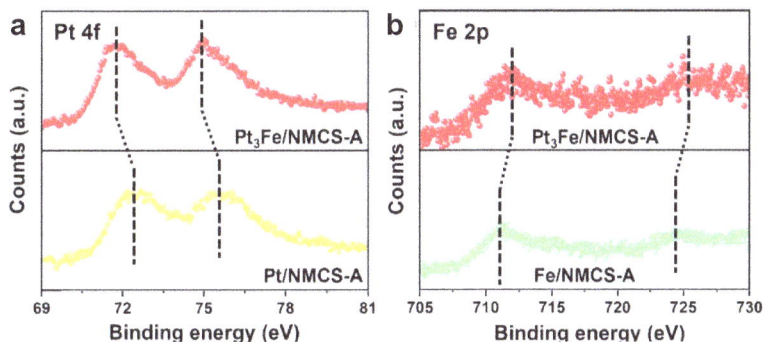

Figure 19. *(a) Pt-4f and (b) Fe-2p XPS spectra of the Pt₃Fe/NMCS-A, Pt/NMCS-A, and Fe/NMCS-A. Reproduced with permission from ref [88]. Copyright 2023, Wiley-VCH.*

A comprehensive evaluation of the HER activity of Pt$_3$Fe/NMCS-A was conducted in acidic, alkaline, and neutral electrolytes, as illustrated in **Figure 20**. In 0.5 M H$_2$SO$_4$ solution, Pt$_3$Fe/NMCS-A exhibited a remarkably low overpotential of only 13 mV to generate the current density of 10 mA cm^{-2}. Notably, the Tafel slope measured for Pt$_3$Fe/NMCS-A was as small as 21 mV dec^{-1}, indicating a Volmer-Tafel pathway for H$_2$ evolution (H$^+$ + e$^-$ + * → H*; H* + H* → H$_2$ ↑). The small Tafel slope reveals the abundance of active centers in Pt$_3$Fe/NMCS-A, providing robust support for its ultrafast reaction kinetics. In alkaline and neutral electrolytes, Pt$_3$Fe/NMCS-A also exhibited outstanding HER activity. In 1 M KOH, the overpotential was only 29 mV to achieve a current density of 10 mA cm^{-2}, with a Tafel slope of 50 mV dec^{-1}. This indicates that Pt$_3$Fe/NMCS-A promotes H$_2$O dissociation and facilitates reaction kinetics. Furthermore, in a 1 M PBS solution, Pt$_3$Fe/NMCS-A exhibited an overpotential of 48 mV to attain a current density of 10 mA cm^{-2}, with a Tafel slope of 58 mV dec^{-1}. These remarkable results highlight the exceptional HER performance of Pt$_3$Fe/NMCS-A and underscore the potential as a highly effective catalyst for hydrogen evolution across a broad range of pH environments.

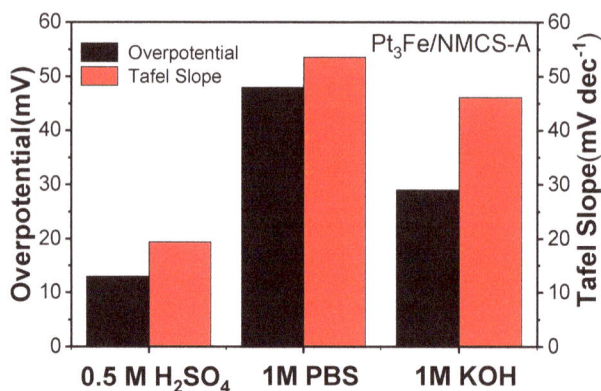

Figure 20. *The overpotential and Tafel slope of Pt₃Fe/NMCS-A measured in various electrolytes with different pH values. Reproduced with permission from ref [88]. Copyright 2023, Wiley-VCH.*

To gain a deeper understanding of the strong electronic interactions between Pt and Fe and to elucidate the outstanding HER activity of the Pt₃Fe/NMCS-A catalyst in universal pH conditions, density functional theory (DFT) calculations were employed for further analysis. Models of Pt₃Fe alloy, Pt metal, and Fe metal were used to represent Pt₃Fe/NMCS-A, Pt/NMCS-A, and Fe/NMCS-A, respectively. Taking the acidic HER as an example, the hydrogen evolution process on the Pt₃Fe/NMCS-A catalyst was found to follow the Volmer-Tafel pathway, as indicated by the small Tafel slope of 21 mV dec^{-1}. The mechanism involves the adsorption of H$^+$ cations on Pt sites through the Volmer step, followed by the combination of H* intermediates with the adsorbed H$^+$ through the Tafel step, resulting in the generation of H₂. This mechanism is schematically illustrated in **Figure 21**.

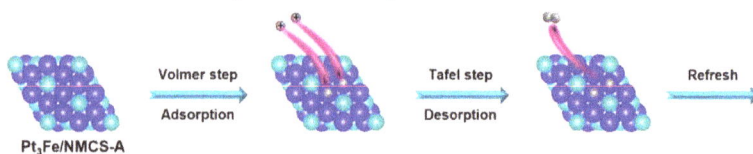

Figure 21. *HER process on the Pt₃Fe/NMCS-A catalyst in acidic condition. Reproduced with permission from ref [88]. Copyright 2023, Wiley-VCH.*

The calculated Gibbs free energy ΔG_{H*} values for different adsorption configurations on Pt₃Fe/NMCS-A (H adsorbed on Pt), Pt₃Fe/NMCS-A (H adsorbed on Fe), Pt/NMCS-A, and Fe/NMCS-A are as follows: −0.18, 0.20, −0.21, and 0.56 eV, respectively, as shown in **Figure 22a**. Positive ΔG_{H*} values indicate weak adsorption of the H* intermediate on the Fe site, which is unfavorable for subsequent H₂ generation. This suggests that Pt₃Fe/NMCS-A exhibits higher HER activity at the Pt sites. Bader charge analysis further revealed strong electronic interactions between Pt and Fe, demonstrating the distinctive electron-accepting characteristics of Pt. The cumulative electron number on Pt (n) increased from 0.19 (without H adsorption) to 0.23 e$^-$ after the H adsorption, as shown in **Figure 22b**. Additionally, the electron structure of Pt-$5d$ orbitals was modulated by Fe atoms, as shown in **Figure 22c**. The Pt$_3$Fe/NMCS-A exhibited lower ϵ_d values of Pt-$5d$ orbitals (-2.45 eV) compared to Pt/NMCS-A (-2.09 eV). Notably, the ϵ_d values of Pt-$5d$ orbitals shifted towards more negative values (-2.46 eV) in Pt$_3$Fe/NMCS-A, further confirming the promoted electron redistribution and weakened Pt-H interactions in Pt$_3$Fe alloy. Furthermore, the co-adsorption energies ($E_{(H^*+^*OH)}$) of H* and *OH intermediates for different materials are illustrated in **Figure 22d**. Pt$_3$Fe/NMCS-A exhibited a more negative $E_{(H^*+^*OH)}$ value (-0.97 eV) compared to Pt/NMCS-A and Fe/NMCS-A. This indicates that the alloying of Fe with Pt significantly promotes H$_2$O dissociation and facilitates the adsorption of H* and *OH intermediates on Pt and Fe sites, respectively. Pt$_3$Fe/NMCS-A possesses a strong OH connection, thereby reducing the barrier for the dissociation of H$_2$O and enhancing HER activity in alkaline and neutral electrolytes.

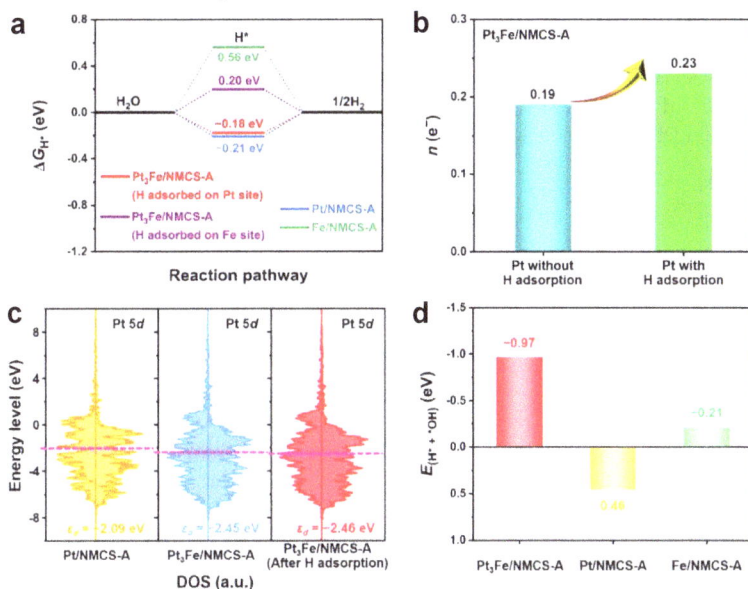

Figure 22. (a) ΔG_{H^*} of the Pt$_3$Fe/NMCS-A, Pt/NMCS-A, and Fe/NMCS-A. (b) n of the Pt$_3$Fe/NMCS-A before and after H adsorption. (c) Pt-5d orbitals and ϵd of the Pt/NMCS-A, Pt$_3$Fe/NMCS-A, and Pt$_3$Fe/NMCS-A after H adsorption. (d) $E_{(H^*+^*OH)}$ of the Pt$_3$Fe/NMCS-A and Pt/NMCS-A. Reproduced with permission from ref [88]. Copyright 2023, Wiley-VCH.

The present study has successfully synthesized the Pt$_3$Fe/NMCS-A catalyst, which has demonstrated remarkable electrocatalytic performance in pH-universal HER application. The alloying of Pt and Fe in the Pt$_3$Fe/NMCS-A catalyst leads to a reduction in the ϵ_d values of the

Pt-5d orbitals, resulting in decreased adsorption energy of H intermediates on the Pt$_3$Fe/NMCS-A catalyst surface. Moreover, the presence of shared adsorption sites for H* and *OH intermediates on Pt and Fe enhances the electrocatalytic performance of Pt$_3$Fe/NMCS-A. this allows for lower energy barriers during the dissociation of H$_2$O into H intermediates, facilitating efficient H$_2$ generation under both alkaline and neutral conditions. The findings highlight the significant application potential of alloy catalysts for electrocatalytic hydrogen production across a wide pH range.

Feng et al. [93] have made significant progress in advancing the understanding of alloy materials for pH-universal hydrogen production. They proposed a rational modification strategy that involves both bulk and surface electronic structure modification of a Ru-based catalyst using carbon quantum dots (CDs) as the carrier, leading to an efficient and pH-adaptable performance in HER. The composition and crystal structure of the as-prepared catalysts were investigated using XRD analysis. The XRD signals of the Ru@CDs catalyst, as shown in **Figure 23a-b**, exhibit good agreement with the standard hexagonal Ru (JCPDS card No.06-0663). The diffraction peaks appear broad, indicating the formation of small-sized hexagonal Ru particles. Notably, these diffraction peaks slightly shift to higher angles compared to those of pure hexagonal Ru and are positioned between the locations of standard hexagonal Ru and hexagonal Co (JCPDS card No.05-0727) in the case of the RuCo@CDs catalyst. This observation indicates the successful fabrication of highly crystalline hexagonal alloyed RuCo nanoparticles.

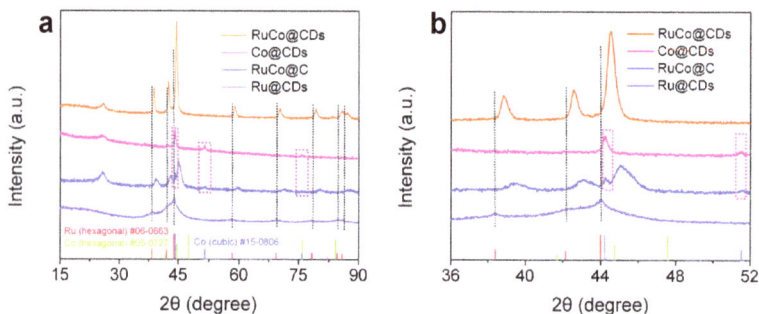

Figure 23. *(a) XRD patterns and (b) magnified pattern of the RuCo@CDs, Co@CDs, RuCo@C and Ru@CDs catalysts. Reproduced with permission from ref [93]. Copyright 2020, the Royal Society of Chemistry.*

XPS was employed to study the oxidation states and electronic structure of the catalysts. The Ru-3p spectrum of the RuCo@CDs catalyst (**Figure 24a**) displays the presence of two chemical states, which are metallic Ru (Ru^0) peaks at 461.7 eV and 483.8 eV and oxidized Ru species (Ru^{4+}) peaks at 463.9 eV and 485.4 eV. Notably, the Ru^{4+} peaks in the RuCo@CDs catalyst shift towards lower binding energy compared to those in the Ru@CDs catalyst, while the Ru^0 peak remains unchanged. This shift indicates electron transfer from Co to Ru, which can be attributed to the higher electronegativity of Ru. The Co-2p spectrum in **Figure 24b** shows that the high-valence Co species in RuCo@CDs shift towards higher binding energy compared to those in Co@CDs, indicating a significant change in the electron structure after alloying. These observations suggest notable differences in the crystal and electronic structure of the RuCo@CDs catalyst. Specifically, the incorporation of Co into the Ru lattice and the interface incorporation with CDs lead to strong electronic coupling interactions among Co, Ru, and CDs. This establishes an electron transfer pathway for Co-Ru-CDs, resulting in an increased content of high oxidation states.

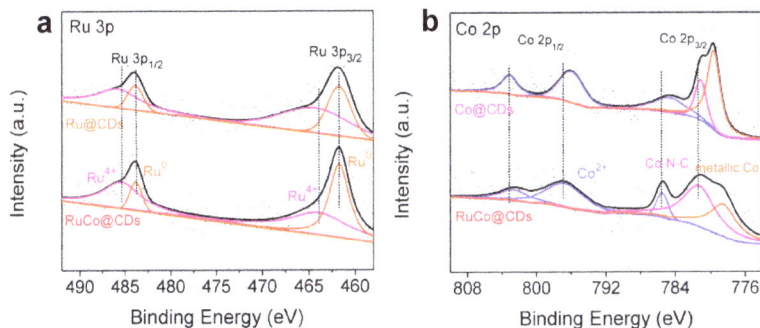

Figure 24. *The high-resolution XPS spectra of (a) Ru-3p and (b) Co-2p of the RuCo@CDs, Ru@CDs and Co@CDs catalysts. Reproduced with permission from ref [93]. Copyright 2020, the Royal Society of Chemistry.*

The HER activity of the RuCo@CDs catalyst was examined using a three-electrode system in electrolytes with varying pH values. For comparison, the polarization curves were determined under the same conditions for Co@CDs, Ru@CDs, RuCo@C, and Pt/C catalysts. The results show that the RuCo@CDs catalyst exhibits the highest HER activity in 1.0 M KOH electrolyte, requiring a low overpotential of only 11 mV to produce a current density of 10 mA cm^{-2}. This overpotential is significantly lower than that of Ru@CDs (70 mV), RuCo@C (50 mV), Co@CDs (168 mV), and commercial Pt/C catalyst (34 mV), as shown in **Figure 25a**. These findings indicate that the intrinsic HER active sites are associated with the Ru sites in RuCo@CDs. The substitution of Co for Ru and the incorporation of CDs have a synergistic effect, accelerating the kinetics of water dissociation. The polarization curves measured in 0.5 M H$_2$SO$_4$ and 1.0 M PBS show that the RuCo@CDs catalyst exhibits optimal performance, with overpotentials of 51 mV and 67 mV, respectively, to afford a current density of 10 mA cm^{-2}. These overpotentials significantly outperform the other catalysts, as indicated in **Figure 25b-c**. Consistently, the performance order of the catalysts in acidic and neutral media is RuCo@CDs → RuCo@C → Ru@CDs → Co@CDs (**Figure 25d**). These results highlight the outstanding performance of the RuCo@CDs catalyst across different pH conditions.

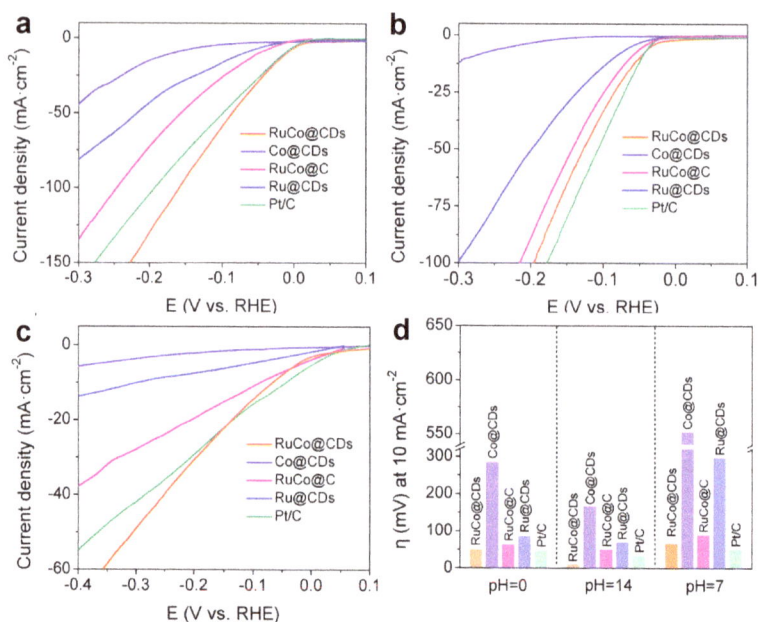

Figure 25. *The polarization curves in (a) 1.0 M KOH, (b) 0.5 M H₂SO₄, (c) 1.0 M PBS solution, and (d) Overpotential comparison of the RuCo@CDs, Co@CDs, RuCo@C, Ru@CDs, and Pt/C catalysts. Reproduced with permission from ref [93]. Copyright 2020, the Royal Society of Chemistry.*

The DFT calculations were conducted to gain further insights into the excellent electrocatalytic activity of the RuCo@CDs catalyst. The crystal structure and energy of the RuCo alloy bulk structure were studied using the pure hexagonal structure of metal Ru as the reference. To simulate the experimentally synthesized RuCo@CDs composite catalyst, a theoretical model of RuCo clusters was constructed based on the RuCo-(002) structure and deposited on a single-layer graphene oxide doped with nitrogen, represented as RuCo@GNO (**Figure 26a**). The nitrogen-doped graphene oxide represents the incorporated CDs with exposed nitrogen and oxygen edge sites in a graphene-like structure. The alloying effects of Ru and Co were initially studied using the Ru-(002) and RuCo-(002) models. The dissociation energy (E_d) of H_2O on the top site (T_{Ru} site) of the original Ru-(002) surface was calculated to be 0.627 eV, as shown in **Figure 26b**. Additionally, the E_d values of H_2O on the Co-doped Ru-(002) surface were also calculated to explore the doping effect of Co. Thus, the E_d values of H_2O on the T_{Ru} and T_{Co} sites on the RuCo-(002) surface were determined to be 0.475 and 0.316 eV, respectively (**Figure 26b**). Both values are smaller than that on the original Ru-(002) surface (0.627 eV). The result indicates that RuCo alloying is favorable for promoting water dissociation, thereby enhancing the electrocatalytic activity in various electrolytes.

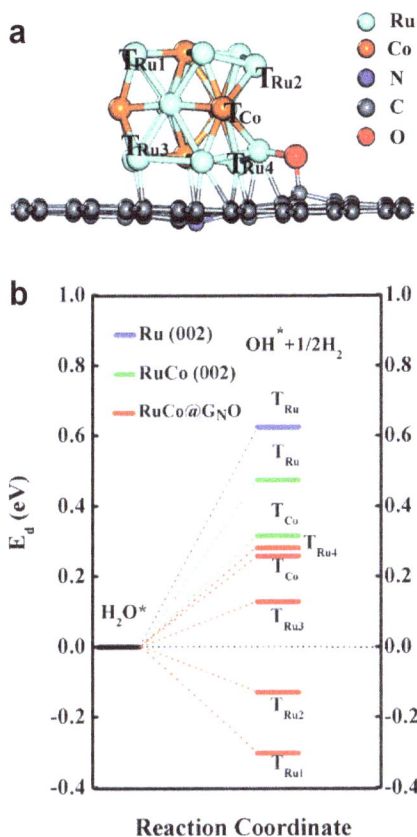

Figure 26. (a) Side view of the RuCo@GNO model, where the obtained adsorption sites are presented. (b) The calculated dissociation energy (E_d) of water at different adsorption sites on the Ru (002), RuCo (002), and RuCo@GNO surfaces. Reproduced with permission from ref [93]. Copyright 2020, the Royal Society of Chemistry.

The strategy of alloying in constructing HER catalysts offers significant advantages for applications across a wide pH range. This approach enhances the overall catalytic performance by leveraging synergistic interactions and electronic tuning. The unique characteristics of alloy catalysts enable them to adapt to different pH conditions, showcasing excellent catalytic activity in acidic, alkaline, and neutral environments. Consequently, alloy catalysts provide a feasible solution for diverse applications.

4.   Conclusion and Prospects

In conclusion, the development of catalysts capable of efficiently driving the HER across a wide pH range is of paramount importance for diverse applications in energy conversion and storage. The exceptional performance of these catalysts in HER across a wide pH range stems from their ability to expedite reaction kinetics and facilitate efficient hydrogen evolution.   The implementation of specific construction strategies, such as doping, heterostructure construction, and alloying, has yielded substantial advancements and promising prospects in this field.

However, persist in the design and synthesis of HER catalysts suitable for wide pH range applications.   Attaining long-term stability, minimizing catalyst degradation, and optimizing catalyst loading are key areas that demand further attention.   Furthermore, comprehending the precise mechanisms governing catalytic processes and identifying active sites within catalyst structures continue to be ongoing research pursuits.   In the future development of efficient HER electrocatalysts across a wide pH range, several key directions should be focused on.

Firstly, efforts can be directed towards further enhancing the catalytic activity and stability of the catalysts to adapt to more extreme or complex acidic and alkaline conditions.   This can be achieved by optimizing the electronic conductivity of the catalyst, increasing the density of surface reaction sites, and designing more stable catalyst structures.   These improvements will result in higher efficiency of hydrogen evolution across a broad range of pH values.

Secondly, sustainability and ease of material preparation are crucial research priorities.   It is important to identify and design catalysts using abundant and cost-effective raw materials. Coupled with the development of straightforward and feasible synthetic methods, this approach will contribute to cost reduction in production and promote the practical application of HER electrocatalysts across a wide pH range.

Lastly, the integration of hydrogen energy with other energy conversion and storage technologies holds promising prospects.   By applying efficient HER electrocatalysts across a wide pH range to renewable energy sources, such as solar and wind energy, a more comprehensive energy system can be established.   This integration has the potential to enhance overall energy utilization efficiency and contribute to the development of a sustainable and clean energy infrastructure.

References

[1]   J. Mahmood, F. Li, S.-M. Jung, M.S. Okyay, I. Ahmad, S.-J. Kim, N. Park, H.Y. Jeong, J.-B. Baek, An efficient and pH-universal ruthenium-based catalyst for the hydrogen evolution reaction, Nat. Nanotechnol., 12 (2017) 441-446. https://doi.org/10.1038/nnano.2016.304

[2]   Y. Zheng, Y. Jiao, M. Jaroniec, S.Z. Qiao, Advancing the Electrochemistry of the Hydrogen-Evolution Reaction through Combining Experiment and Theory, Angew. Chem. Int. Ed., 54 (2015) 52-65. https://doi.org/10.1002/anie.201407031

[3]   S.M. Abu, M.A. Hannan, P.J. Ker, M. Mansor, S.K. Tiong, T.M.I. Mahlia, Recent progress in electrolyser control technologies for hydrogen energy production: A patent

landscape analysis and technology updates, J. Energy Storage, 72 (2023) 108773. https://doi.org/10.1016/j.est.2023.108773

[4] D. Meng, L. Wei, J. Shi, Q. Jiang, J. Tang, A review of enhanced electrocatalytic composites hydrogen/oxygen evolution based on quantum dot, J. Ind. Eng. Chem., 121 (2023) 27-39. https://doi.org/10.1016/j.jiec.2023.01.014

[5] Y. Shi, B. Zhang, Recent advances in transition metal phosphide nanomaterials: Synthesis and applications in hydrogen evolution reaction, Chem. Soc. Rev., 45 (2016) 1529-1541. https://doi.org/10.1039/C5CS00434A

[6] X. Tian, P. Zhao, W. Sheng, Hydrogen Evolution and Oxidation: Mechanistic Studies and Material Advances, Adv. Mater., 31 (2019) 1808066. https://doi.org/10.1002/adma.201808066

[7] G. Zhao, K. Rui, S.X. Dou, W. Sun, Heterostructures for electrochemical hydrogen evolution reaction: A review, Adv. Funct. Mater., 28 (2018) 1803291. https://doi.org/10.1002/adfm.201803291

[8] X.-Z. Fan, Q.-Q. Pang, Recent Research Advances in Ruthenium-Based Electrocatalysts for Water Electrolysis Across the pH-Universal Conditions, Energy Technol., 10 (2022) 2200655. https://doi.org/10.1002/ente.202200655

[9] J. Staszak-Jirkovský, C.D. Malliakas, P.P. Lopes, N. Danilovic, S.S. Kota, K.-C. Chang, B. Genorio, D. Strmcnik, V.R. Stamenkovic, M.G. Kanatzidis, Design of active and stable Co-Mo-S_x chalcogels as pH-universal catalysts for the hydrogen evolution reaction, Nat. Mater., 15 (2016) 197-203. https://doi.org/10.1038/nmat4481

[10] T. Liu, P. Li, N. Yao, G. Cheng, S. Chen, W. Luo, Y. Yin, CoP-Doped MOF-Based Electrocatalyst for pH-Universal Hydrogen Evolution Reaction, Angew. Chem. Int. Ed., 58 (2019) 4679-4684. https://doi.org/10.1002/anie.201901409

[11] L. Su, J. Chen, F. Yang, P. Li, Y. Jin, W. Luo, S. Chen, Electric-Double-Layer Origin of the Kinetic pH Effect of Hydrogen Electrocatalysis Revealed by a Universal Hydroxide Adsorption-Dependent Inflection-Point Behavior, J. Am. Chem. Soc., 145 (2023) 12051-12058. https://doi.org/10.1021/jacs.3c01164

[12] Z. Pu, S. Wei, Z. Chen, S. Mu, Flexible molybdenum phosphide nanosheet array electrodes for hydrogen evolution reaction in a wide pH range, Appl. Catal. B-Environ., 196 (2016) 193-198. https://doi.org/10.1016/j.apcatb.2016.05.027

[13] K. Singh, B.-S. Lou, J.-L. Her, S.-T. Pang, T.-M. Pan, Super Nernstian pH response and enzyme-free detection of glucose using sol-gel derived RuO_x on PET flexible-based extended-gate field-effect transistor, Sensor. Actuat. B-Chem., 298 (2019) 126837. https://doi.org/10.1016/j.snb.2019.126837

[14] G.J. Samuels, T.J. Meyer, An electrode-supported oxidation catalyst based on ruthenium (IV). pH" encapsulation" in a polymer film, J. Am. Chem. Soc., 103 (1981) 307-312. https://doi.org/10.1021/ja00392a010

[15] J. Wu, W. Zheng, Y. Chen, Factors affecting the cathode/electrolyte interfacial pH

change during water reduction: A simulation study, Int. J. Hydrogen Energ., 47 (2022) 18597-18605. https://doi.org/10.1016/j.ijhydene.2022.04.035

[16] B. Nyvad, N. Takahashi, Integrated hypothesis of dental caries and periodontal diseases, J. Oral Microbiol., 12 (2020) 1710953. https://doi.org/10.1080/20002297.2019.1710953

[17] W. Xianhong, S. Zhou, Z. Wang, J. Liu, W. Pei, Y. Pengju, J. Zhao, J. Qiu, Engineering Multifunctional Collaborative Catalytic Interface Enabling Efficient Hydrogen Evolution in All pH Range and Seawater, Adv. Energy Mater., 9 (2019) 1901333. https://doi.org/10.1002/aenm.201901333

[18] Z. Zhou, Z. Pei, L. Wei, S. Zhao, X. Jian, Y. Chen, Electrocatalytic hydrogen evolution under neutral pH conditions: Current understandings, recent advances, and future prospects, Energy Environ. Sci., 13 (2020) 3185-3206. https://doi.org/10.1039/D0EE01856B

[19] A. Han, H. Chen, H. Zhang, Z. Sun, P. Du, Ternary metal phosphide nanosheets as a highly efficient electrocatalyst for water reduction to hydrogen over a wide pH range from 0 to 14, J. Mater. Chem. A, 4 (2016) 10195-10202. https://doi.org/10.1039/C6TA02297A

[20] W.S. Chai, J.Y. Cheun, P.S. Kumar, M. Mubashir, Z. Majeed, F. Banat, S.-H. Ho, P.L. Show, A review on conventional and novel materials towards heavy metal adsorption in wastewater treatment application, J. Clean. Prod., 296 (2021) 126589. https://doi.org/10.1016/j.jclepro.2021.126589

[21] J.N. Hansen, H. Prats, K.K. Toudahl, N. Mørch Secher, K. Chan, J. Kibsgaard, I. Chorkendorff, Is There Anything Better than Pt for HER?, ACS Energy Lett., 6 (2021) 1175-1180. https://doi.org/10.1021/acsenergylett.1c00246

[22] Q. Xu, J. Zhang, H. Zhang, L. Zhang, L. Chen, Y. Hu, H. Jiang, C. Li, Atomic heterointerface engineering overcomes the activity limitation of electrocatalysts and promises highly-efficient alkaline water splitting, Energy Environ. Sci., 14 (2021) 5228-5259. https://doi.org/10.1039/D1EE02105B

[23] F. Liu, C. Shi, X. Guo, Z. He, L. Pan, Z.F. Huang, X. Zhang, J.J. Zou, Rational design of better hydrogen evolution electrocatalysts for water splitting: A review, Adv. Sci., 9 (2022) 2200307. https://doi.org/10.1002/advs.202200307

[24] C. Tang, M.-M. Titirici, Q. Zhang, A review of nanocarbons in energy electrocatalysis: Multifunctional substrates and highly active sites, J. Energy Chem., 26 (2017) 1077-1093. https://doi.org/10.1016/j.jechem.2017.08.008

[25] Z. Pu, T. Liu, G. Zhang, Z. Chen, D.S. Li, N. Chen, W. Chen, Z. Chen, S. Sun, General synthesis of transition-metal-based carbon-group intermetallic catalysts for efficient electrocatalytic hydrogen evolution in wide pH range, Adv. Energy Mater., 12 (2022) 2200293. https://doi.org/10.1002/aenm.202200293

[26] R. Du, W. Jin, R. Hübner, L. Zhou, Y. Hu, A. Eychmüller, Engineering Multimetallic Aerogels for pH-Universal HER and ORR Electrocatalysis, Adv. Energy Mater., 10

(2020) 1903857. https://doi.org/10.1002/aenm.201903857

[27]   W. Zhai, Y. Ma, D. Chen, J.C. Ho, Z. Dai, Y. Qu, Recent progress on the long-term stability of hydrogen evolution reaction electrocatalysts, InfoMat, 4 (2022) e12357. https://doi.org/10.1002/inf2.12357

[28]   M. Li, M. Luo, Z. Xia, Y. Yang, Y. Huang, D. Wu, Y. Sun, C. Li, Y. Chao, W. Yang, W. Yang, Y. Yu, S. Guo, Modulating the surface segregation of PdCuRu nanocrystals for enhanced all-pH hydrogen evolution electrocatalysis, J. Mater. Chem. A, 7 (2019) 20151-20157. https://doi.org/10.1039/C9TA06861A

[29]   X. Peng, C. Pi, X. Zhang, S. Li, K. Huo, P. Chu, Recent Progress of Transition Metal Nitrides for Efficient Electrocatalytic Water Splitting, Sustain. Energ. Fuels, 3 (2019) 366-381. https://doi.org/10.1039/C8SE00525G

[30]   Y. Men, P. Li, F. Yang, G. Cheng, S. Chen, W. Luo, Nitrogen-doped CoP as robust electrocatalyst for high-efficiency pH-universal hydrogen evolution reaction, Appl. Catal. B-Environ., 253 (2019) 21-27. https://doi.org/10.1016/j.apcatb.2019.04.038

[31]   X. Tian, P. Zhao, W. Sheng, Hydrogen evolution and oxidation: mechanistic studies and material advances, Adv. Mater., 31 (2019) 1808066. https://doi.org/10.1002/adma.201808066

[32]   Y. Wang, Q. Lu, F. Li, D. Guan, Y. Bu, Atomic-scale configuration enables fast hydrogen migration for electrocatalysis of acidic hydrogen evolution, Adv. Funct. Mater., 33 (2023) 2213523. https://doi.org/10.1002/adfm.202213523

[33]   A. Lasia, Mechanism and kinetics of the hydrogen evolution reaction, Int. J. Hydrogen Energ., 44 (2019) 19484-19518. https://doi.org/10.1016/j.ijhydene.2019.05.183

[34]   L. Zhang, Y. Zheng, J. Wang, Y. Geng, B. Zhang, J. He, J. Xue, T. Frauenheim, M. Li, Ni/Mo Bimetallic-Oxide-Derived Heterointerface-Rich Sulfide Nanosheets with Co-Doping for Efficient Alkaline Hydrogen Evolution by Boosting Volmer Reaction, Small, 17 (2021) 2006730. https://doi.org/10.1002/smll.202006730

[35]   F. Bao, E. Kemppainen, I. Dorbandt, R. Bors, F. Xi, R. Schlatmann, R. van de Krol, S. Calnan, Understanding the hydrogen evolution reaction kinetics of electrodeposited nickel-molybdenum in acidic, near-neutral, and alkaline conditions, ChemElectroChem, 8 (2021) 195-208. https://doi.org/10.1002/celc.202001436

[36]   M. Lao, P. Li, Y. Jiang, H. Pan, S.X. Dou, W. Sun, From fundamentals and theories to heterostructured electrocatalyst design: An in-depth understanding of alkaline hydrogen evolution reaction, Nano Energy, 98 (2022) 107231. https://doi.org/10.1016/j.nanoen.2022.107231

[37]   C. Hu, L. Zhang, J. Gong, Recent progress made in the mechanism comprehension and design of electrocatalysts for alkaline water splitting, Energy Environ. Sci., 12 (2019) 2620-2645. https://doi.org/10.1039/C9EE01202H

[38]   J. Huang, J. Han, T. Wu, K. Feng, T. Yao, X. Wang, S. Liu, J. Zhong, Z. Zhang, Y. Zhang, B. Song, Boosting Hydrogen Transfer during Volmer Reaction at Oxides/Metal

Nanocomposites for Efficient Alkaline Hydrogen Evolution, ACS Energy Lett., 4 (2019) 3002-3010. https://doi.org/10.1021/acsenergylett.9b02359

[39]   F.-T. Tsai, Y.-T. Deng, C.-W. Pao, J.-L. Chen, J.-F. Lee, K.-T. Lai, W.-F. Liaw, The HER/OER mechanistic study of an FeCoNi-based electrocatalyst for alkaline water splitting, J. Mater. Chem. A, 8 (2020) 9939-9950. https://doi.org/10.1039/D0TA01877E

[40]   S. Anantharaj, S. Noda, V.R. Jothi, S. Yi, M. Driess, P.W. Menezes, Strategies and perspectives to catch the missing pieces in energy-efficient hydrogen evolution reaction in alkaline media, Angew. Chem. Int. Ed., 60 (2021) 18981-19006. https://doi.org/10.1002/anie.202015738

[41]   Z. Zhou, Z. Pei, L. Wei, S.-L. Zhao, X. Jian, Y. Chen, Electrocatalytic Hydrogen Evolution under Neutral pH Conditions: Current Understandings, Recent Advances, and Future Prospects, Energy Environ. Sci., 13 (2020) 3185-3206. https://doi.org/10.1039/D0EE01856B

[42]   S. Xie, Y. Yan, S. Lai, J. He, Z. Liu, B. Gao, M. Javanbakht, X. Peng, P.K. Chu, Ni^{3+}-enriched nickel-based electrocatalysts for superior electrocatalytic water oxidation, Appl. Surf. Sci., 605 (2022) 154743. https://doi.org/10.1016/j.apsusc.2022.154743

[43]   X. Wu, S. Zhou, Z. Wang, J. Liu, W. Pei, P. Yang, J. Zhao, J. Qiu, Engineering multifunctional collaborative catalytic interface enabling efficient hydrogen evolution in all pH range and seawater, Adv. Energy Mater., 9 (2019) 1901333. https://doi.org/10.1002/aenm.201901333

[44]   S. Yuan, Z. Pu, H. Zhou, J. Yu, I.S. Amiinu, J. Zhu, Q. Liang, J. Yang, D. He, Z. Hu, A universal synthesis strategy for single atom dispersed cobalt/metal clusters heterostructure boosting hydrogen evolution catalysis at all pH values, Nano Energy, 59 (2019) 472-480. https://doi.org/10.1016/j.nanoen.2019.02.062

[45]   Y. Cao, Roadmap and direction toward high-performance MoS_2 hydrogen evolution catalysts, ACS Nano, 15 (2021) 11014-11039. https://doi.org/10.1021/acsnano.1c01879

[46]   T. Liang, Y. Liu, Y. Cheng, F. Ma, Z. Dai, Scalable Synthesis of a MoS_2/Black Phosphorus Heterostructure for pH-Universal Hydrogen Evolution Catalysis, ChemCatChem, 12 (2020) 2840-2848. https://doi.org/10.1002/cctc.202000139

[47]   Z. Sun, L. Lin, M. Yuan, H. Yao, Y. Deng, B. Huang, H. Li, G. Sun, J. Zhu, Mott–Schottky heterostructure induce the interfacial electron redistribution of MoS_2 for boosting pH-universal hydrogen evolution with Pt-like activity, Nano Energy, 101 (2022) 107563. https://doi.org/10.1016/j.nanoen.2022.107563

[48]   F.-Y. Chen, Z.-Y. Wu, Z. Adler, H. Wang, Stability challenges of electrocatalytic oxygen evolution reaction: From mechanistic understanding to reactor design, Joule, 5 (2021) 1704-1731. https://doi.org/10.1016/j.joule.2021.05.005

[49]   Y. Zhang, F. Gao, H. You, Z. Li, B. Zou, Y. Du, Recent advances in one-dimensional noble-metal-based catalysts with multiple structures for efficient fuel-cell electrocatalysis, Coord. Chem. Rev., 450 (2022) 214244.

https://doi.org/10.1016/j.ccr.2021.214244

[50] Y.-P.L.Q.-T.Y.H.-J.F. Tao Zhang, Alkaline Seawater Electrolysis at Industrial Level: Recent Progress and Perspective, J. Electrochem., 28 (2022) 2214006. https://dx.doi.org/10.13208/j.electrochem.2214006

[51] K. Wu, F. Chu, Y. Meng, K. Edalati, Q. Gao, W. Li, H.-J. Lin, Cathodic corrosion activated Fe-based nanoglass as a highly active and stable oxygen evolution catalyst for water splitting, J. Mater. Chem. A, 9 (2021) 12152-12160. https://doi.org/10.1039/D1TA00769F

[52] J.C. Miramontes, C. Gaona Tiburcio, E. García Mata, M.Á. Esneider Alcála, E. Maldonado-Bandala, M. Lara-Banda, D. Nieves-Mendoza, J. Olguín-Coca, P. Zambrano-Robledo, L.D. López-León, F. Almeraya Calderón, Corrosion Resistance of Aluminum Alloy AA2024 with Hard Anodizing in Sulfuric Acid-Free Solution, Materials, 15 (2022) 6401. https://doi.org/10.3390/ma15186401

[53] Z. Shi, X. Wang, J. Ge, C. Liu, W. Xing, Fundamental understanding of the acidic oxygen evolution reaction: mechanism study and state-of-the-art catalysts, Nanoscale, 12 (2020) 13249-13275. https://doi.org/10.1039/D0NR02410D

[54] L. An, C. Wei, M. Lu, H. Liu, Y. Chen, G. Scherer, A. Fisher, P. Xi, Z. Xu, C.-H. Yan, Recent Development of Oxygen Evolution Electrocatalysts in Acidic Environment, Adv. Mater., 33 (2021) 2006328. https://doi.org/10.1002/adma.202006328

[55] S. Xie, X. Zhang, P. Xu, B. Hatcher, Y. Liu, L. Ma, S.N. Ehrlich, S. Hong, F. Liu, Effect of surface acidity modulation on Pt/Al$_2$O$_3$ single atom catalyst for carbon monoxide oxidation and methanol decomposition, Catal. Today, 402 (2022) 149-160. https://doi.org/10.1016/j.cattod.2022.03.028

[56] S. Rojas-Carbonell, K. Artyushkova, A. Serov, C. Santoro, I. Matanovic, P. Atanassov, Effect of pH on the activity of platinum group metal-free catalysts in oxygen reduction reaction, ACS Catal., 8 (2018) 3041-3053. https://doi.org/10.1021/acscatal.7b03991

[57] S.L. Fereja, P. Li, J. Guo, Z. Fang, Z. Zhang, Z. Zhuang, X. Zhang, K. Liu, W. Chen, W-Doped MoP Nanospheres as Electrocatalysts for pH-Universal Hydrogen Evolution Reaction, ACS Appl. Nano Mater., 4 (2021) 5992-6001. https://doi.org/10.1021/acsanm.1c00850

[58] Y. Xu, R. Wang, J. Wang, J. Li, T. Jiao, Z. Liu, Facile fabrication of molybdenum compounds (Mo$_2$C, MoP and MoS$_2$) nanoclusters supported on N-doped reduced graphene oxide for highly efficient hydrogen evolution reaction over broad pH range, Chem. Eng. J., 417 (2021) 129233. https://doi.org/10.1016/j.cej.2021.129233

[59] M. Qin, L. Chen, H. Zhang, M. Humayun, Y. Fu, X. Xu, X. Xue, C. Wang, Achieving highly efficient pH-universal hydrogen evolution by Mott-Schottky heterojunction of Co$_2$P/Co$_4$N, Chem. Eng. J., 454 (2023) 140230. https://doi.org/10.1016/j.cej.2022.140230

[60] Z. Wang, W. Xu, K. Yu, Y. Feng, Z. Zhu, 2D heterogeneous vanadium compound interfacial modulation enhanced synergistic catalytic hydrogen evolution for full pH

range seawater splitting, Nanoscale, 12 (2020) 6176-6187.
https://doi.org/10.1039/D0NR00207K

[61]   B. Zhang, J. Yao, J. Liu, T. Zhang, H. Wan, H. Wang, Reducing the pH dependence of
hydrogen evolution kinetics via surface reactivity diversity in medium-entropy alloys,
EES Catal., 1 (2023) 1017-1024. https://doi.org/10.1039/D3EY00157A

[62]   D. Zhang, H. Zhao, X. Wu, Y. Deng, Z. Wang, Y. Han, H. Li, Y. Shi, X. Chen, S. Li,
Multi-site electrocatalysts boost pH-universal nitrogen reduction by high-entropy alloys,
Adv. Funct. Mater., 31 (2021) 2006939. https://doi.org/10.1002/adfm.202006939

[63]   D. Xu, S.-N. Zhang, J.-S. Chen, X.-H. Li, Design of the synergistic rectifying interfaces
in Mott–Schottky catalysts, Chem. Rev., 123 (2022) 1-30.
https://doi.org/10.1021/acs.chemrev.2c00426

[64]   Y. He, S. Liu, C. Priest, Q. Shi, G. Wu, Atomically dispersed metal–nitrogen–carbon
catalysts for fuel cells: advances in catalyst design, electrode performance, and durability
improvement, Chem. Soc. Rev., 49 (2020) 3484-3524.
https://doi.org/10.1039/C9CS00903E

[65]   Y. Hou, J. Lv, W. Quan, Y. Lin, Z. Hong, Y. Huang, Strategies for Electrochemically
Sustainable H_2 Production in Acid, Adv. Sci., 9 (2022) 2104916.
https://doi.org/10.1002/advs.202104916

[66]   C. Pariya, K.N. Jayaprakash, A. Sarkar, Alkene metathesis: new developments in catalyst
design and application, Coord. Chem. Rev., 168 (1998) 1-48.
https://doi.org/10.1016/S0010-8545(97)00066-0

[67]   Y. Gao, Q. Wang, G. Ji, A. Li, J. Niu, Doping strategy, properties and application of
heteroatom-doped ordered mesoporous carbon, RSC Adv., 11 (2021) 5361-5383.
https://doi.org/10.1039/D0RA08993A

[68]   W. Xiao, L. Zhang, D. Bukhvalov, Z. Chen, Z. Zou, L. Shang, X. Yang, D. Yan, F. Han,
T. Zhang, Hierarchical ultrathin carbon encapsulating transition metal doped MoP
electrocatalysts for efficient and pH-universal hydrogen evolution reaction, Nano Energy,
70 (2020) 104445. https://doi.org/10.1016/j.nanoen.2020.104445

[69]   J. Wang, W. Fang, Y. Hu, Y. Zhang, J. Dang, Y. Wu, B. Chen, H. Zhao, Z. Li, Single
atom Ru doping 2H-MoS$_2$ as highly efficient hydrogen evolution reaction electrocatalyst
in a wide pH range, Appl. Catal. B-Environ., 298 (2021) 120490.
https://doi.org/10.1016/j.apcatb.2021.120490

[70]   S.M. El-Refaei, P.A. Russo, N. Pinna, Recent advances in multimetal and doped
transition-metal phosphides for the hydrogen evolution reaction at different pH values,
ACS Appl. Mater. Interfaces, 13 (2021) 22077-22097.
https://doi.org/10.1021/acsami.1c02129

[71]   A. Zhang, Y. Liang, H. Zhang, Z. Geng, J. Zeng, Doping regulation in transition metal
compounds for electrocatalysis, Chem. Soc. Rev., 50 (2021) 9817-9844.
https://doi.org/10.1039/D1CS00330E

[72] Y. Zheng, Y. Chen, X. Yue, S. Huang, Heteroatom Doping of Molybdenum Carbide Boosts pH-Universal Hydrogen Evolution Reaction, ACS Sustain. Chem. Eng., 8 (2020) 10284-10291. https://doi.org/10.1021/acssuschemeng.0c03311

[73] R. Zhang, X. Du, S. Li, J. Guan, Y. Fang, X. Li, Y. Dai, M. Zhang, Application of heteroatom doping strategy in electrolyzed water catalytic materials, J. Electroanal. Chem., 921 (2022) 116679. https://doi.org/10.1016/j.jelechem.2022.116679

[74] Q. Song, J. Li, S. Wang, J. Liu, X. Liu, L. Pang, H. Li, H. Liu, Enhanced electrocatalytic performance through body enrichment of Co-based bimetallic nanoparticles in situ embedded porous N-doped carbon spheres, Small, 15 (2019) 1903395. https://doi.org/10.1002/smll.201903395

[75] W. Zhang, X. Jiang, Z. Dong, J. Wang, N. Zhang, J. Liu, G.-R. Xu, L. Wang, Porous Pd/NiFeO$_x$ Nanosheets Enhance the pH-Universal Overall Water Splitting, Adv. Funct. Mater., 31 (2021) 2107181. https://doi.org/10.1002/adfm.202107181

[76] D. Zheng, L. Yu, W. Liu, X. Dai, X. Niu, W. Fu, W. Shi, F. Wu, X. Cao, Structural advantages and enhancement strategies of heterostructure water-splitting electrocatalysts, Cell Rep. Phys. Sci., 2 (2021) 100443. https://doi.org/10.1016/j.xcrp.2021.100443

[77] N. Yao, R. Meng, F. Wu, Z. Fan, G. Cheng, W. Luo, Oxygen-Vacancy-Induced CeO$_2$/Co$_4$N heterostructures toward enhanced pH-Universal hydrogen evolution reactions, Appl. Catal. B-Environ., 277 (2020) 119282. https://doi.org/10.1016/j.apcatb.2020.119282

[78] Z. Lin, K. Li, Y. Tong, W. Wu, X. Cheng, H. Wang, P. Chen, P. Diao, Engineering Coupled NiS$_x$-WO$_{2.9}$ Heterostructure as pH-Universal Electrocatalyst for Hydrogen Evolution Reaction, ChemSusChem, 16 (2023) e202201985. https://doi.org/10.1002/cssc.202201985

[79] T. Liu, L. Bai, N. Tian, J. Liu, Y. Zhang, H. Huang, Interfacial engineering in two-dimensional heterojunction photocatalysts, Int. J. Hydrogen Energ., 48 (2023) 12257-12287. https://doi.org/10.1016/j.ijhydene.2022.12.121

[80] W. Liu, X. Wang, F. Wang, K. Du, Z. Zhang, Y. Guo, H. Yin, D. Wang, A durable and pH-universal self-standing MoC-Mo$_2$C heterojunction electrode for efficient hydrogen evolution reaction, Nat. Commun., 12 (2021) 6776. https://doi.org/10.1038/s41467-021-27118-6

[81] M. Martino, C. Ruocco, E. Meloni, P. Pullumbi, V. Palma, Main hydrogen production processes: An overview, Catalysts, 11 (2021) 547. https://doi.org/10.3390/catal11050547

[82] Z. Cheng, Y. Xiao, W. Wu, X. Zhang, Q. Fu, Y. Zhao, L. Qu, All-pH-tolerant in-plane heterostructures for efficient hydrogen evolution reaction, ACS Nano, 15 (2021) 11417-11427. https://doi.org/10.1021/acsnano.1c01024

[83] J. Wang, T. Liao, Z. Wei, J. Sun, J. Guo, Z. Sun, Heteroatom-doping of non-noble metal-based catalysts for electrocatalytic hydrogen evolution: An electronic structure tuning strategy, Small Methods, 5 (2021) 2000988. https://doi.org/10.1002/smtd.202000988

[84]   X. Peng, S. Xie, S. Xiong, R. Li, P. Wang, X. Zhang, Z. Liu, L. Hu, B. Gao, P. Kelly, P.K. Chu, Ultralow-voltage hydrogen production and simultaneous Rhodamine B beneficiation in neutral wastewater, J. Energy Chem., 81 (2023) 574-582. https://doi.org/10.1016/j.jechem.2023.03.022

[85]   M. Kim, M.A.R. Anjum, M. Choi, H.Y. Jeong, S.H. Choi, N. Park, J.S. Lee, Covalent 0D-2D Heterostructuring of Co_9S_8-MoS_2 for Enhanced Hydrogen Evolution in All pH Electrolytes, Adv. Funct. Mater., 30 (2020) 2002536. https://doi.org/10.1002/adfm.202002536

[86]   R.-Q. Yao, Y.-T. Zhou, H. Shi, Q.-H. Zhang, L. Gu, Z. Wen, X.-Y. Lang, Q. Jiang, Nanoporous palladium-silver surface alloys as efficient and pH-universal catalysts for the hydrogen evolution reaction, ACS Energy Lett., 4 (2019) 1379-1386. https://doi.org/10.1021/acsenergylett.9b00845

[87]   B. Pang, X. Liu, T. Liu, T. Chen, X. Shen, W. Zhang, S. Wang, T. Liu, D. Liu, T. Ding, Laser-assisted high-performance PtRu alloy for pH-universal hydrogen evolution, Energy Environ. Sci., 15 (2022) 102-108. https://doi.org/10.1039/D1EE02518J

[88]   P. Kuang, Z. Ni, B. Zhu, Y. Lin, J. Yu, Modulating the d-Band Center Enables Ultrafine Pt_3Fe Alloy Nanoparticles for pH-Universal Hydrogen Evolution Reaction, Adv. Mater., 35 (2023) 2303030. https://doi.org/10.1002/adma.202303030

[89]   Z. Cui, W. Jiao, Z. Huang, G. Chen, B. Zhang, Y. Han, W. Huang, Design and Synthesis of Noble Metal-Based Alloy Electrocatalysts and Their Application in Hydrogen Evolution Reaction, Small, 19 (2023) 2301465. https://doi.org/10.1002/smll.202301465

[90]   M. Du, X. Li, H. Pang, Q. Xu, Alloy electrocatalysts, EnergyChem, 5 (2023) 100083. https://doi.org/10.1016/j.enchem.2022.100083

[91]   D. Wu, K. Kusada, T. Yamamoto, T. Toriyama, S. Matsumura, I. Gueye, O. Seo, J. Kim, S. Hiroi, O. Sakata, On the electronic structure and hydrogen evolution reaction activity of platinum group metal-based high-entropy-alloy nanoparticles, Chem. Sci., 11 (2020) 12731-12736. https://doi.org/10.1039/D0SC02351E

[92]   T. Zhang, A.G. Walsh, J. Yu, P. Zhang, Single-atom alloy catalysts: structural analysis, electronic properties and catalytic activities, Chem. Soc. Rev., 50 (2021) 569-588. https://doi.org/10.1039/D0CS00844C

[93]   T. Feng, G. Yu, S. Tao, S. Zhu, R. Ku, R. Zhang, Q. Zeng, M. Yang, Y. Chen, W. Chen, W. Chen, B. Yang, A highly efficient overall water splitting ruthenium-cobalt alloy electrocatalyst across a wide pH range via electronic coupling with carbon dots, J. Mater. Chem. A, 8 (2020) 9638-9645. https://doi.org/10.1039/D0TA02496A

Materials Research Forum LLC
https://doi.org/10.21741/9781644903070

CHAPTER 10

Catalysts for Oxygen Evolution Reaction

Xiang Peng*

Hubei Key Laboratory of Plasma Chemistry and Advanced Materials, Engineering Research Center of Phosphorus Resources Development and Utilization of Ministry of Education, School of Materials Science and Engineering, Wuhan Institute of Technology, Wuhan 430205, China

xpeng@wit.edu.cn

Abstract

The oxygen evolution reaction (OER) is a complex process involving the breaking of O-H bonds and the formation of O-O bonds through a four-electron transfer mechanism. Efficient catalysts are necessary to enhance the slower reaction kinetics and improve the energy conversion efficiency in water splitting. This chapter provides a concise overview of the reaction mechanism involved in OER catalysis, followed by a discussion on the correlation between the structure and properties of various catalysts.

Keywords

Oxygen Evolution Reaction, Hydrogen Evolution Reaction, Water Splitting, Catalyst, Nanomaterials

1.   Introduction

Electrocatalytic water splitting involves two half-cell reactions: the hydrogen evolution reaction (HER) occurring at the anode and the oxygen evolution reaction (OER) occurring at the cathode. The OER is a complex process involving the breaking of O-H bonds and the formation of O-O bonds through a four-electron transfer mechanism. Efficient catalysts are necessary to enhance the slower reaction kinetics and improve the energy conversion efficiency in water splitting [1]. Currently, noble metal oxides such as IrO_2 and RuO_2 are considered benchmark catalysts for OER. However, their limited reserves, high costs, and poor stability impose significant limitations on their feasibility for large-scale applications [2, 3]. Therefore, there is an urgent need to develop transition metal-based OER catalysts that are abundant, cost-effective, and stable, thus enabling the widespread adoption of water-splitting technology for large-scale hydrogen production.

This chapter provides a concise overview of the reaction mechanism involved in OER catalysis, followed by a discussion on the correlation between the structure and properties of various catalysts. By understanding the fundamental principles underlying OER catalysis and exploring the relationship between catalyst structure and performance, researchers aim to identify promising transition metal-based catalysts with enhanced activity, durability, and cost-effectiveness for efficient and sustainable hydrogen production.

2.   Mechanism of the Oxygen Evolution Reaction

The production of high-purity hydrogen through electrocatalytic water splitting is of critical importance in addressing the current crisis and environmental challenges.   Within this technology, the OER plays a significant role.   The OER is a slow kinetic process that involves the coupling of four electrons and protons, resulting in a high thermodynamic energy barrier [4-6].   Therefore, achieving faster kinetics of water-splitting reactions and improving energy conversion efficiency heavily relies on the utilization of efficient and stable OER catalysts. Research indicates that different catalysts exhibit variations in the mechanism of the OER under alkaline conditions.   In an alkaline environment, water undergoes oxidation to produce O* and OH* ions.   Both reaction mechanisms involve the adsorption and conversion of three intermediate species (OH*, O*, OOH*) [7, 8].   Two well-established OER mechanisms have been proposed based on thermodynamic and kinetic theoretical models.   The first mechanism is the conventional adsorbate evolution mechanism (AEM) [9, 10], as shown in **Figure 1**.   In this mechanism, the OER proceeds through the adsorption and oxidation of OH* intermediates, leading to the release of oxygen gas.

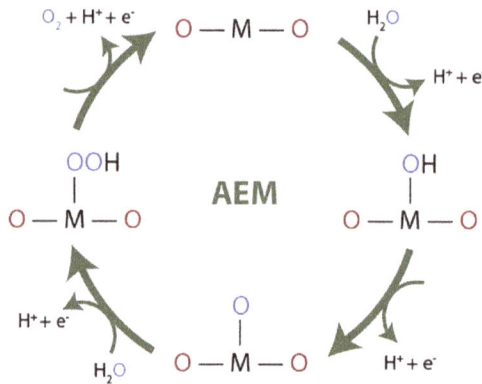

Figure 1. The conventional adsorbate evolution OER mechanism. Reproduced with permission from ref [11]. Copyright 2021, Cell Press.

The second mechanism is the lattice oxygen activation mechanism (LOM) [12, 13], as shown in **Figure 2**.   In this mechanism, the OER involves the activation of lattice oxygen from the catalyst material itself, resulting in the release of oxygen gas.

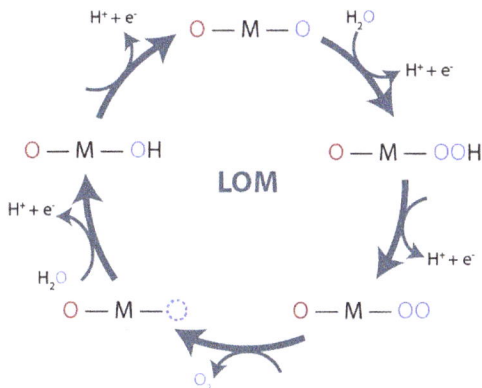

Figure 2. *The lattice oxygen activation OER mechanism. Reproduced with permission from ref [11]. Copyright 2021, Cell Press.*

Understanding these different mechanisms and their corresponding reaction pathways is crucial for designing and optimizing OER catalysts. By elucidating the fundamental principles governing the OER and exploring the variations in reaction mechanisms, researchers aim to develop efficient and stable catalysts that can enhance the kinetics and overall performance of water-splitting reactions, ultimately contributing to the production of high-purity hydrogen. Both catalytic OER mechanisms indeed involve a four-electron transfer process, and their reaction pathways can be described as follows:

The overall reaction for the OER in alkaline conditions is given by:

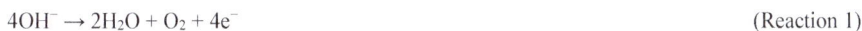

$$4OH^- \rightarrow 2H_2O + O_2 + 4e^- \qquad \text{(Reaction 1)}$$

In the anodic OER equation, the following series of reactions take place [14, 15]:

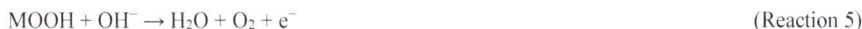

$$M + OH^- \rightarrow MOH + e^- \qquad \text{(Reaction 2)}$$

$$MOH + OH^- \rightarrow H_2O + MO + e^- \qquad \text{(Reaction 3)}$$

$$MO + OH^- \rightarrow MOOH + e^- \qquad \text{(Reaction 4)}$$

$$MOOH + OH^- \rightarrow H_2O + O_2 + e^- \qquad \text{(Reaction 5)}$$

In these reactions, M represents the active site, typically the metal atoms, present in the catalyst. These reaction pathways illustrate the stepwise process of the four-electron transfer involved in the OER and demonstrate the role of the active site in facilitating the oxidation of hydroxide ions and the formation of oxygen gas during the OER.

3.   Catalysts for Oxygen Evolution Reaction

3.1.   Zero-Dimensional Nanoparticles

Zero-dimensional (0D) nanoparticles often referred to as nanoclusters or quantum dots, possess unique properties due to their small size and high surface-to-volume ratio, which have gained significant attention as promising electrocatalysts for the OER [16, 17]. The small size and high surface area of 0D nanoparticles provide a large number of active sites for OER [18-20]. The unique electronic properties arising from quantum confinement effects can facilitate charge transfer and reaction kinetics [21]. These characteristics contribute to improved catalytic activity compared to bulk materials. Furthermore, the size and composition of 0D nanoparticles can be precisely controlled during synthesis. By tuning the nanoparticle size, the electronic structure and surface reactivity can be optimized for enhanced OER activity [22]. Additionally, varying the composition and doping of the nanoparticles can further enhance catalytic performance [23]. 0D nanoparticles can also be integrated with suitable support materials to enhance their catalytic performance further [24-26]. Support materials can provide structural stability, facilitate charge transport, and improve mass transfer. Exploring different support materials and optimizing their interactions with 0D nanoparticles can lead to synergistic effects and improved OER activity.

For example, Peng et al. [27] proposed a novel strategy involving rhizobium-like metal clusters anchored to a porous TiN nanowires array on carbon cloth (CC) to enhance the charge transport and improve the catalytic activity for the OER. The preparation procedure for the metal clusters supported by the porous TiN nanowires array is outlined as follows:

Firstly, a TiN/CC nanowires skeleton was prepared through a traditional hydrothermal reaction, followed by high-temperature nitridation. The resulting TiN/CC structure exhibited a porous and dense arrangement of nanowires, surrounded by carbon fibers, as shown in **Figure 3a** in the SEM images, where the inset highlights the structure.

Subsequently, the TiN/CC structure was immersed in a solution of nickel oleate (Ni-OA) at a concentration of 1 mg mL^{-1}. This process facilitated the adsorption of Ni-OA onto the surface of the porous TiN nanowires. Finally, the Ni-OA adsorbed TiN/CC structure was annealed at 700 °C in an N_2/H_2 gas environment for 90 minutes, resulting in the formation of Ni/TiN/CC.

Scanning electron microscopy (SEM) images depicted in **Figure 3b-d** illustrate the progressive increase in the amount of loaded nanoparticles through repetitive immersion and drying cycles. In **Figure 3b**, it can be observed that the nanowires in Ni/TiN/CC-1 were not completely covered by nanoparticles. However, by increasing the immersion/drying cycle to 3 (Ni/TiN/CC-3), uniform coverage of nanoparticles on the nanowires was achieved (**Figure 3c**). On the other hand, excessive repetition of the process, as seen in 5 cycles of Ni/TiN/CC-5, led to nanowire entanglement due to the high quantity of loaded nanoparticles, as indicated in **Figure 3d**.

It is important to note that while loaded nanoparticles serve as active electrocatalytic materials, an excessive amount of active material can have adverse effects, such as impeding charge transfer and mass transport, consequently hindering the catalytic capability. Therefore, achieving an optimal loading of nanoparticles is crucial for enhancing catalytic performance.

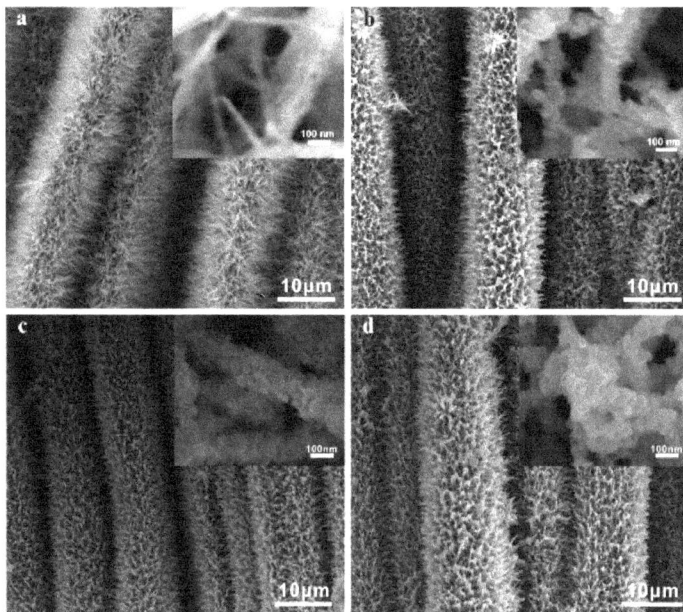

Figure 3. SEM images of (a) TiN/CC, (b) Ni/TiN/CC-1, (c) Ni/TiN/CC-3, and (d) Ni/TiN/CC-5, where the digitals represent the immersion/drying cycles of the TiN/CC skeleton in the Ni-OA solution. The insets are the high-magnification images. Reproduced with permission from ref [27] Copyright 2021, Elsevier.

X-ray diffraction (XRD) pattern of the Ni/TiN/CC samples, as shown in **Figure 4a,** indicates strong diffraction peaks corresponding to cubic TiN (JCPDS card No. 87-0628), which is consistent with previous observations [28]. Additionally, a weak signal of metallic Ni (JCPDS card No. 04-0850) is observed, indicating the presence of metallic Ni in the catalyst. The high conductivity of both TiN and metallic Ni is expected to facilitate charge transfer and catalytic kinetics during the catalytic reactions. Moreover, the metallic Ni surface undergoes conversion into high-valence Ni-based (oxy)hydroxyl compounds, which can enhance the catalytic properties [29]. This conversion likely occurs due to the interaction between the metallic Ni and the TiN substrate. The high-resolution X-ray photoelectron spectroscopy (XPS) N-$2p$ spectrum, as shown in **Figure 4b**, provides additional insights. The spectrum shows a peak at 853 eV, corresponding to metallic Ni (Ni^0), while peaks between 855 and 860 eV represent Ni^{2+} and Ni^{3+} of nickel oxides, respectively [30, 31]. This observation confirms the presence of different oxidation states of nickel on the catalyst surface.

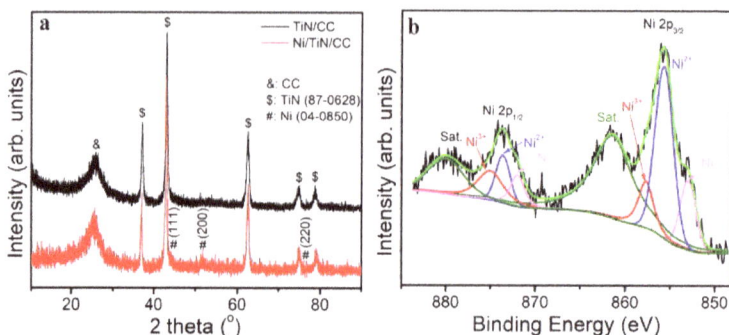

Figure 4. (a) XRD patterns of TiN/CC and Ni/TiN/CC. (b) High-resolution XPS Ni-2p spectrum of the Ni/TiN/CC catalyst. Reproduced with permission from ref [27]. Copyright 2021, Elsevier.

The OER performance of the synthesized series Ni/TiN/CC catalysts was investigated in 1.0 M KOH aqueous solution using a three-electrode system. **Figure 5a** demonstrates the phenomenon of "reconstruction" during the OER, where the oxidation of Ni^{2+} to NiOOH is observed within the potential range of 1.4 to 1.5 V *vs.* RHE (reversible hydrogen electrode). Among the different Ni/TiN/CC catalysts investigated, Ni/TiN/CC-3 exhibited remarkable performance, generating a current density of 10 mA cm^{-2} with an overpotential of only 360 mV. This overpotential was significantly lower than that of Ni/TiN/CC-1, Ni/TiN/CC-5, and TiN/CC catalysts, indicating superior catalytic activity. Furthermore, the kinetics of the OER were evaluated by the Tafel slope. **Figure 5b** illustrates that the Ni/TiN/CC-3 catalyst displayed the smallest Tafel slope, measuring only 133 mV dec^{-1}. A smaller Tafel slope indicates faster reaction kinetics during the OER. The electrochemical impedance spectroscopy (EIS) results shown in **Figure 5c** further confirmed the enhanced OER kinetics of the Ni/TiN/CC-3 catalyst. The catalyst exhibited a smaller charge-transfer resistance, indicating improved charge-transfer efficiency. The electrochemically active surface area (ECSA) of the Ni/TiN/CC catalysts was studied and evaluated using the double layer capacitance (C_{dl}) in the non-faradic region. **Figure 5d** shows that the Ni/TiN/CC-3 catalyst exhibited the largest C_{dl}, measuring 15 mF cm^{-2}. A larger C_{dl} indicates a greater number of exposed active sites for electrochemical reactions, thus reflecting better catalytic performance.

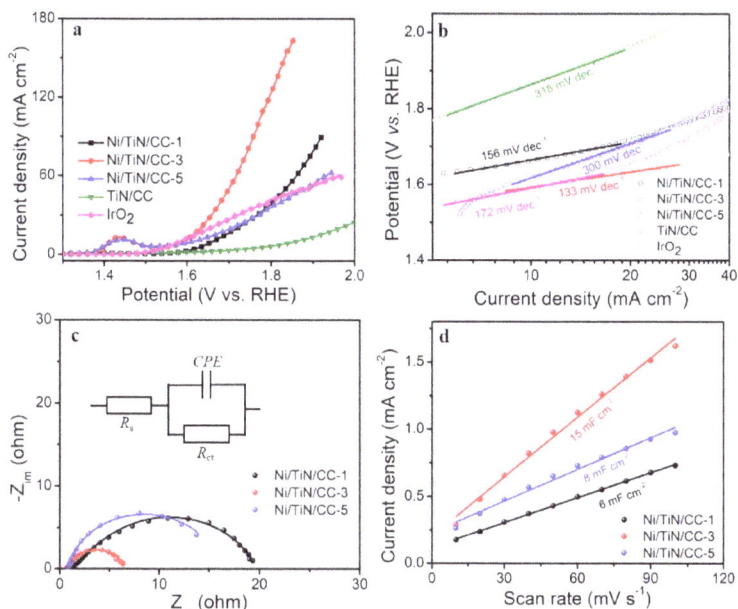

Figure 5. *Electrochemical properties of Ni/TiN/CC-1, Ni/TiN/CC-3, and Ni/TiN/CC-5 catalysts. (a) Polarization curves, (b) Tafel slopes, (c) EIS spectra with the equivalent circuit in the inset, and (d) ECSA. Reproduced with permission from ref [27]. Copyright 2021, Elsevier.*

The results of the study suggest that the Ni cluster in the Ni/TiN/CC-3 catalyst exhibits outstanding activity for the OER. The interaction between the Ni cluster and the TiN substrate is beneficial in promoting the OER. However, excessive loading of the Ni species on the TiN/CC skeleton can lead to material overlap, which reduces the exposure of active sites and slows down charge transfer. Consequently, the excess loading of active material negatively affects the OER performance.

Notably, the Ni/TiN/CC-3 catalyst also exhibited excellent stability. **Figure 6a** shows that the current density remained stable during continuous operation for more than 20 hours, indicating the catalyst's robustness. Furthermore, **Figure 6b** reveals that the morphology of the catalyst did not exhibit significant changes after the long-term stability testing. The high-resolution XPS Ni-2*p* spectra of the Ni/TiN/CC-3 catalyst after the long-term test, as shown in **Figure 6c**, indicated the disappearance of the metallic Ni species. This disappearance is attributed to the surface reconstruction of the Ni species during the OER process. This study presents an effective strategy for preparing highly efficient and stable metal cluster catalysts for electrochemical water splitting, exemplified by the excellent activity and stability of the Ni/TiN/CC-3 catalyst.

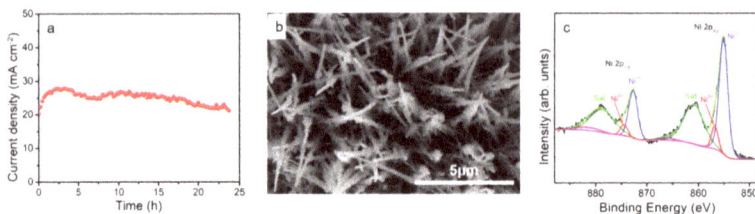

***Figure 6**. (a) Stability test. (b) SEM image and (c) XPS Ni-2p spectra of the Ni/TiN/CC-3 electrocatalyst after the stability test. Reproduced with permission from ref [27]. Copyright 2021, Elsevier.*

3.2.   One-Dimensional Nanomaterials

One-dimensional (1D) nanomaterials, such as nanotubes [32-34], nanowires [35-37], nanofibers [38, 39], and nanorods [40] offer several advantages over other materials due to their unique geometrical morphology. These benefits include superior mechanical properties, high specific surface area, fast electron transport paths, and facilitation of mass transfer [41, 42]. As a result, 1D nanomaterials have a wide range of potential applications in electrochemistry [27, 32, 43, 44], photocatalysis [45, 46], supercapacitor [47, 48], and other fields [49-52]. In recent years, there has been increasing research focus on the study of 1D catalysts, leading to enhanced stability and catalytic activity.

Wang et al. [53] conducted a comprehensive analysis of the catalytic activity of Co-based catalysts by combining theoretical predictions with experimental investigations. They developed models of the Co (111) surface, the Co_2C (111) surface, and the Co-Co_2C heterostructure to perform theoretical predictions, as depicted in **Figure 7a**. The density of states (DOS) results in **Figure 7b** show that the DOS of the Co-Co_2C interface near the Fermi level is higher than that of the pure Co and Co_2C surfaces. This indicates that the coupling of Co and Co_2C at the heterostructure interface beneficially modifies the electronic structure [54, 55]. In addition, the Gibbs free energy (ΔG) was calculated using density functional theory (DFT) to gain a deeper understanding of the heterostructure construction on catalytic performance. Specifically, a four-step reaction mechanism was employed to study the OER process in alkaline media, considering the adsorption of *O, *OH, and *OOH as intermediates during the OER. The theoretical models are illustrated in **Figure 7c**. The calculated free energy diagram in **Figure 7d** shows that the rate-determining step (RDS) for the pure Co catalyst is the fourth step involving the forming of O_2 gas from the *OOH intermediate. However, for the pure Co_2C surface and the heterostructured Co-Co_2C interface, the RDS steps are the third step involving the conversion of *O into the *OOH group. Furthermore, the RDS at the heterostructured Co-Co_2C interface exhibits the smallest free energy of 3.43 eV, indicating that the heterointerface leads to a thermodynamics favorable process for the OER [56, 57].

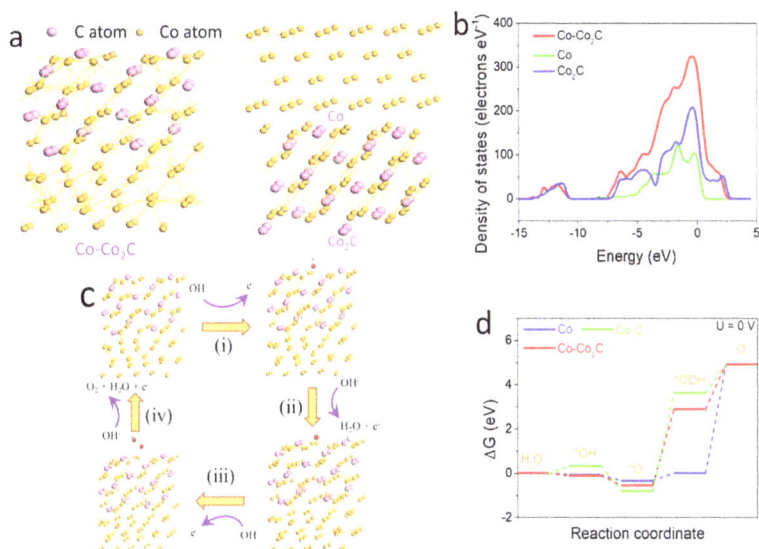

Figure 7. *(a) Constructed models of Co (111) surface, Co_2C (111) surface, and heterostructured Co-Co_2C interface. (b) The calculated DOS results of Co, Co_2C, and heterostructured Co-Co_2C models. (c) The proposed four-step mechanism for OER under an alkaline medium by adsorbing *OH, *O and *OOH intermediates. (d) Calculated free energy diagram of OER intermediates at U = 0 V. Reproduced with permission from ref [53]. Copyright 2021, Elsevier.*

Based on the theoretical prediction, the researchers prepared the Co_2C nanowires modified with superficial Co nanoparticles on CC (Co-Co_2C/CC). The synthesis of the Co-Co_2C/CC heterostructure catalysts involved a two-step process, as indicated in **Figure 8**. First, a hydrothermal route was employed to produce the $Co(OH)_2$/CC precursor. Then, an annealing treatment was conducted using melamine as the carbon source under a 5% H_2/Ar flow [58]. During the annealing treatment, the as-prepared $Co(OH)_2$/CC precursor reacted with melamine, resulting in the formation of Co_2C nanowires. Simultaneously, the reduction environment created by the H_2/Ar flow, along with the relatively high reaction temperature, facilitated the formation of metallic Co nanoparticles on the surface of Co_2C nanowires.

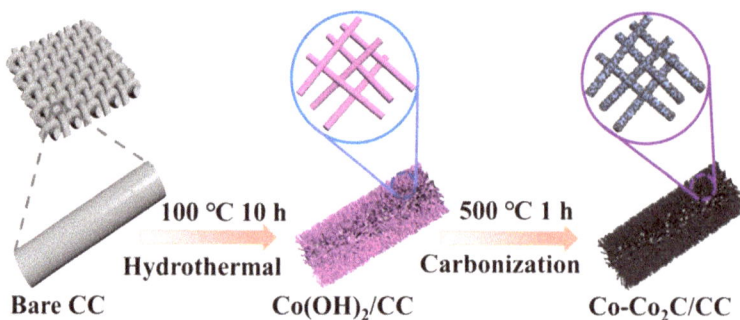

Figure 8. Schematic illustration of the synthetic procedure for the Co-Co₂C/CC electrocatalyst. Reproduced with permission from ref [53]. Copyright 2021, Elsevier.

In **Figure 9a**, the XRD pattern demonstrates diffraction peaks that can be attributed to the cubic Co phase (indexed to JCPDS card No. 15-0806) and the orthorhombic Co_2C phase (indexed to JCPDS card No. 65-1457). This indicates the presence of both Co and Co_2C in the Co-Co₂C/CC heterostructure catalyst. The SEM images of Co-Co₂C/CC show the nanowire arrays grow uniformly on the CC skeleton, as shown in **Figure 9b-c**. This indicates a well-controlled synthesis process. Moreover, the high-magnification SEM image in **Figure 9d** provides a closer look at the surface of the nanowires, showing that a substantial number of nanoparticles have been modified on the nanowire surface. This modification suggests the presence of both nanowires and nanoparticles in the Co-Co₂C/CC heterostructure catalyst. The combination of nanowires and nanoparticles in the catalyst structure may lead to a synergistic effect, enhanced surface area, improved charge transfer, and optimized active site exposure, all of which contribute to improved catalytic performance.

The TEM images presented in **Figure 9e-f** provide further confirmation of the coupling properties between the nanowires and nanoparticles in the Co-Co₂C/CC heterostructure catalyst. The high-resolution TEM (HR-TEM) image in **Figure 9g** reveals the heterointerface between the two phases, with lattice spacings of 0.205 nm and 0.211 nm attributed to the Co_2C (111) planes and metallic Co (200) planes, respectively. These TEM and HR-TEM results provide direct evidence of the composite structure composed of metallic Co nanoparticles and Co_2C nanowires. The elemental mappings shown in **Figure 9h-j** demonstrate the homogeneous distribution of carbon (C) elements in the nanowires. Elemental Co is observed on both the nanowires and nanoparticles, indicating their presence in the Co-Co₂C heterostructure. XPS analysis further confirms the presence of Co^0, Co^{2+}, and Co-C species in the Co-Co₂C/CC heterostructure. The peak shift of Co in the Co-Co₂C heterostructure compared to pure Co/CC and Co₂C/CC suggests electronic interactions between the two species at the heterointerface.

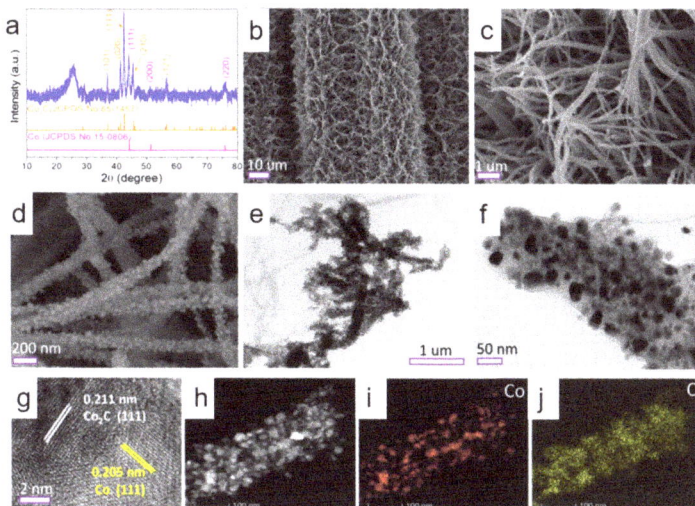

Figure 9. *(a) XRD patterns of Co-Co₂C/CC. (b-d) SEM images of Co-Co₂C/CC with different magnifications. (e-f) TEM images and (g) HR-TEM image of the heterostructured Co-Co₂C catalyst. (h-j) HAADF image and elemental mappings of Co and C for the heterostructured Co-Co₂C catalyst. Reproduced with permission from ref [53]. Copyright 2021, Elsevier.*

To validate the theoretical predictions and further investigate the catalytic performance, the OER performance of the Co-Co$_2$C/CC catalyst was studied in a 1.0 M KOH electrolyte at room temperature. Comparative groups, including bare CC, Co$_2$C/CC, and commercial RuO$_2$ coated on CC, were also evaluated. The self-supporting construction of the Co-Co$_2$C/CC catalyst allowed for direct use as a working electrode without additional binders.

Figure 10a illustrates the OER performance of the different catalysts. The bare CC substrate exhibits negligible OER activity, indicating that the CC substrate itself does not contribute significantly to the activity of the Co-Co$_2$C/CC catalyst. In contrast, the composite Co-Co$_2$C/CC heterostructure catalyst demonstrates the best OER catalytic activity, achieving a current density of 10 mA cm^{-2} with an overpotential of only 261 mV. This overpotential is lower than that of Co$_2$C/CC and RuO$_2$ on CC, indicating the improved catalytic performance resulting from the nanoparticle/nanowire heterogeneous interface construction. **Figure 10b** compares the catalytic activity of the Co-Co$_2$C/CC catalyst for OER with other reported OER electrocatalysts in alkaline media. The Co-Co$_2$C/CC catalyst shows comparable performance to these catalysts, further highlighting its potential as an efficient OER electrocatalyst.

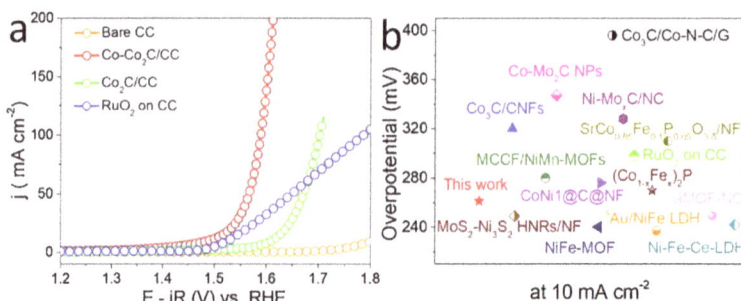

Figure 10. (a) Polarization curves for Co-Co$_2$C/CC, RuO$_2$ on CC, Co$_2$C/CC, and bare CC measured in 1 M KOH with a scan rate of 5 mV s^{-1}. (b) Comparison of the overpotentials of different catalysts at the current density of 10 mA cm^{-2}. Reproduced with permission from ref [53]. Copyright 2021, Elsevier.

The Tafel slope of the Co-Co$_2$C/CC catalyst in OER is determined to be 58.8 mV dec^{-1}, which is significantly smaller than that of Co$_2$C/CC (96.8 mV dec^{-1}) and even RuO$_2$ on CC (64.5 mV dec^{-1}), as shown in **Figure 11a.** The result indicates the Co-Co$_2$C/CC catalyst exhibits fast kinetics in the OER process. A smaller Tafel slope suggests a more efficient electrochemical reaction and a higher catalytic activity. To further investigate the kinetics and electron transport capability of the catalysts, EIS measurements were conducted. **Figure 11b** presents the results, showing the charge transfer resistance (R_{ct}) of the different catalysts. The Co-Co$_2$C/CC catalyst displays a lower R_{ct} compared to both the Co$_2$C/CC and bare CC. This indicates a significant improvement in kinetics and charge transfer efficiency for the Co-Co$_2$C/CC catalyst. The lower charge transfer resistance suggests that the Co-Co$_2$C/CC catalyst facilitates faster electron transfer during the OER process, contributing to its enhanced catalytic performance.

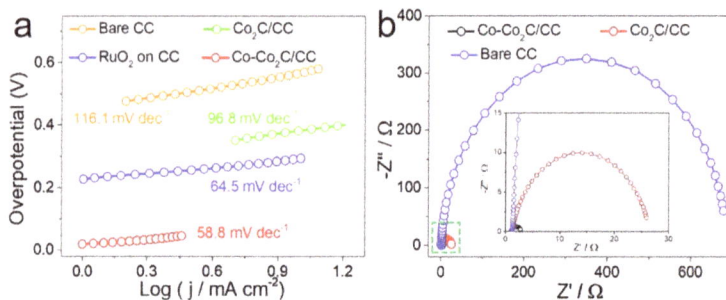

Figure 11. (a) Tafel plots and (b) EIS plots of Co-Co$_2$C/CC, Co$_2$C/CC, and bare CC recorded at an overpotential of 320 mV with the inset showing an enlarged image of the green dashed frame. Reproduced with permission from ref [53]. Copyright 2021, Elsevier.

In **Figure 12a**, the turnover frequency (TOF) is calculated and plotted against overpotential. The Co-Co$_2$C/CC catalyst exhibits larger TOF values compared to the Co$_2$C/CC catalyst across the entire range. This indicates that the active sites in the Co-Co$_2$C/CC catalyst possess a higher intrinsic activity than those in the Co$_2$C/CC catalyst. The higher TOF values further support the superior catalytic performance of the Co-Co$_2$C/CC catalyst in the OER process.

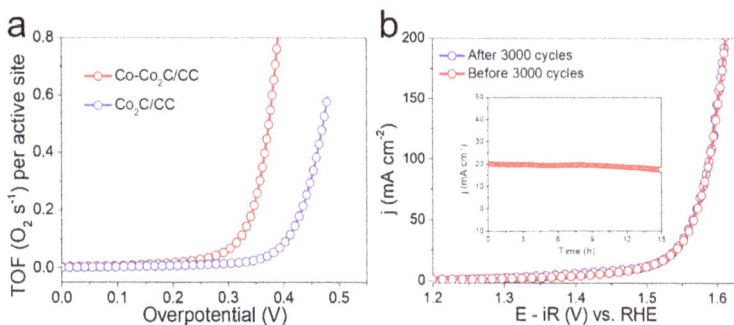

Figure 12. (a) Calculated TOF values for Co-Co$_2$C/CC and Co$_2$C/CC catalysts. (b) Polarization curves for the Co-Co$_2$C/CC catalyst before and after 3000 cycles of CV scanning. The inset in (b) displays the current density-time curve of the Co-Co$_2$C/CC catalyst during the long-term OER operation. Reproduced with permission from ref [53]. Copyright 2021, Elsevier.

Figure 12b provides additional evidence of the stability and durability of the Co-Co$_2$C/CC catalyst. The polarization curve of the Co-Co$_2$C/CC catalyst shows negligible loss after cycling through 3000 continuous cyclic voltammetry (CV) scans compared to the initial curve. This indicates that the catalyst maintains its catalytic activity without significant degradation or loss of performance during repeated cycling. Moreover, the chronoamperometric (CA) curve shown in the inset of **Figure 12b** demonstrates the excellent long-term stability of the Co-Co$_2$C/CC catalyst. The CA curve reveals that the Co-Co$_2$C/CC catalyst retains its OER activity over a continuous testing period of 15 hours, with negligible degradation in current density. This suggests that the Co-Co$_2$C/CC catalyst exhibits long-term durability and demonstrates its potential for practical applications requiring stable and continuous OER performance.

Doan et al. [59] developed a novel composite electrocatalyst composed of CuO nanowires (CuO NWs) with a shell of aerobically doped cobalt phosphide (O-doped Co$_2$P). The fabrication process involved a simple and cost-effective method, resulting in vertically grown composite nanowire catalysts on a conductive framework, as illustrated in **Figure 13**. To synthesize the catalyst, the researchers first prepared a precursor called Co(OH)$_2$/Cu(OH)$_2$ NWs/CF using a solution-based method. Subsequently, the precursor was subjected to air annealing, leading to the formation of Co$_3$O$_4$/CuO NWs/CF. During this step, the Co(OH)$_2$/Cu(OH)$_2$ NWs/CF precursor was heated in the presence of air. Finally, the Co$_3$O$_4$/CuO NWs/CF underwent heat treatment in an argon environment, resulting in the formation of the O-doped Co$_2$P/CuO NWs/CF catalyst. This heat treatment step facilitated the conversion of the Co$_3$O$_4$/CuO structure into O-doped Co_2P. The morphology characterization results clearly demonstrated that the O-doped Co_2P nanolayer completely enveloped the CuO NWs, forming a core-shell nanostructure.

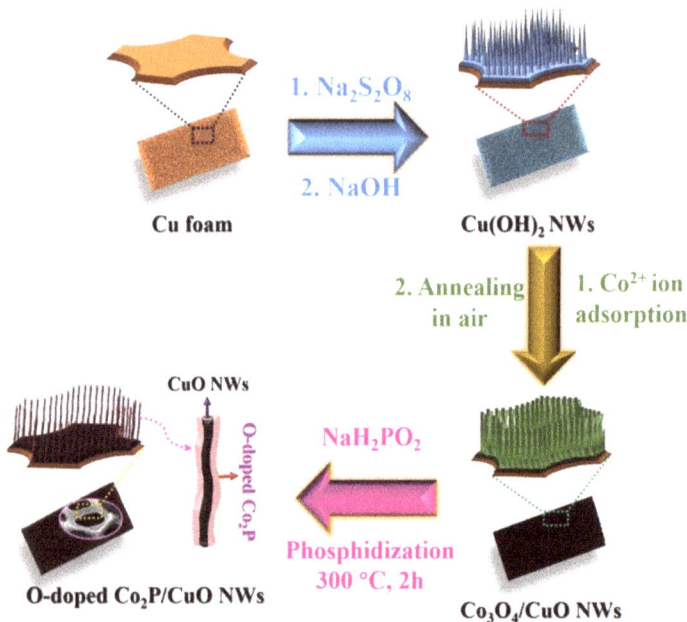

Figure 13. *Schematic illustration of the fabrication of the O-doped Co_2P/CuO NWs/CF hybrid. Reproduced with permission from ref [59]. Copyright 2020, Elsevier.*

The synthesized materials were analyzed using XPS. A comparison of the Cu-$2p$ spectra between CuO NWs, Co_3O_4/CuO NWs, and O-doped Co_2P/CuO NWs revealed significant shifts in the latter two spectra, indicating the formation of an electronic interaction between the Co-based nanolayer and the CuO core [60]. Moreover, successful phosphating led to the formation of O-doped Co_2P/CuO NWs. The incorporation of a small amount of oxygen (O) into the Co_2P structure was found to promote the electron transfer process and adjust the Gibbs free energy, enhancing the catalytic activity by modifying the adsorption properties of reactant molecules, intermediates, and products on the catalyst surface [61].

The electrocatalytic performance of the synthetic catalysts in OER was investigated using a typical three-electrode system in an N_2-saturated 1.0 M KOH electrolyte. Doan et al. [59] observed that the O-doped Co_2P/CuO NWs catalyst exhibited a maximum double-layer capacitance (C_{dl}) of 39.6 mF cm^{-2}. This catalyst required an overpotential of only 270 mV to achieve a current density of 10 mA cm^{-2}, with a Tafel slope of 74.4 mV dec^{-1}. Additionally, the O-doped Co_2P/CuO NWs/CF catalyst demonstrated excellent stability. These findings suggest that the novel O-doped Co_2P/CuO NWs catalyst, with its core-shell structure, exhibits modified physicochemical properties, exposes more active sites, and facilitates the electron transfer process, leading to improved electrocatalytic performance in the OER.

3.3.   Two-Dimensional Nanomaterials

Two-dimensional (2D) nanomaterials have attracted significant attention due to their unique structure and exceptional physical and chemical properties [62-64]. The unsaturated edge sites present in 2D materials exhibit high electrochemical activity, making them promising candidates for electrocatalysis applications [65, 66]. However, the catalytic activity of 2D nanomaterials is often limited by their abundant basal sites, which are less active in electrocatalysis. Additionally, the high surface energy of nano-sheets can lead to their aggregation, hindering the exposure of active sites and resulting in decreased catalytic performance. To overcome these challenges, various strategies such as heteroatom doping, interface structure engineering, and defect engineering have been employed to optimize the electronic structure of 2D nanomaterials and enhance their catalytic properties.

Xie et al. [67] proposed an efficient strategy for preparing a nickel-based nanosheet electrocatalyst enriched with Ni^{3+} and oxygen vacancies (Ov) by doping Fe atoms into NiO nanosheets (Fe/NiO/CC). This catalyst demonstrated significant potential for OER. The Fe/NiO/CC electrocatalyst was synthesized through a simple solution method followed by chemical adsorption. Initially, vertically aligned $Ni(OH)_2$ nanosheets ($Ni(OH)_2$/CC) were prepared on a CC substrate using a solution-based approach, followed by an annealing treatment to stabilize the NiO/CC nanostructures and improve their adhesion to the CC substrate. Fe atoms were introduced into the NiO nanosheets through sonication in a $FeSO_4$ solution, which modulated the electronic structure, charge distribution, and coordination of active sites, including Ni and Fe atoms, for efficient OER.

The SEM images of NiO/CC and Fe/NiO/CC are presented in **Figure 14a-b**, illustrating the nanosheet array structure of both samples. The nanosheets are vertically grown on the CC substrate. Importantly, the introduction of Fe^{2+} cations into NiO/CC through chemical adsorption does not disrupt the original array structure of NiO. Typically, the morphology of nanosheet arrays vertically grown on CC exposes larger active sites and electrolyte interfaces, creating favorable conditions to enhance catalytic activity.

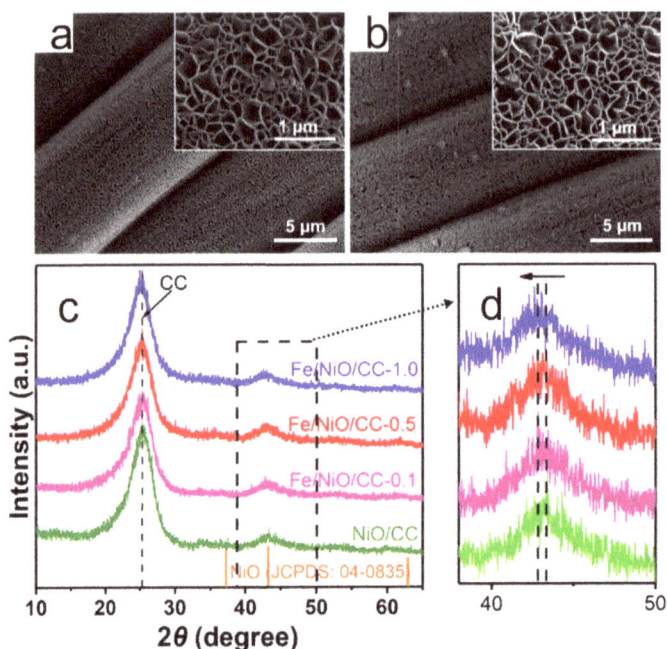

Figure 14*. SEM images of (a) NiO/CC and (b) Fe/NiO/CC. (c) XRD patterns of Fe/NiO/CC doped with different amounts of Fe. (d) XRD patterns in the 2θ range of 38–50°. Reproduced with permission from ref [67]. Copyright 2022, Elsevier.*

Additionally, the structure and crystallinity of the aforementioned samples were characterized using XRD, as depicted in **Figure 14c**. The main characteristic peaks of both sample sets correspond to NiO (JCPDS card No. 04-0835), with the exception of the peak at approximately 25° originating from the CC substrate. Notably, compared to pure NiO, the introduction of Fe^{2+} cations induces a slight shift of the characteristic peak towards smaller angles. Additionally, with an increase in Fe^{2+} content from 0.1 to 1.0 mL, the prominent diffraction peak at $2\theta=43.3°$ notably shifts towards smaller angles, as illustrated in **Figure 14d**. This phenomenon can be attributed to the introduction of Fe^{2+}, which leads to disorder in the NiO lattice structure and internal microstrains. Such modifications facilitate charge transfer and enhance the electrochemical activity of the catalyst. These results suggest that the introduction of Fe^{2+} causes expansion and distortion in the original NiO lattice, potentially accelerating charge transfer and improving electrocatalytic activity.

To elucidate the mechanism of the NiO/CC and Fe/NiO/CC catalysts during the OER, their reconstruction processes were systematically investigated. **Figure 15a** presents the polarization curves of the NiO/CC catalyst during the initial scanning cycles, indicating that it stabilizes after approximately 20 cycles. In contrast, the Fe/NiO/CC catalyst achieves stable reconstruction within 10 cycles, as depicted in **Figure 15b**. Furthermore, the current density for the reconstruction of the Fe/NiO/CC catalyst is lower than that of the NiO/CC catalyst. **Figure 15c** illustrates the peak potentials (ranging from 1.3 to 1.6 V *vs.* RHE) for the reconstruction of the NiO/CC and Fe/NiO/CC catalysts during the initial scanning cycles. The Fe-doped catalyst requires a lower potential and less energy consumption for surface reconstruction. These findings suggest that Fe doping facilitates the reconstruction process and improves energy efficiency during this process.

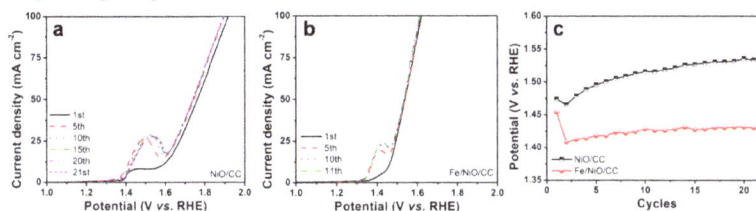

Figure 15. Polarization curves obtained in different stages from (a) NiO/CC and (b) Fe/NiO/CC catalysts during reconstruction. (c) Peak potentials of NiO/CC and Fe/NiO/CC catalysts for different scanning cycles. Reproduced with permission from ref [67]. Copyright 2022, Elsevier.

To further analyze the effect of Fe^{2+} doping on the surface electronic structure of Ni and Fe atoms during the reconstruction, the surface chemistry of the catalysts was characterized before and after reconstruction. Figure **16a-b** displays the high-resolution XPS Ni-$2p$ spectra of the fresh NiO/CC and Fe/NiO/CC catalysts. The binding energies of Ni^{2+} and Ni^{3+} are located at 855.3/856.4 eV ($2p_{3/2}$) and 873/874.6 eV ($2p_{1/2}$), respectively. The $Ni^{3+}/(Ni^{3+} + Ni^{2+})$ ratio of the fresh NiO/CC and Fe/NiO/CC catalysts was calculated to be 0.32 and 0.33, respectively. After successive linear sweep voltammetry scans for surface reconstruction, the elemental composition of the catalysts remained unchanged. However, the contents of Fe^{3+} and Ni^{3+} exhibited significant variations. In the NiO/CC catalyst, the surface concentration of Ni^{3+} increased from 0.32 to 0.39 after reconstruction, as shown in **Figure 16c**, indicating a 7% increase in Ni^{3+} concentration over the initial 20 cycles of reconstruction. In the Fe/NiO/CC catalyst, the Ni^{3+} concentration increased from 0.33 to 0.54, as depicted in **Figure 16c**, indicating a 21% increase in Ni^{3+} concentration during the initial 10 cycles of surface reconstruction. The increase in Ni^{3+} in the Fe/NiO/CC catalyst was three times greater than that in the NiO/CC catalyst, despite the reconstruction process consisting of half the number of scanning cycles. Consequently, the reconstruction kinetics of the catalyst increased fivefold after Fe incorporation, and the amount of Ni^{3+} in the final product increased, indicating the generation of a Ni^{3+}-enriched surface in the actual catalyst for OER.

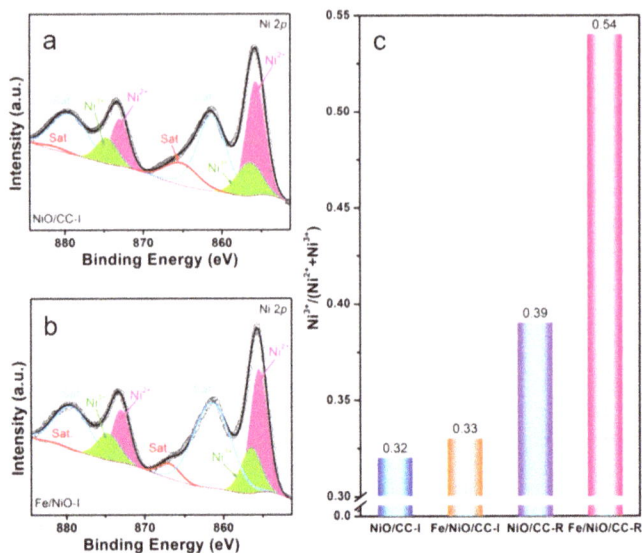

Figure 16. *High-resolution Ni-2p spectra of (a) NiO/CC and (b) Fe/NiO/CC before reconstruction. (c) Ni^{3+} concentrations in NiO/CC and Fe/NiO/CC before and after reconstruction. Reproduced with permission from ref [67]. Copyright 2022, Elsevier.*

Furthermore, the $Fe^{3+}/(Fe^{2+} + Fe^{3+})$ ratio increased from 0.27 for the fresh sample to 0.43 for the reconstructed sample, indicating a 16% increase in Fe^{3+} concentration, as depicted in **Figure 17a-b**. This observation highlights that the introduction of iron not only accelerates the reconstruction process of NiO but also enhances the Ni^{3+} content. Consequently, the incorporation of iron has a positive effect on improving the OER properties of the catalyst.

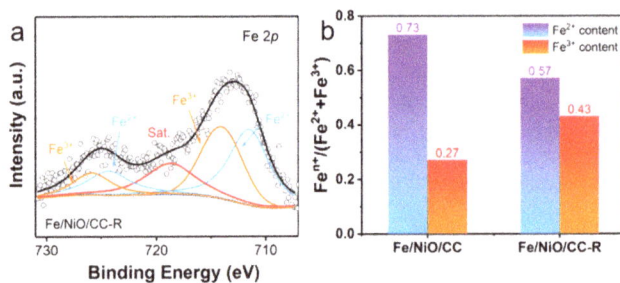

Figure 17. *(a) High-resolution XPS Fe-2p spectrum of Fe/NiO/CC after reconstruction. (b) Fe^{2+} and Fe^{3+} concentrations in the Fe/NiO/CC catalyst before (Fe/NiO/CC) and after (Fe/NiO/CC-R) the reconstruction. Reproduced with permission from ref [67]. Copyright 2022, Elsevier.*

The OER characteristics of the NiO/CC and Fe/NiO/CC catalysts after reconstruction were investigated in 1.0 M KOH using a three-electrode system. **Figure 18a** presents the polarization curves of the catalysts. The Fe/NiO/CC-0.5 catalyst requires an overpotential of only 288 mV to achieve an oxygen evolution current density of 100 mA cm^{-2}, which is significantly lower than that of NiO/CC (434 mV) and even lower than that of the commercial IrO$_2$ electrocatalyst. This demonstrates the superior activity of Fe/NiO/CC-0.5 compared to Fe/NiO/CC-0.1 and Fe/NiO/CC-1.0, highlighting the importance of the doped Fe concentration, as shown in **Figure 18b**. The overpotentials of the NiO/CC and Fe/NiO/CC-0.5 catalysts are compared in **Figure 18c**, further indicating the higher catalytic activity of the sample after Fe doping. **Figure 18d** displays the Tafel slopes of the series of Fe/NiO/CC catalysts compared to NiO/CC. The Fe/NiO/CC-0.5 catalyst exhibits a smaller Tafel slope of 72.6 mV dec^{-1}, which is only half that of NiO/CC (146.4 mV dec^{-1}). This confirms the accelerated OER kinetics achieved through Fe doping.

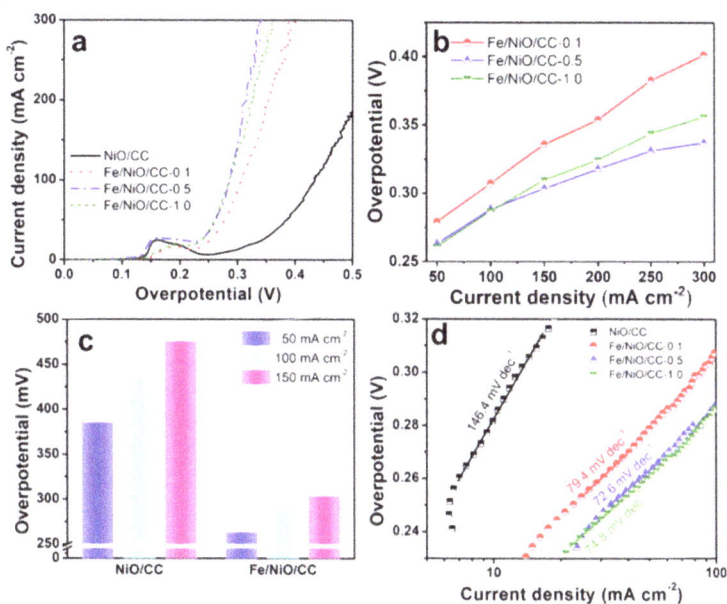

***Figure 18**. (a) Polarization curves. (b) Overpotentials required for the Fe/NiO/CC-n (n = 0.1, 0.5, and 1.0) catalysts to generate different current densities. (c) Overpotentials required to produce different current densities for the NiO/CC and Fe/NiO/CC catalysts. (d) Tafel slopes. Reproduced with permission from ref [67]. Copyright 2022, Elsevier.*

To investigate the reaction kinetics in OER, electrochemical impedance spectroscopy (EIS) was performed. An equivalent circuit model consisting of a series solution resistance (R_s), a constant phase element (CPE), and a R_{ct} was derived, as depicted in **Figure 19a**. Compared to the large charge transfer resistance observed for pure NiO/CC, the Fe/NiO/CC catalyst exhibits a smaller charge transfer resistance. This indicates higher charge transfer efficiency and faster reaction kinetics in the Fe/NiO/CC catalyst. The ECSA of the Fe/NiO/CC catalyst was measured to be 3.17 mF cm^{-2}, which is 11 times higher than that of the pristine NiO/CC (0.26 mF cm^{-2}), as shown in **Figure 19b**. This indicates that the Fe/NiO/CC catalyst has a significantly larger electrochemically active surface area, which can contribute to enhanced catalytic performance.

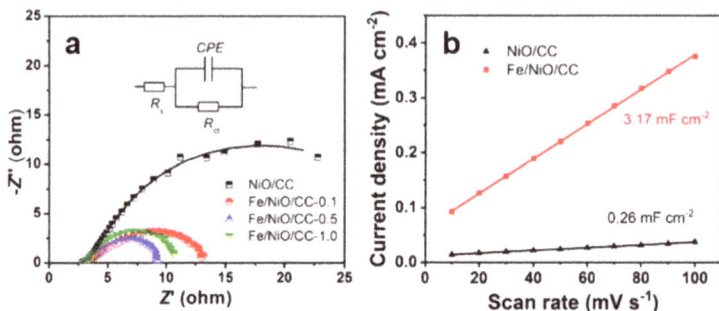

***Figure 19**. (a) EIS and (b) ECSA of the catalysts. Reproduced with permission from ref [67]. Copyright 2022, Elsevier.*

Long-term stability is a crucial factor for the commercial application of OER catalysts. The Fe/NiO/CC catalyst demonstrates excellent activity during continuous assessment for 50 hours at a constant overpotential of approximately 300 mV. This implies outstanding stability in OER, further highlighting the potential of the Fe/NiO/CC catalyst for practical applications.

In the study conducted by Huang et al. [68], they investigated the mechanism by which Fe ions in alkaline electrolytes promote the oxygen evolution activity of nickel-based catalysts. They prepared Ni(OH)$_2$ nanosheet arrays (NSAs) on nickel foam (NF) using a simple ultrasonic method, which served as the working electrode. The effect of different concentrations of Fe(III) in 1.0 M KOH solutions as the electrolyte on the catalytic activity of OER was studied.

To examine the impact of Fe(III) ions in the electrolyte on the performance of the Ni(OH)$_2$ NSAs catalyst, the OER properties were measured using a three-electrode system in purified KOH and unpurified KOH solutions with varying concentrations of Fe(III). The results indicated that the OER catalytic activity increased with the increase in Fe(III) concentration in the electrolyte. The Ni(OH)$_2$ catalyst exhibited the highest OER activity in the electrolyte with a Fe(III) concentration of 500 μM. At a potential of 1.6 V *vs.* RHE, the current density substantially increased from 36 to 1052 mA cm^{-2}. The onset potential of the Ni(OH)$_2$ catalyst in the Fe(III)-containing KOH electrolyte was 1.45 V *vs.* RHE, significantly lower than the onset potential in the pure KOH electrolyte (1.60 V *vs.* RHE), as shown in **Figure 20a**. Additionally, the Tafel slope of the Ni(OH)$_2$ catalyst in the Fe(III)-containing KOH electrolyte was measured to be 51.9 mV dec^{-1}, which was much smaller than the Tafel slope in the pure KOH electrolyte (108.2 mV dec^{-1}). These findings demonstrated that the adsorption of OH$^-$ was relatively easier and strongly catalyzed by Fe(III), leading to an improvement in the OER activity of the Ni(OH)$_2$

NSAs catalyst [69]. Therefore, the presence of Fe(III) ions in the electrolyte enhances the catalytic performance of the Ni-based catalysts in OER.

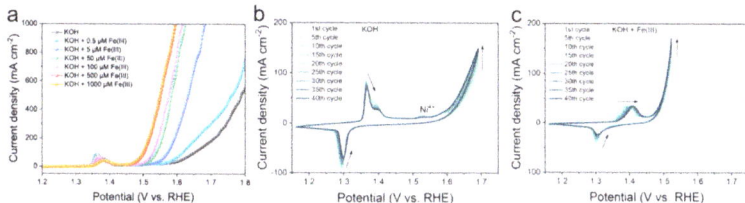

Figure 20. *(a) Polarization curves, CV profiles for 40 cycles of the Ni(OH)₂ in (b) KOH and (c) KOH + Fe(III) at a scanning rate of 5 mV s⁻¹. Reproduced with permission from ref [68]. Copyright 2023, Elsevier.*

In order to elucidate the mechanism behind the enhanced catalytic activity of the Ni(OH)$_2$ NSAs catalyst for OER in the presence of Fe(III), in situ EIS analysis was conducted at various potentials. In the high-frequency region of the EIS plots, the phase angle value measured in the Fe(III)-containing KOH electrolyte was lower than that in the pure KOH electrolyte, indicating faster reaction kinetics in the presence of Fe(III) [70]. The addition of Fe(III) to the electrolyte facilitates the fast transfer of electrically active species (OH⁻), thereby promoting the OER process [69, 71, 72].

In the Nyquist plots, the R_{ct} decreases with applied potentials. The Ni(OH)$_2$ catalyst exhibits a smaller semicircle at all potentials in the Fe(III)-containing KOH electrolyte, indicating smaller charge transfer resistance and faster reaction kinetics facilitated by the presence of Fe(III). Additionally, CV curves provide information on phase transition and OER activity based on the redox features of Ni. The polarization curves exhibit a wide anodic peak at approximately 1.37 V *vs.* RHE corresponding to the transition of Ni^{2+} to Ni^{3+}, as shown in **Figure 20b-c**. The oxygen evolution current density increases with the number of CV scanning cycles, as more Ni^{2+} species are converted into Ni^{3+}/Ni^{4+} species on the surface of Ni(OH)$_2$.

However, the redox peak observed in the Fe(III)-containing KOH electrolyte is positively shifted compared to the peaks measured in the pure KOH electrolyte. The introduction of Fe(III) into the electrolyte hinders the transition of Ni^{2+} to Ni^{3+}, thus eliminating the peaks corresponding to the oxidation of Ni^{2+} to Ni^{3+}/Ni^{4+}. It is well-known that Fe(III) enhances the formation of intermediates via the following route: Fe(III) → (Ni)Fe^{x+} → (Ni)Fe-OH → (Ni)Fe-O → (Ni)Fe-OOH → Fe(III), to produce the oxygen gas (**Figure 21**). Furthermore, the Fe(III) in the electrolyte generates the Fe-NiOOH/Ni-FeOOH, leading to the production of oxygen gas. Furthermore, Fe(III) in the electrolyte generates Fe-NiOOH/Ni-FeOOH interfaces in situ on the surface of the Ni(OH)$_2$ catalyst, facilitating the OER process. The stable chemistry of the Ni species in the catalyst contributes to its robust OER capability.

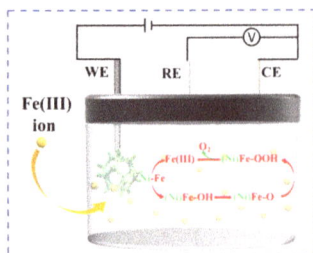

Figure 21. *Proposed Fe(III)-catalyzed mechanism of OER on Ni(OH)$_2$ in alkaline electrolyte. WE: working electrode, RE: reference electrode, CE: counter electrode. Reproduced with permission from ref [68]. Copyright 2023, Elsevier.*

In a separate study by Peng et al. [73], they reported the in-situ segregation of metallic cobalt nanoparticles on vanadium nitride (VN) nanosheets, forming a nanoparticle/nanosheet hierarchical structure known as Co/VN. This novel composite enhances the rapid charge transfer, resulting in improved OER activity and enhanced stability. The Co/VN nanosheet catalyst was prepared through a traditional hydrothermal reaction followed by nitridation in an ammonia atmosphere. The hydrothermal reaction was employed to prepare the Co$_2$V$_2$O$_7$ nanosheet precursors. Metallic Co nanoparticles were formed and uniformly embedded into the nanosheet via a high-temperature in-situ phase segregation strategy. Simultaneously, the vanadium species were reduced to nitride, forming the VN nanosheet, as illustrated in **Figure 22a**. The formation of the nanoparticle/nanosheet hierarchical structure provides a large number of stably exposed active sites, contributing to the enhanced OER activity of the catalyst.

In **Figure 22b**, SEM images of the hydrothermal product of Co$_2$V$_2$O$_7$ with a nanosheet structure and smooth surface are shown. However, after high-temperature nitridation in an ammonia atmosphere (**Figure 22c**), the surface of the nanosheets becomes rough and embedded with numerous nanoparticles, indicating the separation of nanoparticles from the nanosheets and the formation of a porous structure after ammonia nitridation. XRD analysis confirms the presence of a composite of VN and metallic Co based on the diffraction peak in the XRD pattern. TEM and HR-TEM results further reveal that the nanosheets are composed of VN, while the nanoparticles correspond to metallic Co. The OER performance of the Co/VN catalyst is evaluated through polarization curves, as depicted in **Figure 22d**. The results demonstrate that the Co/VN catalyst requires an overpotential of 320 mV to achieve a current density of 10 mA cm^{-2}, surpassing the performance of other reported cobalt-based catalysts for OER.

The superior OER performance of the hierarchical Co/VN catalyst can be attributed to the in-situ segregation of metallic Co nanoparticles on the conductive VN nanosheets after ammonia nitridation. This segregation results in the highly dispersed metallic Co nanoparticles, which greatly enhance the OER activity. This was further confirmed by the ECSA analysis, which revealed that the hierarchical Co/VN catalyst exhibits the largest C_{dl}. Additionally, the excellent stability of the hierarchical Co/VN catalyst is attributed to the strong binding between the metallic Co nanoparticles and the conductive VN substrate, which originates from the in-situ phase separation process.

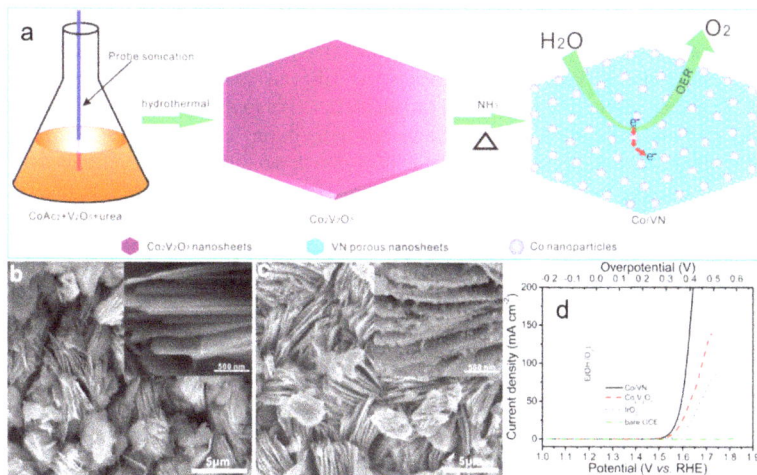

Figure 22. *(a) Schematic illustration of the preparation procedures of the porous Co/VN nanosheet catalyst for high-performance OER. SEM images of (b) hydrothermal product $Co_2V_2O_7$ and (c) Co/VN with insets showing the corresponding high-magnification images. (d) Polarization curves of Co/VN, $Co_2V_2O_7$, and commercial IrO_2 catalysts compared to bare GCE. Reproduced with permission from ref [73]. Copyright 2017, Elsevier.*

3.4. Three-Dimensional Nanomaterials

Nanomaterials typically refer to materials with sizes ranging from 1 nm to 100 nm in at least one dimension. Three-dimensional (3D) nanomaterials, on the other hand, are not nano-sized materials themselves but are constructed using low-dimensional nanomaterials as building blocks. Despite their larger size, 3D nanomaterials retain similar physicochemical properties as their low-dimensional counterparts [74-76]. The unique physicochemical properties of 3D nanomaterials, including a porous structure, high specific surface area, large surface/volume ratio, and numerous surface active sites, often contribute significantly to enhanced catalytic efficiency for OER [77-79].

Feng et al. [80] reported a cost-effective, highly active, and extremely durable 3D intercalated NiFe/C array grown on NF using a generic method. The synthesis process involved the growth of Fe-doped Ni(OH)$_2$ nanosheets, referred to as (Fe)Ni(OH)$_2$, on nickel foam via a hydrothermal method. Subsequently, glucose was uniformly distributed on the (Fe)Ni(OH)$_2$ nanosheets, followed by high-temperature calcination to produce the NiFe/C arrays. This unique 3D structure exhibits high electrical conductivity and a large specific surface area, providing a greater number of active sites for catalysis. Additionally, the nanopores present on the outer

Materials Research Forum LLC
https://doi.org/10.21741/9781644903070

carbon layer effectively enhance the transport of OH⁻ ions and fast electron transfer, resulting in high catalytic activity.

Furthermore, the in-situ growth of the 3D structure on the nickel foam enhances the conductivity of the catalyst and strengthens its binding to the substrate. The sandwich structure, as observed in the SEM image in **Figure 23a**, consists of NiFe nanoparticles encapsulated within a graphitized carbon layer. The graphitized carbon layer serves to stabilize the catalyst, preventing agglomeration and detachment of the NiFe nanoparticles during electrochemical operations. As a result, the NiFe/C catalyst exhibits excellent long-term durability in OER applications.

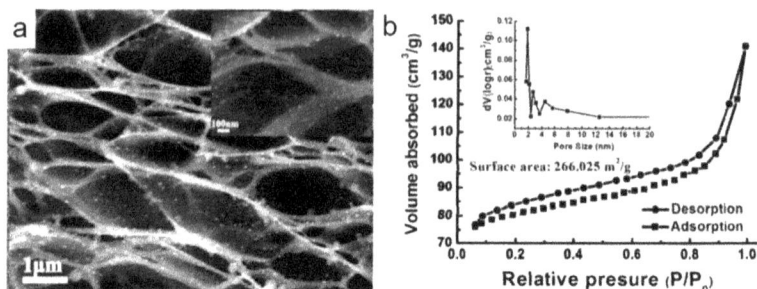

Figure 23. (a) SEM image of 3D sandwiched NiFe/C arrays with inset showing the images obtained at higher magnification. (b) BET isotherm of 3D sandwiched NiFe/C arrays. Reproduced with permission from ref [80]. Copyright 2016, American Chemical Society.

The as-prepared NiFe/C arrays possess a substantial specific surface area of approximately $266.025 \ m^2 \ g^{-1}$, as determined by nitrogen adsorption and desorption isotherms (**Figure 23b**). The Barrett-Joyner-Halenda (BJH) analysis of the catalyst reveals a pore size distribution ranging from 1.8 to 4.5 nm, with a concentration of pores at around 1.9 nm, as shown in the inset of **Figure 23b**. This large specific surface area provides greater exposure to active sites, while the porous structure expands the volume during successive redox reactions and facilitates the rapid diffusion of OH⁻ ions, thereby contributing to excellent OER activity.

Electrochemical results demonstrate that the NiFe/C catalyst with sandwich structures exhibits superior catalytic activity compared to NiFe nanoparticles and $(Fe)Ni(OH)_2$ nanosheets (**Figure 24a-b**). The NiFe/C arrays catalyst demonstrates an initial potential of 1.43 V *vs*. RHE in OER, accompanied by a small Tafel slope of 30 mV dec⁻¹. These results highlight the outstanding catalytic activity and fast kinetics of the NiFe/C arrays catalyst for OER. Additionally, the sandwich-structured NiFe/C arrays catalyst exhibits excellent stability. These findings further confirm that the unique 3D structure, characterized by high conductivity and a large specific surface area, facilitates the OER process and contributes to the catalyst's performance.

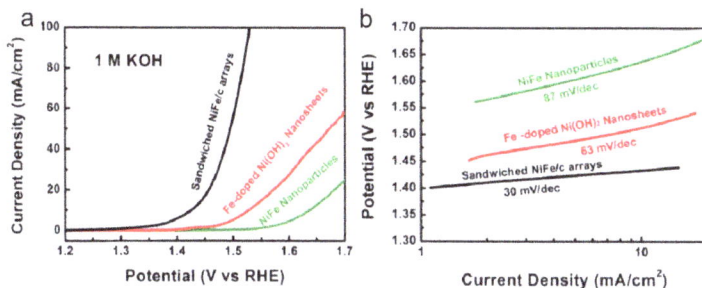

Figure 24 (a) Polarization curves and (b) Tafel plots of the 3D sandwiched NiFe/C arrays, Fe-doped Ni(OH)$_2$ nanosheets, and NiFe nanoparticles. Reproduced with permission from ref [80]. Copyright 2016, American Chemical Society.

Deng et al. [81] conducted a study on the development of efficient OER electrocatalysts by precisely regulating the vacancies of C≡N (V$_{C≡N}$) in NiFe Prussian blue analogues (PBAs), as shown in **Figure 25**. The morphology of the sample was observed using SEM. The NiFe PBAs precursor exhibited a 3D structure with a smooth surface, uniformly distributed on the conductive CC substrate. However, after the heat treatment, the surface of the NiFe PBAs gradually became rough and porous. To confirm the formation of V$_{C≡N}$ after heat treatment, Raman scattering spectra were obtained. The spectra revealed a gradual decrease in the peaks corresponding to Fe^{3+}–C≡N–Ni^{2+} and Fe^{2+}–C≡N–Ni^{2+}, indicating the formation of V$_{C≡N}$. This characterization technique confirmed the successful regulation of V$_{C≡N}$ content in the NiFe PBAs through the heat treatment process.

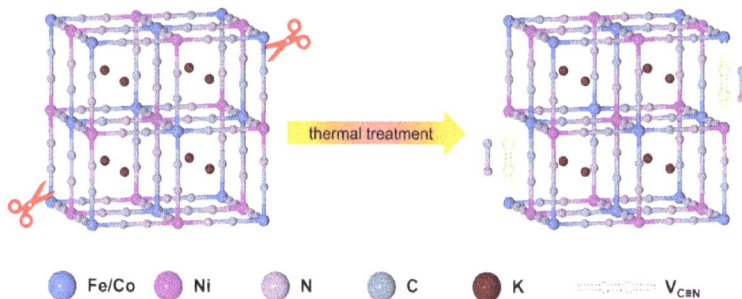

Figure 25. C≡N vacancy engineering of PBAs. Reproduced with permission from ref [81]. Copyright 2023, Elsevier.

The surface chemical state of the NiFe PBAs was analyzed using XPS. The XPS results revealed that as the heat treatment temperature increased, the concentration of Ni(III) decreased, while the concentration of Fe(III) increased on the surface of the catalyst. This indicates the presence of electronic interaction between Ni and Fe atoms after heat treatment, with electron transfer occurring from Fe to Ni. The Fe sites with higher valence states have a stronger ability to adsorb OH⁻ ions, promoting the formation of active FeOOH species and playing a crucial role in enhancing the OER properties of the catalyst. High-resolution XPS spectra of the N-1s region indicated that the concentration of $V_{C\equiv N}$ increased with the temperature. At a temperature of 300 °C, nearly 80% of the surface of the NiFe PBAs exhibited C≡N defects. The precisely controlled $V_{C\equiv N}$ content in PBAs-based catalysts plays a significant role in enhancing catalytic activity and reaction kinetics [82].

The electrochemical properties of the NiFe PBAs catalyst were evaluated using a three-electrode system in a 1.0 M KOH electrolyte. The PBA-250 catalyst, which had a deficiency of C≡N, displayed a lower overpotential compared to the NiFe PBA precursor, indicating improved catalytic activity in OER through the introduction of $V_{C\equiv N}$ into the NiFe PBAs catalyst. By applying an overpotential of only 270 mV, a current density of 50 mA cm⁻² was achieved, along with a small Tafel slope of 53 mV dec⁻¹, as shown in **Figure 26a-b**. Additionally, the PBA-250 catalyst exhibited faster reaction kinetics and a larger ECSA compared to other catalysts, as indicated in **Figure 26c**. The enhanced electrochemical performance was attributed to the optimized electronic structure and surface reconstruction of Fe active sites achieved by the moderate $V_{C\equiv N}$ content in the NiFe PBAs catalyst. However, when there were excessive C≡N vacancies, as shown in the PBA-300 catalyst, the 3D structure of the catalyst was damaged, leading to blockage of electron transfer and inferior OER performance.

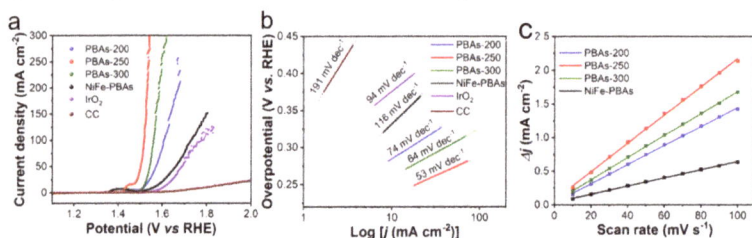

Figure 26. *Electrochemical characteristics of the electrocatalysts: (a) Polarization curves, (b) Tafel plots, and (c) ECSA. Reproduced with permission from ref [81]. Copyright 2023, Elsevier.*

Xu et al. [83] presented a hybrid nanostructure consisting of a hollow CoS core and CeOₓ nanoparticles precisely modulated on its surface. The synthesis process involved the preparation of the hollow CoS core from ZIF-67 nanoparticles using a solution method. Subsequently, CeOₓ nanoparticles were grown in situ on the surface of the hollow CoS core through a hydrothermal method. This resulted in the formation of a 3D hollow CeOₓ/CoS hybrid nanostructure, as depicted in **Figure 27**. Generally, Co-based metal-organic frameworks (MOFs) exhibit a large surface area and abundant pore structures. The SEM images also revealed that the surface of the CeOₓ/CoS nanostructure displayed a highly porous morphology, which provides numerous accessible sites for electrocatalysis. This porous structure is beneficial for enhancing the catalytic activity [84].

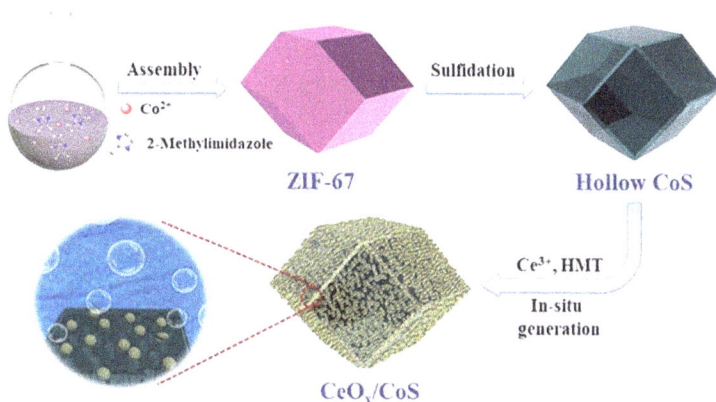

Figure 27. *Schematically illustration of the fabrication process of the hybrid nanostructure of CeO$_x$/CoS. Reproduced with permission from ref [83]. Copyright 2021, Wiley-VCH.*

The XPS analysis provided insights into the electronic interaction between CoS and CeO$_x$ in the 3D CeO$_x$/CoS catalyst. The Co-3p spectra of the CeO$_x$/CoS catalyst exhibited a negative shift compared to pure CoS. Similarly, the Ce-3d spectrum showed a positive shift compared to pure CeO$_2$. These findings suggest the presence of electron coupling and transfer in the 3D CeO$_x$/CoS catalyst [85]. Moreover, the introduction of Ce into the catalyst led to an increase in the Co^{2+}/Co^{3+} molar ratio in the CeO$_x$/CoS catalyst [86]. This is attributed to the redox transition of Ce^{3+} to Ce^{4+}, which promotes the rapid diffusion of oxygen through the formation of oxygen vacancies. Consequently, a protective CeO$_x$ thin layer forms on the surface of CoS, effectively inhibiting the oxidation and corrosion of CoS. This mechanism contributes to the improved catalytic activity and kinetics of the OER [87]. Furthermore, the optimized electronic structure of the CeO$_x$/CoS hybrid structure resulted in impressive electrochemical performance. The catalyst achieved a current density of 10 mA cm^{-2} at a low overpotential of 269 mV, accompanied by a small Tafel slope of 50 mV dec^{-1} in OER. These results demonstrate the significantly enhanced catalytic activity of the 3D CeO$_x$/CoS catalyst in OER.

3.5. Others

In recent years, researchers have focused on exploring amorphous materials as promising catalysts for the OER. Amorphous materials offer advantages over crystalline materials due to their increased exposure to active sites and enhanced flexibility in the presence of coordination unsaturation [88, 89]. Various types of amorphous materials have been investigated, including amorphous transition metal compounds [90-93], amorphous alloys [94-96], amorphous-crystalline core-shell structures [97-99], and amorphous-crystalline heterostructure [100, 101].

For example, Duan et al. [90] reported the synthesis of an amorphous NiFeMo oxide catalyst through rapid co-precipitation. The synthesis procedure involved mixing a high concentration of Ni/Fe chloride precursor in water, followed by the rapid addition of a 1.0 M Na$_2$MoO$_4$·2H$_2$O aqueous solution under ultrasonic conditions. This resulted in the formation of a milky yellow emulsion within just 2 minutes. The obtained NiFeMo oxide was subsequently freeze-dried. The formation of amorphous NiFeMo oxides (a-NiFeMo) was confirmed by SEM and TEM images. The results demonstrated that the amorphous NiFeMo oxides exhibited rapid surface self-reconstruction, leading to the formation of an oxygen vacancy-rich layer and displaying significant OER activity.

The a-NiFeMo catalyst exhibited an onset potential of only 1.45 V *vs.* RHE in OER, obtained in a 0.1 M KOH electrolyte under O_2 saturation. To achieve a current density of 10 mA cm^{-2}, the catalyst required an overpotential of 280 mV, with a small Tafel slope of 49 mV dec^{-1}. Additionally, the a-NiFeMo catalyst displayed a low R_{ct} of only 32 Ω. Long-term stability tests confirmed the preservation of the amorphous structure, indicating excellent OER properties including low overpotential, small Tafel slope, fast reaction kinetics, and long-term durability.

In another study, Cai et al. [94] successfully synthesized amorphous nickel-iron alloy compounds at room temperature. The XRD pattern exhibited a broad diffraction peak centered at approximately 45°, indicating the amorphous nature of the nickel-iron alloy compound. The morphology and structure were further confirmed by TEM and high-angle annular dark-field imaging-scanning transmission electron microscopy (HAADF-STEM), as shown in **Figure 28a-b**. The amorphous nature of the product was also supported by the diffraction rings observed in the selected area electron diffraction (SAED) pattern (inset of **Figure 28a**).

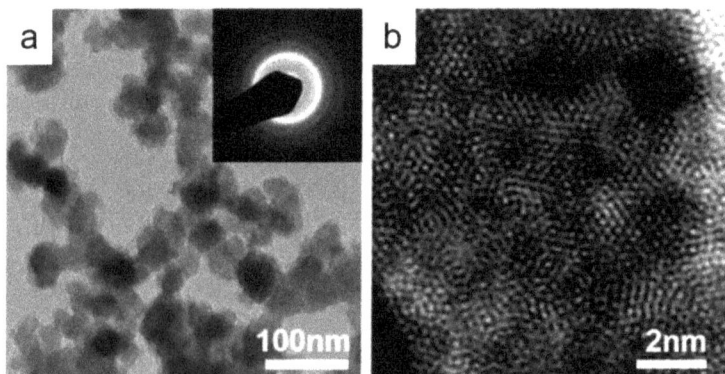

Figure 28. (a) TEM with inset showing the SAED pattern and (b) HAADF-STEM images of the amorphous NiFe alloy catalyst. Reproduced with permission from ref [94]. Copyright 2020, American Chemical Society.

The electrochemical properties of the amorphous NiFe alloy compound in the OER were evaluated using a rotating disk electrode. Among the catalysts tested, the amorphous nickel-iron alloy catalyst with a Ni/Fe atomic ratio of 3:1 demonstrated outstanding catalytic activity in OER. It required an overpotential of only 242 mV to achieve a current density of 10 mA cm^{-2}, which was 100 mV lower than that of the crystalline catalyst. To assess the durability of the amorphous NiFe alloy catalyst during the OER process, the concentrations of nickel and iron in the electrolyte were measured before and after stability tests using inductively coupled plasma-optical emission spectrometry (ICP-OES). The results showed that the concentrations of Ni and Fe in the electrolyte after the long-term tests were below the detection limit of ICP-OES (0.1 < ppm), indicating excellent durability of the amorphous NiFe alloy catalyst. In addition, isotope (^{18}O) labelling was employed to investigate OH adsorption during the OER process. The intensity of the $H_2^{18}O$ signal of the amorphous catalyst was two orders of magnitude higher than that of the crystalline catalyst under the same test conditions. This suggests that the amorphous catalyst exposed more active sites, enabling a higher level of OH adsorption.

4.   Conclusion and Prospects

In conclusion, this chapter has highlighted the importance of catalysts for the OER and discussed various types of catalyst materials and their performance in OER. The OER is a critical reaction in electrocatalytic hydrogen production and plays a vital role in renewable energy conversion and storage systems. Further research and development in the field of catalysts for high-performance OER hold promising prospects.

(i) Catalyst design and optimization: Researchers can explore the design and optimization of catalyst materials at the atomic and molecular levels. This includes studying the effects of composition, structure, and morphology on catalytic activity, stability, and selectivity. Tailoring catalysts using advanced synthesis techniques, such as atomic layer deposition, sol-gel methods, and self-assembly, can lead to the development of highly active and durable catalysts.

(ii) Bifunctional catalysts: Developing bifunctional catalysts that can simultaneously catalyze both the OER and the HER is important for overall water splitting efficiency. Designing catalysts with optimized interfaces, hybrid materials, or heterostructures can enhance catalytic activity and facilitate efficient charge transfer between different reaction sites.

(iii) Catalyst stability and durability: Ensuring long-term stability and durability of catalysts is crucial for practical applications. Researchers can investigate degradation mechanisms, surface passivation, and ion transport phenomena to mitigate catalyst degradation and improve long-term performance. Protective coatings, encapsulation techniques, and catalyst support materials can also be explored to enhance stability.

(iv) Computational modelling and machine learning: The use of computational modelling and machine learning approaches can aid in catalyst discovery, design, and optimization. High-throughput computational screening can accelerate the identification of promising catalyst candidates and guide experimental efforts. Machine learning techniques can also help in understanding structure-activity relationships and predicting catalytic performance.

(v) Catalyst integration and device engineering: The integration of catalysts into practical device architectures, such as electrolyzers and photoelectrochemical cells, is a crucial aspect of research. Optimizing catalyst-electrode interfaces, exploring novel electrode materials, and developing efficient reactor designs are important for enhancing overall system performance.

References

[1]   H. Dau, C. Limberg, T. Reier, M. Risch, S. Roggan, P. Strasser, The mechanism of water oxidation: From electrolysis via homogeneous to biological catalysis, ChemCatChem, 2 (2010) 724-761. https://doi.org/10.1002/cctc.201000126

[2]   R.H. Zhang, N. Dubouis, M. Ben Osman, W. Yin, M.T. Sougrati, D.A.D. Corte, D. Giaume, A. Grimaud, A Dissolution/Precipitation Equilibrium on the Surface of Iridium-Based Perovskites Controls Their Activity as Oxygen Evolution Reaction Catalysts in Acidic Media, Angew. Chem. Int. Ed., 58 (2019) 4571-4575. https://doi.org/10.1002/anie.201814075

[3]   X. Peng, Y. Yan, X. Jin, C. Huang, W. Jin, B. Gao, P.K. Chu, Recent advance and prospectives of electrocatalysts based on transition metal selenides for efficient water splitting, Nano Energy, 78 (2020) 105234. https://doi.org/10.1016/j.nanoen.2020.105234

[4]   D.Y. Kuo, J.K. Kawasaki, J.N. Nelson, J. Kloppenburg, G. Hautier, K.M. Shen, D.G. Schlom, J. Suntivich, Influence of Surface Adsorption on the Oxygen Evolution Reaction on $IrO_2(110)$, J. Am. Chem. Soc., 139 (2017) 3473-3479. https://doi.org/10.1021/jacs.6b11932

[5]   F.-Y. Chen, Z.-Y. Wu, Z. Adler, H. Wang, Stability challenges of electrocatalytic oxygen evolution reaction: From mechanistic understanding to reactor design, Joule, 5 (2021) 1704-1731. https://doi.org/10.1016/j.joule.2021.05.005

[6]   D. Vikraman, S. Hussain, I. Rabani, A. Feroze, M. Ali, Y.-S. Seo, S.-H. Chun, J. Jung, H.-S. Kim, Engineering $MoTe_2$ and Janus SeMoTe nanosheet structures: First-principles roadmap and practical uses in hydrogen evolution reactions and symmetric supercapacitors, Nano Energy, 87 (2021) 106161. https://doi.org/10.1016/j.nanoen.2021.106161

[7]   T. Reier, H.N. Nong, D. Teschner, R. Schlögl, P. Strasser, Electrocatalytic Oxygen Evolution Reaction in Acidic Environments-Reaction Mechanisms and Catalysts, Adv. Energy Mater., 7 (2016) 1601275. https://doi.org/10.1002/aenm.201601275

[8]   A.J. Tkalych, H.L. Zhuang, E.A. Carter, A Density Functional $+U$ Assessment of Oxygen Evolution Reaction Mechanisms on β-NiOOH, ACS Catal., 7 (2017) 5329-5339. https://doi.org/10.1021/acscatal.7b00999

[9]   M.D. Bhatt, J.Y. Lee, Theoretical insights into the mechanism of oxygen evolution reaction (OER) on pristine $BiVO_4(001)$ and $BiVO_4(110)$ surfaces in acidic medium both in the gas and solution (water) phases, Nanotechnology, 32 (2021) 335401. https://doi.org/10.1088/1361-6528/abfcfd

[10]   X. Li, Y. Sun, Q. Wu, H. Liu, W. Gu, X. Wang, Z. Cheng, Z. Fu, Y. Lu, Optimized Electronic Configuration to Improve the Surface Absorption and Bulk Conductivity for Enhanced Oxygen Evolution Reaction, J. Am. Chem. Soc., 141 (2019) 3121-3128. https://doi.org/10.1021/jacs.8b12299

[11]   X. Xie, L. Du, L. Yan, S. Park, Y. Qiu, J. Sokolowski, W. Wang, Y. Shao, Oxygen Evolution Reaction in Alkaline Environment: Material Challenges and Solutions, Adv. Funct. Mater., 32 (2022) 2110036. https://doi.org/10.1002/adfm.202110036

[12]   Z.F. Huang, S. Xi, J. Song, S. Dou, X. Li, Y. Du, C. Diao, Z.J. Xu, X. Wang, Tuning of lattice oxygen reactivity and scaling relation to construct better oxygen evolution electrocatalyst, Nat. Commun., 12 (2021) 3992. https://doi.org/10.1038/s41467-021-24182-w

[13]   A. Grimaud, O. Diaz-Morales, B. Han, W.T. Hong, Y.L. Lee, L. Giordano, K.A. Stoerzinger, M.T.M. Koper, Y. Shao-Horn, Activating lattice oxygen redox reactions in metal oxides to catalyse oxygen evolution, Nat. Chem., 9 (2017) 457-465. https://doi.org/10.1038/nchem.2695

[14]   J.S. Kim, B. Kim, H. Kim, K. Kang, Recent Progress on Multimetal Oxide Catalysts for the Oxygen Evolution Reaction, Adv. Energy Mater., 8 (2018) 1702774. https://doi.org/10.1002/aenm.201702774

[15]   Q. Zhao, Z. Yan, C. Chen, J. Chen, Spinels: Controlled Preparation, Oxygen Reduction/Evolution Reaction Application, and Beyond, Chem. Rev., 117 (2017) 10121-10211. https://doi.org/10.1021/acs.chemrev.7b00051

[16]   S. Pedireddy, H.K. Lee, W.W. Tjiu, I.Y. Phang, H.R. Tan, S.Q. Chua, C. Troadec, X.Y. Ling, One-step synthesis of zero-dimensional hollow nanoporous gold nanoparticles with enhanced methanol electrooxidation performance, Nat. Commun., 5 (2014) 4947. https://doi.org/10.1038/ncomms5947

[17]   A. Yang, K. Su, S. Wang, Y. Wang, X. Qiu, W. Lei, Y. Tang, Self-stabilization of zero-dimensional PdIr nanoalloys at two-dimensional manner for boosting their OER and HER performance, Appl. Sur. Sci., 510 (2020) 145408. https://doi.org/10.1016/j.apsusc.2020.145408

[18]   U.P. Suryawanshi, U.V. Ghorpade, D.M. Lee, M. He, S.W. Shin, P.V. Kumar, J.S. Jang, H.R. Jung, M.P. Suryawanshi, J.H. Kim, Colloidal Ni_2P Nanocrystals Encapsulated in Heteroatom-Doped Graphene Nanosheets: A Synergy of 0D@2D Heterostructure Toward Overall Water Splitting, Chem. Mater., 33 (2020) 234-245. https://doi.org/10.1021/acs.chemmater.0c03543

[19]   D. Meng, L. Wei, J. Shi, Q. Jiang, J. Tang, A review of enhanced electrocatalytic composites hydrogen/oxygen evolution based on quantum dot, J. Ind. Eng. Chem., 121 (2023) 27-39. https://doi.org/10.1016/j.jiec.2023.01.014

[20]   R. Gui, H. Jin, Z. Wang, J. Li, Black phosphorus quantum dots: synthesis, properties, functionalized modification and applications, Chem. Soc. Rev., 47 (2018) 6795-6823. https://doi.org/10.1039/C8CS00387D

[21]   L. Tian, Z. Li, P. Wang, X. Zhai, X. Wang, T. Li, Carbon quantum dots for advanced electrocatalysis, J. Energy Chem., 55 (2021) 279-294. https://doi.org/10.1016/j.jechem.2020.06.057

[22]   M. Kuang, P. Han, L. Huang, N. Cao, L. Qian, G. Zheng, Electronic Tuning of Co, Ni-Based Nanostructured (Hydr)oxides for Aqueous Electrocatalysis, Adv. Funct. Mater., 28 (2018) 1804886. https://doi.org/10.1002/adfm.201804886

[23]   Y. Wang, W. Nong, N. Gong, T. Salim, M. Luo, T.L. Tan, K. Hippalgaonkar, Z. Liu, Y. Huang, Tuning Electronic Structure and Composition of FeNi Nanoalloys for Enhanced Oxygen Evolution Electrocatalysis via a General Synthesis Strategy, Small, 18 (2022) 2203340. https://doi.org/10.1002/smll.202203340

[24]   Y. Lu, D. Fan, Z. Chen, W. Xiao, C. Cao, X. Yang, Anchoring Co_3O_4 nanoparticles on MXene for efficient electrocatalytic oxygen evolution, Sci. Bull., 65 (2020) 460-466. https://doi.org/10.1016/j.scib.2019.12.020

[25]   J. Li, X. Xu, B. Zhang, W. Hou, S. Lv, Y. Shi, Controlled synthesis and fine-tuned interface of NiS nanoparticles/Bi_2WO_6 nanosheets heterogeneous as electrocatalyst for oxygen evolution reaction, Appl. Surf. Sci., 526 (2020) 146718. https://doi.org/10.1016/j.apsusc.2020.146718

[26]   H.-J. Liu, N. Yu, X.-Q. Yuan, H.-Y. Zhao, X.-Y. Zhang, Y.-M. Chai, B. Dong, 0D–2D Schottky heterostructure coupling of FeS nanosheets and Co_9S_8 nanoparticles for long-term industrial-level water oxidation, Nano Res., 16 (2022) 5929-5937. https://doi.org/10.1007/s12274-022-5176-7

[27]   X. Peng, X. Jin, N. Liu, P. Wang, Z. Liu, B. Gao, L. Hu, P.K. Chu, A high-performance electrocatalyst composed of nickel clusters encapsulated with a carbon network on TiN nanaowire arrays for the oxygen evolution reaction, Appl. Surf. Sci., 567 (2021) 150779. https://doi.org/10.1016/j.apsusc.2021.150779

[28]   X. Peng, A.M. Qasim, W. Jin, L. Wang, L. Hu, Y. Miao, W. Li, Y. Li, Z. Liu, K. Huo, K.-y. Wong, P.K. Chu, Ni-doped amorphous iron phosphide nanoparticles on TiN nanowire arrays: An advanced alkaline hydrogen evolution electrocatalyst, Nano Energy, 53 (2018) 66-73. https://doi.org/10.1016/j.nanoen.2018.08.028

[29]   B.R. Wygant, K. Kawashima, C.B. Mullins, Catalyst or Precatalyst? The Effect of Oxidation on Transition Metal Carbide, Pnictide, and Chalcogenide Oxygen Evolution Catalysts, ACS Energy Lett., 3 (2018) 2956-2966. https://doi.org/10.1021/acsenergylett.8b01774

[30]   L. Trotochaud, J.K. Ranney, K.N. Williams, S.W. Boettcher, Solution-cast metal oxide thin film electrocatalysts for oxygen evolution, J. Am. Chem. Soc., 134 (2012) 17253-17261. https://doi.org/10.1021/ja307507a

[31]   L. Hu, X. Zeng, X. Wei, H. Wang, Y. Wu, W. Gu, L. Shi, C. Zhu, Interface engineering for enhancing electrocatalytic oxygen evolution of NiFe LDH/NiTe heterostructures, Appl. Catal. B: Environ., 273 (2020) 119014. https://doi.org/10.1016/j.apcatb.2020.119014

[32]   X. Wu, Y. Yang, T. Zhang, B. Wang, H. Xu, X. Yan, Y. Tang, CeO_x-Decorated Hierarchical $NiCo_2S_4$ Hollow Nanotubes Arrays for Enhanced Oxygen Evolution

Reaction Electrocatalysis, ACS Appl. Mater. Interfaces, 11 (2019) 39841-39847. https://doi.org/10.1021/acsami.9b12221

[33]   J. Xia, J. Zhao, B. Huang, L. Xu, M. Luo, J. Wang, F. Luo, Y. Du, C.H. Yan, Efficient Optimization of Electron/Oxygen Pathway by Constructing Ceria/Hydroxide Interface for Highly Active Oxygen Evolution Reaction, Adv. Funct. Mater., 30 (2020) 1908367. https://doi.org/10.1002/adfm.201908367

[34]   J.X. Feng, S.H. Ye, H. Xu, Y.X. Tong, G.R. Li, Design and Synthesis of FeOOH/CeO$_2$ Heterolayered Nanotube Electrocatalysts for the Oxygen Evolution Reaction, Adv. Mater., 28 (2016) 4698-4703. https://doi.org/10.1002/adma.201600054

[35]   G. Zhang, B. Wang, J. Bi, D. Fang, S. Yang, Constructing ultrathin CoP nanomeshes by Er-doping for highly efficient bifunctional electrocatalysts for overall water splitting, J. Mater. Chem. A, 7 (2019) 5769-5778. https://doi.org/10.1039/C9TA00530G

[36]   S.M.N. Jeghan, G. Lee, One-dimensional hierarchical nanostructures of NiCo$_2$O$_4$, NiCo$_2$S$_4$ and NiCo$_2$Se$_4$ with superior electrocatalytic activities toward efficient oxygen evolution reaction, Nanotechnology, 31 (2020) 295405. https://doi.org/10.1088/1361-6528/ab8667

[37]   P. Chen, K. Xu, Z. Fang, Y. Tong, J. Wu, X. Lu, X. Peng, H. Ding, C. Wu, Y. Xie, Metallic Co$_4$N Porous Nanowire Arrays Activated by Surface Oxidation as Electrocatalysts for the Oxygen Evolution Reaction, Angew. Chem. Int. Ed., 54 (2015) 14710-14714. https://doi.org/10.1002/anie.201506480

[38]   Z. Li, K.H. Xue, J. Wang, J.G. Li, X. Ao, H. Sun, X. Song, W. Lei, Y. Cao, C. Wang, Cation and Anion Co-doped Perovskite Nanofibers for Highly Efficient Electrocatalytic Oxygen Evolution, ACS Appl. Mater. Interfaces, 12 (2020) 41259-41268. https://doi.org/10.1021/acsami.0c10045

[39]   D. Hu, R. Wang, P. Du, G. Li, Y. Wang, D. Fan, X. Pan, Electrospinning Ru doped Co$_3$O$_4$ porous nanofibers as promising bifunctional catalysts for oxygen evolution and oxygen reduction reactions, Ceram. Int., 48 (2022) 6549-6555. https://doi.org/10.1016/j.ceramint.2021.11.202

[40]   Z. Zhang, B. He, L. Chen, H. Wang, R. Wang, L. Zhao, Y. Gong, Boosting Overall Water Splitting via FeOOH Nanoflake-Decorated PrBa$_{0.5}$Sr$_{0.5}$Co$_2$O$_{5+\delta}$ Nanorods, ACS Appl. Mater. Interfaces, 10 (2018) 38032-38041. https://doi.org/10.1021/acsami.8b12372

[41]   X. Wang, Z. Li, J. Shi, Y. Yu, One-dimensional titanium dioxide nanomaterials: nanowires, nanorods, and nanobelts, Chem. Rev., 114 (2014) 9346-9384. https://doi.org/10.1021/cr400633s

[42]   L. Zhang, H. Zhao, S. Xu, Q. Liu, T. Li, Y. Luo, S. Gao, X. Shi, A.M. Asiri, X. Sun, Recent Advances in 1D Electrospun Nanocatalysts for Electrochemical Water Splitting, Small Struct., 2 (2020) 2000048. https://doi.org/10.1002/sstr.202000048

[43]   J. Li, G. Zheng, One-Dimensional Earth-Abundant Nanomaterials for Water-Splitting

Electrocatalysts, Adv. Sci., 4 (2017) 1600380. https://doi.org/10.1002/advs.201600380

[44] S.D. Ghadge, O.I. Velikokhatnyi, M.K. Datta, P.M. Shanthi, S. Tan, K. Damodaran, P.N. Kumta, Experimental and Theoretical Validation of High Efficiency and Robust Electrocatalytic Response of One-Dimensional (1D) (Mn,Ir)O_2: 10F Nanorods for the Oxygen Evolution Reaction in PEM-Based Water Electrolysis, ACS Catal., 9 (2019) 2134-2157. https://doi.org/10.1021/acscatal.8b02901

[45] F.X. Xiao, J. Miao, H.B. Tao, S.F. Hung, H.Y. Wang, H.B. Yang, J. Chen, R. Chen, B. Liu, One-dimensional hybrid nanostructures for heterogeneous photocatalysis and photoelectrocatalysis, Small, 11 (2015) 2115-2131. https://doi.org/10.1002/smll.201402420

[46] B. Weng, S. Liu, Z.-R. Tang, Y.-J. Xu, One-dimensional nanostructure based materials for versatile photocatalytic applications, RSC Adv., 4 (2014) 12685–12700. https://doi.org/10.1039/c3ra47910b

[47] A. Jiang, Z. Wang, Q. Li, M. Dong, Ionic Liquid-Assisted Synthesis of Hierarchical One-Dimensional MoP/NPC for High-Performance Supercapacitor and Electrocatalysis, ACS Sustain. Chem. Eng., 8 (2020) 6343-6351. https://doi.org/10.1021/acssuschemeng.0c00238

[48] N.G. Prakash, M. Dhananjaya, A.L. Narayana, D.P.M.D. Shaik, P. Rosaiah, O.M. Hussain, High Performance One Dimensional α-MoO_3 Nanorods for Supercapacitor Applications, Ceram. Int., 44 (2018) 9967-9975. https://doi.org/10.1016/j.ceramint.2018.03.032

[49] J. Vergara-Figueroa, O. Erazo, H. Pesenti, P. Valenzuela, A. Fernández-Pérez, W. Gacitúa, Development of Thin Films from Thermomechanical Pulp Nanofibers of Radiata Pine (Pinus radiata D. Don) for Applications in Bio-Based Nanocomposites, Fibers, 11 (2022) 1. https://doi.org/10.3390/fib11010001

[50] N.K. Swamy, K.N.S. Mohana, M.B. Hegde, A.M. Madhusudana, Fabrication of 1D graphene nanoribbon and malenized linseed oil-based nanocomposite: a highly impervious bio-based anti-corrosion coating material for mild steel, J. Appl. Electrochem., 52 (2022) 1133-1148. https://doi.org/10.1007/s10800-022-01692-z

[51] M. Cordero, C. Ruiz, D.A. Palacio, P. Turunen, A. Rowan, B.F. Urbano, Effect of low aspect ratio one-dimensional nanoparticles on properties of photocrosslinked alginate nanocomposite hydrogels, Int. J. Biol. Macromol. , 204 (2022) 635-643. https://doi.org/10.1016/j.ijbiomac.2022.02.059

[52] Y. Ding, K. Tu, I. Burgert, T. Keplinger, Janus wood membranes for autonomous water transport and fog collection, J. Mater. Chem. A, 8 (2020) 22001-22008. https://doi.org/10.1039/D0TA07544B

[53] P. Wang, J. Zhu, Z. Pu, R. Qin, C. Zhang, D. Chen, Q. Liu, D. Wu, W. Li, S. Liu, J. Xiao, S. Mu, Interfacial engineering of Co nanoparticles/Co_2C nanowires boosts overall water splitting kinetics, Appl. Catal. B: Environ., 296 (2021) 120334.

https://doi.org/10.1016/j.apcatb.2021.120334

[54] W. Chen, Y. Zhang, R. Huang, Y. Zhou, Y. Wu, Y. Hu, K. Ostrikov, Ni–Co hydroxide nanosheets on plasma-reduced Co-based metal–organic nanocages for electrocatalytic water oxidation, J. Mater. Chem. A, 7 (2019) 4950-4959. https://doi.org/10.1039/C9TA00070D

[55] K. Zhu, J. Chen, W. Wang, J. Liao, J. Dong, M.O.L. Chee, N. Wang, P. Dong, P.M. Ajayan, S. Gao, J. Shen, M. Ye, Etching-Doping Sedimentation Equilibrium Strategy: Accelerating Kinetics on Hollow Rh-Doped CoFe-Layered Double Hydroxides for Water Splitting, Adv. Funct. Mater., 30 (2020) 2003556. https://doi.org/10.1002/adfm.202003556

[56] X. Wang, A. Vasileff, Y. Jiao, Y. Zheng, S.Z. Qiao, Electronic and Structural Engineering of Carbon-Based Metal-Free Electrocatalysts for Water Splitting, Adv. Mater., 31 (2019) 1803625. https://doi.org/10.1002/adma.201803625

[57] B. Fei, Z. Chen, J. Liu, H. Xu, X. Yan, H. Qing, M. Chen, R. Wu, Ultrathinning Nickel Sulfide with Modulated Electron Density for Efficient Water Splitting, Adv. Energy Mater., 10 (2020) 2001963. https://doi.org/10.1002/aenm.202001963

[58] X. Zhang, F. Zhou, W. Pan, Y. Liang, R. Wang, General Construction of Molybdenum-Based Nanowire Arrays for pH-Universal Hydrogen Evolution Electrocatalysis, Adv. Funct. Mater., 28 (2018) 1804600. https://doi.org/10.1002/adfm.201804600

[59] T.L. Luyen Doan, D.T. Tran, D.C. Nguyen, H. Tuan Le, N.H. Kim, J.H. Lee, Hierarchical three-dimensional framework interface assembled from oxygen-doped cobalt phosphide layer-shelled metal nanowires for efficient electrocatalytic water splitting, Appl. Catal. B: Environ., 261 (2020) 118268. https://doi.org/10.1016/j.apcatb.2019.118268

[60] Y. Yang, W. Zhang, Y. Xiao, Z. Shi, X. Cao, Y. Tang, Q. Gao, CoNiSe$_2$ heteronanorods decorated with layered-double-hydroxides for efficient hydrogen evolution, Appl. Catal. B: Environ., 242 (2019) 132-139. https://doi.org/10.1016/j.apcatb.2018.09.082

[61] P. Cai, J. Huang, J. Chen, Z. Wen, Oxygen-Containing Amorphous Cobalt Sulfide Porous Nanocubes as High-Activity Electrocatalysts for the Oxygen Evolution Reaction in an Alkaline/Neutral Medium, Angew. Chem. Int. Ed., 56 (2017) 4858-4861. https://doi.org/10.1002/anie.201701280

[62] X. Kong, K. Xu, C. Zhang, J. Dai, S. Norooz Oliaee, L. Li, X. Zeng, C. Wu, Z. Peng, Free-Standing Two-Dimensional Ru Nanosheets with High Activity toward Water Splitting, ACS Catal., 6 (2016) 1487-1492. https://doi.org/10.1021/acscatal.5b02730

[63] M. Liu, K.-A. Min, B. Han, L.Y.S. Lee, Interfacing or Doping? Role of Ce in Highly Promoted Water Oxidation of NiFe-Layered Double Hydroxide, Adv. Energy Mater., 11 (2021) 2101281. https://doi.org/10.1002/aenm.202101281

[64] S.F. Zai, X.Y. Gao, C.C. Yang, Q. Jiang, Ce-Modified $Ni(OH)_2$ Nanoflowers Supported on $NiSe_2$ Octahedra Nanoparticles as High-Efficient Oxygen Evolution Electrocatalyst, Adv. Energy Mater., 11 (2021) 2101266. https://doi.org/10.1002/aenm.202101266

[65] F. Xiao, P. Zhou, R. Weng, P. Yang, W. Tang, L. Liao, Y. Wang, M. Zhao, W. Zhang, P. He, B. Jia, Co-Mn-S nanosheets decorated with CeO_2: A highly active electrocatalyst toward oxygen evolution reaction, J. Alloys Compd., 901 (2022) 163621. https://doi.org/10.1016/j.jallcom.2022.163621

[66] S. Liu, J. Zhu, M. Sun, Z. Ma, K. Hu, T. Nakajima, X. Liu, P. Schmuki, L. Wang, Promoting the hydrogen evolution reaction through oxygen vacancies and phase transformation engineering on layered double hydroxide nanosheets, J. Mater. Chem. A, 8 (2020) 2490-2497. https://doi.org/10.1039/C9TA12768B

[67] S. Xie, Y. Yan, S. Lai, J. He, Z. Liu, B. Gao, M. Javanbakht, X. Peng, P.K. Chu, Ni^{3+}-enriched nickel-based electrocatalysts for superior electrocatalytic water oxidation, Appl. Surf. Sci., 605 (2022) 154743. https://doi.org/10.1016/j.apsusc.2022.154743

[68] Z. Huang, A. Reda Woldu, X. Peng, P.K. Chu, Q.-X. Tong, L. Hu, Remarkably boosted water oxidation activity and dynamic stability at large-current–density of $Ni(OH)_2$ nanosheet arrays by Fe ion association and underlying mechanism, Chem. Eng. J., 477 (2023) 147155. https://doi.org/10.1016/j.cej.2023.147155

[69] J. Feng, M. Chen, P. Zhou, D. Liu, Y.-Y. Chen, B. He, H. Bai, D. Liu, W.F. Ip, S. Chen, D. Liu, W. Feng, J. Ni, H. Pan, Reconstruction optimization of distorted FeOOH/Ni hydroxide for enhanced oxygen evolution reaction, Mater. Today Energy, 27 (2022) 101005. https://doi.org/10.1016/j.mtener.2022.101005

[70] P. Zhou, X. Lv, S. Tao, J. Wu, H. Wang, X. Wei, T. Wang, B. Zhou, Y. Lu, T. Frauenheim, X. Fu, S. Wang, Y. Zou, Heterogeneous-Interface-Enhanced Adsorption of Organic and Hydroxyl for Biomass Electrooxidation, Adv. Mater., 34 (2022) 2204089. https://doi.org/10.1002/adma.202204089

[71] C. Chen, M. Sun, F. Zhang, H. Li, M. Sun, P. Fang, T. Song, W. Chen, J. Dong, B. Rosen, P. Chen, B. Huang, Y. Li, Adjacent Fe Site boosts electrocatalytic oxygen evolution at Co site in single-atom-catalyst through a dual-metal-site design, Energy Environ. Sci., 16 (2023) 1685-1696. https://doi.org/10.1039/D2EE03930C

[72] S. Wang, K. Zhao, Z. Chen, L. Wang, Z. Qi, J. Hao, W. Shi, New insights into cations effect in oxygen evolution reaction, Chem. Eng. J., 433 (2022) 133518. https://doi.org/10.1016/j.cej.2021.133518

[73] X. Peng, L. Wang, L. Hu, Y. Li, B. Gao, H. Song, C. Huang, X. Zhang, J. Fu, K. Huo, P.K. Chu, In situ segregation of cobalt nanoparticles on VN nanosheets via nitriding of $Co_2V_2O_7$ nanosheets as efficient oxygen evolution reaction electrocatalysts, Nano Energy, 34 (2017) 1-7. https://doi.org/10.1016/j.nanoen.2017.02.016

[74] Y. Liu, B. Workalemahu, X. Jiang, The Effects of Physicochemical Properties of Nanomaterials on Their Cellular Uptake In Vitro and In Vivo, Small, 13 (2017) 1701815.

https://doi.org/10.1002/smll.201701815

[75]   Y. Yu, Y. Shi, B. Zhang, Synergetic Transformation of Solid Inorganic-Organic Hybrids into Advanced Nanomaterials for Catalytic Water Splitting, Acc. Chem. Res., 51 (2018) 1711-1721. https://doi.org/10.1021/acs.accounts.8b00193

[76]   N.K. Chaudhari, H. Jin, B. Kim, K. Lee, Nanostructured materials on 3D nickel foam as electrocatalysts for water splitting, Nanoscale, 9 (2017) 12231-12247. https://doi.org/10.1039/C7NR04187J

[77]   Y. Fang, X.Y. Yu, X.W.D. Lou, Formation of Hierarchical Cu-Doped CoSe₂ Microboxes via Sequential Ion Exchange for High-Performance Sodium-Ion Batteries, Adv. Mater., 30 (2018) 1706668. https://doi.org/10.1002/adma.201706668

[78]   K. Karuppasamy, R. Bose, D. Vikraman, S. Ramesh, H.S. Kim, E. Alhseinat, A. Alfantazi, H.-S. Kim, Revealing the effect of various organic ligands on the OER activity of MOF-derived 3D hierarchical cobalt oxide @ carbon nanostructures, J. Alloys Compd., 934 (2023) 167909. https://doi.org/10.1016/j.jallcom.2022.167909

[79]   Y. He, Z. Yin, Z. Wang, H. Wang, W. Xiong, B. Song, H. Qin, P. Xu, G. Zeng, Metal-organic frameworks as a good platform for the fabrication of multi-metal nanomaterials: design strategies, electrocatalytic applications and prospective, Adv. Colloid Interface Sci. , 304 (2022) 102668. https://doi.org/10.1016/j.cis.2022.102668

[80]   Y. Feng, H. Zhang, L. Fang, Y. Mu, Y. Wang, Uniquely Monodispersing NiFe Alloyed Nanoparticles in Three-Dimensional Strongly Linked Sandwiched Graphitized Carbon Sheets for High-Efficiency Oxygen Evolution Reaction, ACS Catal., 6 (2016) 4477-4485. https://doi.org/10.1021/acscatal.6b00481

[81]   W. Deng, B. Xu, Q. Zhao, S. Xie, W. Jin, X. Zhang, B. Gao, Z. Liu, Z. Abd-Allah, P.K. Chu, X. Peng, C≡N vacancy engineering of Prussian blue analogs for the advanced oxygen evolution reaction, J. Environ. Chem. Eng., 11 (2023) 109407. https://doi.org/10.1016/j.jece.2023.109407

[82]   Z.Y. Yu, Y. Duan, J.D. Liu, Y. Chen, X.K. Liu, W. Liu, T. Ma, Y. Li, X.S. Zheng, T. Yao, M.R. Gao, J.F. Zhu, B.J. Ye, S.H. Yu, Unconventional CN vacancies suppress iron-leaching in Prussian blue analogue pre-catalyst for boosted oxygen evolution catalysis, Nat. Commun., 10 (2019) 2799. https://doi.org/10.1038/s41467-019-10698-9

[83]   H. Xu, J. Cao, C. Shan, B. Wang, P. Xi, W. Liu, Y. Tang, MOF-Derived Hollow CoS Decorated with CeOₓ Nanoparticles for Boosting Oxygen Evolution Reaction Electrocatalysis, Angew. Chem. Int. Ed., 57 (2018) 8654-8658. https://doi.org/10.1002/anie.201804673

[84]   L. Yu, H. Hu, H.B. Wu, X.W. Lou, Complex Hollow Nanostructures: Synthesis and Energy-Related Applications, Adv. Mater., 29 (2017) 1604563. https://doi.org/10.1002/adma.201604563

[85]   L. Xu, Q. Jiang, Z. Xiao, X. Li, J. Huo, S. Wang, L. Dai, Plasma-Engraved Co₃O₄

Nanosheets with Oxygen Vacancies and High Surface Area for the Oxygen Evolution Reaction, Angew. Chem. Int. Ed., 55 (2016) 5277-5281. https://doi.org/10.1002/anie.201600687

[86]   B. Feng, I. Sugiyama, H. Hojo, H. Ohta, N. Shibata, Y. Ikuhara, Atomic structures and oxygen dynamics of CeO_2 grain boundaries, Sci. Rep., 6 (2016) 20288. https://doi.org/10.1038/srep20288

[87]   M. Breitwieser, C. Klose, A. Hartmann, A. Büchler, M. Klingele, S. Vierrath, R. Zengerle, S. Thiele, Cerium Oxide Decorated Polymer Nanofibers as Effective Membrane Reinforcement for Durable, High-Performance Fuel Cells, Adv. Energy Mater., 7 (2016) 1602100. https://doi.org/10.1002/aenm.201602100

[88]   S. Anantharaj, S. Noda, Amorphous Catalysts and Electrochemical Water Splitting: An Untold Story of Harmony, Small, 16 (2020) 1905779. https://doi.org/10.1002/smll.201905779

[89]   Y. Zhai, X. Ren, J. Yan, S. Liu, High Density and Unit Activity Integrated in Amorphous Catalysts for Electrochemical Water Splitting, Small Struct., 2 (2020) 2000096. https://doi.org/10.1002/sstr.202000096

[90]   Y. Duan, Z.Y. Yu, S.J. Hu, X.S. Zheng, C.T. Zhang, H.H. Ding, B.C. Hu, Q.Q. Fu, Z.L. Yu, X. Zheng, J.F. Zhu, M.R. Gao, S.H. Yu, Scaled-Up Synthesis of Amorphous NiFeMo Oxides and Their Rapid Surface Reconstruction for Superior Oxygen Evolution Catalysis, Angew. Chem. Int. Ed., 58 (2019) 15772-15777. https://doi.org/10.1002/anie.201909939

[91]   Y. Zhang, F. Gao, D. Wang, Z. Li, X. Wang, C. Wang, K. Zhang, Y. Du, Amorphous/Crystalline Heterostructure Transition-Metal-based Catalysts for High-Performance Water Splitting, Coord. Chem. Rev., 475 (2023) 214916. https://doi.org/10.1016/j.ccr.2022.214916

[92]   Y. Yang, L. Zhuang, T.E. Rufford, S. Wang, Z. Zhu, Efficient water oxidation with amorphous transition metal boride catalysts synthesized by chemical reduction of metal nitrate salts at room temperature, RSC Adv., 7 (2017) 32923-32930. https://doi.org/10.1039/C7RA02558K

[93]   X. Wang, X. Han, R. Du, Z. Liang, Y. Zuo, P. Guardia, J. Li, J. Llorca, J. Arbiol, R. Zheng, A. Cabot, Unveiling the role of counter-anions in amorphous transition metal-based oxygen evolution electrocatalysts, Appl. Catal. B: Environ., 320 (2023) 121988. https://doi.org/10.1016/j.apcatb.2022.121988

[94]   W. Cai, R. Chen, H. Yang, H.B. Tao, H.Y. Wang, J. Gao, W. Liu, S. Liu, S.F. Hung, B. Liu, Amorphous versus Crystalline in Water Oxidation Catalysis: A Case Study of NiFe Alloy, Nano Lett., 20 (2020) 4278-4285. https://doi.org/10.1021/acs.nanolett.0c00840

[95]   K.M. Cole, D.W. Kirk, S.J. Thorpe, In Situ Raman Study of Amorphous and Crystalline Ni-Co Alloys for the Alkaline Oxygen Evolution Reaction, J. Electrochem. Soc., 165 (2018) J3122-J3129. https://doi.org/10.1149/2.0131815jes

[96]   S. Yan, M. Zhong, C. Wang, X. Lu, Amorphous aerogel of trimetallic FeCoNi alloy for highly efficient oxygen evolution, Chem. Eng. J., 430 (2022) 132955. https://doi.org/10.1016/j.cej.2021.132955

[97]   S. Anantharaj, P.N. Reddy, S. Kundu, Core-Oxidized Amorphous Cobalt Phosphide Nanostructures: An Advanced and Highly Efficient Oxygen Evolution Catalyst, Inorg. Chem., 56 (2017) 1742-1756. https://doi.org/10.1021/acs.inorgchem.6b02929

[98]   L. Li, H. Sun, X. Xu, M. Humayun, X. Ao, M.F. Yuen, X. Xue, Y. Wu, Y. Yang, C. Wang, Engineering Amorphous/Crystalline Rod-like Core-Shell Electrocatalysts for Overall Water Splitting, ACS Appl. Mater. Interfaces, 14 (2022) 50783-50793. https://doi.org/10.1021/acsami.2c13417

[99]   H. Sheng, H. Qu, B. Zeng, Y. Li, C. Xia, C. Li, L. Cao, B. Dong, Enriched Fe Doped on Amorphous Shell Enable Crystalline@Amorphous Core-Shell Nanorod Highly Efficient Electrochemical Water Oxidation, Small, 19 (2023) 2300876. https://doi.org/10.1002/smll.202300876

[100]   M. Kuang, J. Zhang, D. Liu, H. Tan, K.N. Dinh, L. Yang, H. Ren, W. Huang, W. Fang, J. Yao, X. Hao, J. Xu, C. Liu, L. Song, B. Liu, Q. Yan, Amorphous/Crystalline Heterostructured Cobalt-Vanadium-Iron (Oxy)hydroxides for Highly Efficient Oxygen Evolution Reaction, Adv. Energy Mater., 10 (2020) 2002215. https://doi.org/10.1002/aenm.202002215

[101]   M. Singh, D.C. Cha, T.I. Singh, A. Maibam, D.R. Paudel, D.H. Nam, T.H. Kim, S. Yoo, S. Lee, A critical review on amorphous-crystalline heterostructured electrocatalysts for efficient water splitting, Mater. Chem. Front., 7 (2023) 6254-6280. https://doi.org/10.1039/D3QM00940H

Electrocatalytic Hydrogen Production: Catalysts and Applications | Materials Research Forum LLC
Materials Research Foundations **165** (2024) | https://doi.org/10.21741/9781644903070

CHAPTER 11

Electrolyzer for Electrocatalytic Hydrogen Production

Xiang Peng*

Hubei Key Laboratory of Plasma Chemistry and Advanced Materials, Engineering Research Center of Phosphorus Resources Development and Utilization of Ministry of Education, School of Materials Science and Engineering, Wuhan Institute of Technology, Wuhan 430205, China

xpeng@wit.edu.cn

Abstract

Electrolyzer devices are crucial equipment for implementing hydrogen production through water electrolysis. They play a significant role in regulating reaction conditions and improving efficiency. Different types of electrolyzer cells, including alkaline electrolyzer cells, membrane alkaline electrolyzer cells, and polymer electrolyte membrane electrolyzer cells, have contributed uniquely to the development of clean hydrogen energy technology. Continuous improvements in the performance of electrolyzer devices can lead to more efficient, economical, and sustainable hydrogen production.

Keywords

Electrolyzer, Alkaline Electrolyzer, Acidic Electrolyzer, Solid Oxide Electrolyzer, Intermediate-Temperature Electrolyzer

1. Introduction

The process of hydrogen production through water electrolysis involves the use of split water into hydrogen gas (H_2) and oxygen gas (O_2). This process consists of two reactions known as the hydrogen evolution reaction (HER) and the oxygen evolution reaction (OER) [1-3]. These reactions occur at the cathode and anode, respectively. During the HER, protons (H^+) and electrons (e^-) derived from water combine to form hydrogen gas H_2. On the other hand, in the OER, water is oxidized to produce oxygen gas. Electrocatalysts play a vital role in enhancing the efficiency of both the HER and OER [4, 5]. They help reduce the barriers to the reaction, facilitate electron transfer, and accelerate reaction kinetics, thereby improving the overall energy efficiency [6, 7]. Various materials, such as metals [8], metal alloys [9, 10], metal compounds [11, 12], and carbon-based materials [13, 14], are commonly employed in the design of electrocatalysts.

Electrolyzer devices are crucial equipment for implementing hydrogen production through water electrolysis. They play a significant role in regulating reaction conditions and improving efficiency [15]. Different types of electrolyzer cells, including alkaline electrolyzer cells [16], membrane alkaline electrolyzer cells [15, 17], and polymer electrolyte membrane electrolyzer cells [18, 19], have contributed uniquely to the development of clean hydrogen energy technology. The design and performance of these devices directly impact hydrogen production kinetics and energy conversion efficiency. Therefore, conducting in-depth research on the construction and optimization of electrolyzer devices is essential for the advancement of clean hydrogen energy production technology. Continuous improvements in the performance of electrolyzer devices can lead to more efficient, economical, and sustainable hydrogen production.

2.   Alkaline Electrolyzers

Under alkaline conditions, the use of an electrolyte, such as sodium hydroxide (NaOH) or potassium hydroxide (KOH), is preferred over acidic conditions due to several advantages. Alkaline electrolyzers play a crucial role in hydrogen production through water electrolysis. The use of an alkaline electrolyte provides higher stability and tolerance for the electrolyzer compared to acidic conditions. This milder environment allows for improved durability and longevity of the electrolyzer system. Employing electrolytes like NaOH and KOH enables efficient catalysis of both HER and OER at relatively lower potentials. This means that less electricity is consumed during the electrolysis process, resulting in higher energy efficiency.

Additionally, alkaline conditions often allow for the use of relatively inexpensive catalysts, such as nickel-based alloys. This contributes to cost reduction and makes alkaline electrolyzers more economically competitive compared to other types of electrolyzers. The economic efficiency of alkaline electrolyzers makes them highly favorable in clean hydrogen technologies. Operating conditions in alkaline electrolyzers are relatively mild, with lower temperature and pressure requirements. This reduces the demands on equipment and manufacturing costs. Moreover, alkaline electrolytes exhibit strong adaptability to different water qualities, which enhances the flexibility of electrolyzers in practical applications, as they can effectively operate with varying water sources.

2.1.   Alkaline Electrolyzer

Alkaline electrolysis, a well-established technology, has been employed for large-scale hydrogen production since the early 20th century, with capacities reaching the MW scale [20]. Alkaline water electrolysis is a mature technology that covers a wide capacity range, from 1 to 1000 Nm^3 h^{-1} and is commercially viable for large-scale hydrogen production, with investment costs ranging between €700 and €1300/kWel [21, 22]. This method operates by circulating a liquid electrolyte, typically a 25–35 wt% aqueous KOH solution, at temperatures between 70 and 90 °C and pressures below 3.2 MPa. Commercial electrolyzers have a rated efficiency, based on the high heating value (HHV), ranging from 65 to 80%, resulting in specific energy consumption of 4.5-5.5 kWh Nm^{-3}. These electrolyzers are designed to have a lifetime exceeding 30 years. Alkaline water electrolysis is widely considered the most favorable technology for large-scale green hydrogen production due to its proven maturity, durability, and relatively low cost [23].

Ren et al. [24] reported a semi-empirical model to describe the behavior of a real-scale alkaline electrolyzer system. The system integrates an alkaline electrolyzer with a nominal production rate of 50 Nm^3 h^{-1}, as shown in **Figure 1**. The power sources could supply up to 250 kW of DC electricity. The electrolyzer consisted of 45 electrolysis cells, with each cell containing an anode, a porous diaphragm, and a cathode housed by bipolar plates. The electrodes had a geometrical area of 1.24 m^2, and their rated current was 2480 A, corresponding to a current density of 2000 A m^{-2}. The electrolyzer could operate at pressures up to 1.6 MPa and temperatures up to 90 °C. The electrolyte used was approximately 30 wt% KOH, which circulated in the system at a flow rate of 2.7 m^3 h^{-1}. During electrolysis, hydrogen and oxygen bubbles generated on the electrodes were carried by the electrolyte into separators, where the gas was separated from the bulk electrolyte by gravity. The bottoms of the separators were connected through a pipe to maintain the electrolyte level and minimize the differential pressure between the cathode and anode sides of the electrolysis cell. Additionally, the liquid electrolyte levels in each separator were continuously monitored to prevent gas crossover through the connecting pipe.

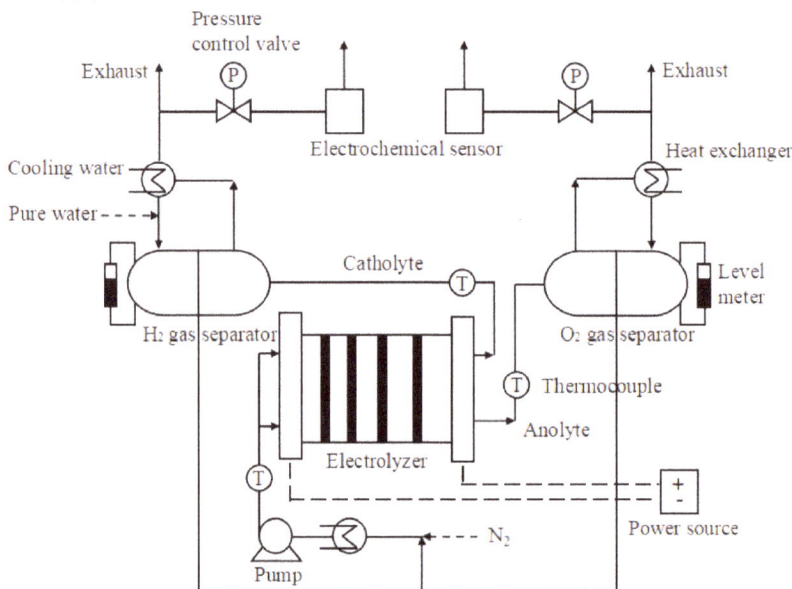

Figure 1. *Schematic diagram of the alkaline water electrolysis system. Reproduced with permission from ref [24]. Copyright 2022, Elsevier.*

The design and optimization of liquid-gas separators and the control of liquid electrolyte levels were crucial for improving the performance and safety of the pressurized water electrolysis system. The purity of the gases produced directly affected the operational safety of the system. The gases exited through the upper exit of the gas-liquid separators and passed through a mist trap and heat exchanger to remove most of the water content. Oxygen was subsequently vented, while hydrogen was sent to the purification unit to reduce humidity and oxygen levels. Online electrochemical sensors were utilized to monitor the purity of the produced hydrogen and oxygen, ensuring quality control and safety monitoring. All operations were carried out to prevent the hydrogen level from approaching the explosion limit of the gas mixture. The system pressure was maintained by controlling the gas release rate through a pneumatic valve installed at the hydrogen output. The pneumatic valve at the oxygen output was adjusted to balance the liquid levels between separators. The valve opening was adjusted using the proportional-integral-differential (PID) control technique.

The heat generated during the electrolysis reaction led to an increase in the temperature of the electrolyte and the produced gases within the electrolyzer. To maintain heat balance, waste heat was extracted by recycling cooling water through a gas and liquid heat exchanger. The gas heat exchanger could condense water vapor impurities before the high-temperature gas from separators enters the next unit. Simultaneously, the liquid heat exchanger ensured that the electrolyte temperature remained stable at the electrolyzer inlet, thereby controlling the electrolysis temperature. Deionized water, with a maximum electrical conductivity of 5 mS cm^{-1}, was automatically introduced into the hydrogen separator based on the reduction in liquid electrolyte level to meet water consumption requirements. Nitrogen was employed to purge the internal volume and pipes both before starting up and after shutdown to prevent potential explosion hazards.

The liquid KOH electrolyte had the potential to react with CO_2 in the air, leading to the formation of K_2CO_3 carbonate precipitation. This carbonate precipitation could obstruct the porous structure of the electrode and diaphragm, insulating electron pathways and contributing to an overall loss of electrical efficiency. To prevent this, the alkaline water electrolysis system maintained a positive pressure nitrogen gas blanket during idle states to prevent CO_2 from infiltrating the lye container. The control unit, developed based on a programmable logic controller (PLC), collected and processed information from various sensors such as gas analyzers, level sensors, pressure transducers, thermocouples, etc. Safe and automated operations were achieved through user-defined parameters and control logic. Alarms and interlocks were triggered if any abnormal operations were detected. In the event of a power outage, the control unit was temporarily supported by power from an uninterruptible power supply (UPS).

2.2.   Anion Exchange Membrane Electrolyzer

The recent technology of anion-exchange membrane (AEM) water electrolysis is still in the early stages of development [25]. In this technology, the diaphragm traditionally used in classical alkaline water electrolysis technology is replaced by a membrane that facilitates the transport of hydroxide ions. This substitution has the potential to reduce gas crossover and cell resistances, similar to proton-exchange membrane (PEM) technology. Additionally, AEM water electrolysis allows for the utilization of less expensive and more abundant catalysts associated with alkaline technology. The primary objective of AEM water electrolysis is to combine the cost-effective materials of the alkaline system with the performance characteristics of PEM water electrolysis. The ultimate goal is to develop an efficient and low-cost electrolyzer suitable for storing energy from renewables in the form of hydrogen.

The AEM water electrolyzer consists of a membrane electrode assembly (MEA), which includes an anode for the OER, a cathode for the HER, anode/cathode bipolar plates, and the anion exchange membrane [26]. **Figure 2** illustrates the membrane electrode assembly and the catalyst electrode layer, where the ionomer and catalyst come into contact. The catalyst facilitates the conduction of electrons, while the ionomer, if present, may provide a pathway for the transportation of hydroxide ions as well [27].

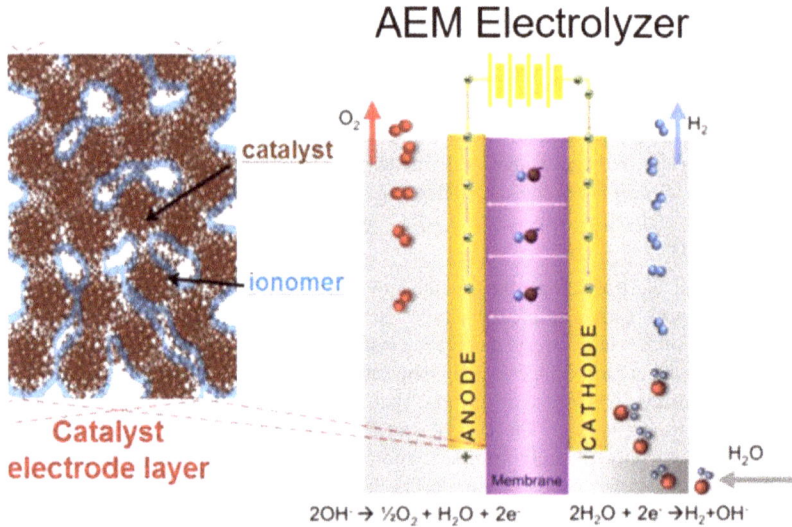

Figure 2. *Schematic of the membrane electrode assembly and catalyst electrode layer for AEM electrolyzer. Reproduced with permission from ref [28]. Copyright 2019, Elsevier.*

AEM electrolysis offers a significant advantage by allowing for the utilization of platinum group metal (PGM)-free electrocatalysts for both the HER and OER in separate reaction chambers. This capability contributes to reducing the capital cost of AEM water electrolysis. However, there are current challenges in AEM catalyst development, including the optimization of chemical composition, stability, and overall activity when integrated into the AEM system [29]. PGM-free electrocatalysts typically exhibit relatively low mass-specific activity compared to noble metal-based catalysts, leading to higher catalyst loading on the MEA and increased ohmic resistance losses.

The HER in the AEM electrolysis requires additional energy to break the robust H-O covalent bonds in water, which is considered a crucial rate-determining step [30]. The rational design of non-PGM electrocatalysts for HER in AEM systems plays a vital role in the future of hydrogen production. Some studies have reported superior performance of AEM electrolyzers using non-PGM HER catalysts compared to those based on PGM materials, further demonstrating the reliable capability of the AEM approach to efficiently hydrogen production.

Ni-based hybrids are commonly employed as HER catalysts in AEM electrolysis; however, their catalytic activity is reported to be inferior to that of PGM materials. For example, Ni nano-powders with a mass loading of 2 mg cm^{-2} have been utilized as a HER catalyst in AEM electrolysis systems, demonstrating a current density of 100 mA cm^{-2} at an applied voltage of 1.99 V and a temperature of 44 °C [31]. Xiao et al. [32] reported a water electrolysis electrolyzer that utilizes alkaline polymer catalysts and non-precious metal catalysts, working only with pure water. The MEA is fabricated by sandwiching a self-crosslinking quaternary ammonia polysulfone (xQAPS) membrane between a NiFe anode and a NiMo cathode, both impregnated with xQAPS ionomer. The sponge-like structure of the NiMo alloy is believed to expose a much higher specific surface area. Within this assembly, OH$^-$ is oxidized to O$_2$, and H$_2$O is reduced to H$_2$. The polymeric electrolyte membrane transports OH$^-$ rather than H$^+$, and unlike alkaline water electrolyzers, the reactant is pure water. The anode and cathode have a high catalyst loading of 40 mg cm^{-2}, which led to an AEM cell performance of 400 mA cm^{-2} at a cell voltage of 1.8-1.85 V under 70 °C in the pure water-electrolyte. Faid et al. [33] synthesized Ni$_{0.9}$Mo$_{0.1}$ nanosheets and found that they presented excellent cathode performance comparable to Pt nanoparticles in an AEM cell. The NiMo supported on Vulcan X72 carbon (NiMo/X72) achieved a current density of 1 A cm^{-2} at a voltage of 1.9 V in 1 M KOH, compared to the voltage of 1.8 V of the Pt particle catalyst to reach the same current density, suggesting a small performance gap between the NiMo alloy and Pt-PGM catalysts.

Nickel/iron (NiFe)-based materials have been found to exhibit excellent oxygen evolution reaction (OER) activity across various applications [34]. Their performance in AEM cells has also been evaluated. NiFe-based layered double hydroxides (LDHs), known for their excellent OER properties, have been widely studied [35]. A strategy involving particle size reduction was proposed to enhance the performance of NiFe-LDHs based on an inverse relationship between lateral size and effective surface area. This increase in the lateral parameter was hypothesized to significantly boost hydroxyl ions in the feed. Another advantage was the reduction in the film thickness of the catalyst layer, minimizing ohmic losses on the MEA. With this approach, a one-spot spontaneous gelation-deflocculation method, as illustrated in **Figure 3**, was introduced to produce ultra-fine NiFe-LDHs particles with a small lateral size of less than 10 nm [36].

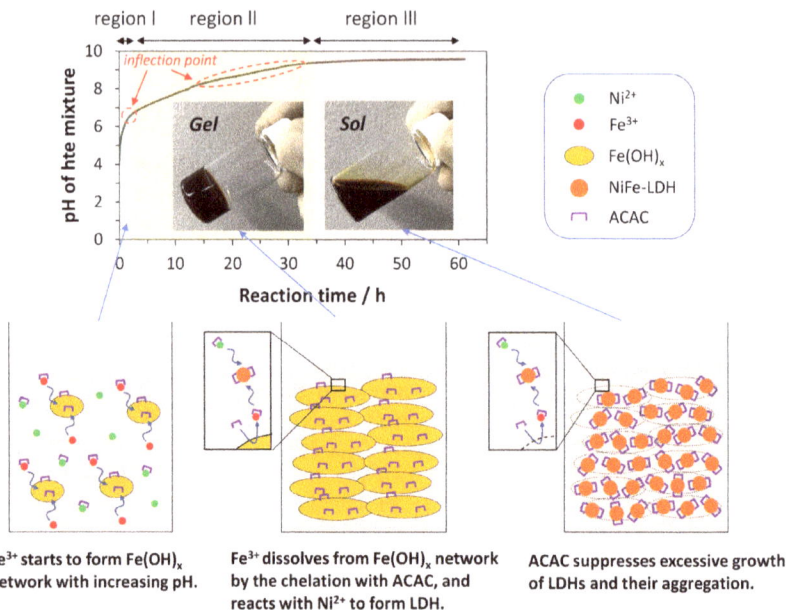

Fe^{3+} starts to form Fe(OH)$_x$ network with increasing pH.

Fe^{3+} dissolves from Fe(OH)$_x$ network by the chelation with ACAC, and reacts with Ni^{2+} to form LDH.

ACAC suppresses excessive growth of LDHs and their aggregation.

Figure 3. Time course of the pH of the reaction mixture during NiFe-LDH synthesis and schematic illustration of the synthesis mechanism of NiFe-LDH. The inset shows photographs of the reaction mixture taken at the corresponding periods. Reproduced with permission from ref [36]. Copyright 2020, American Chemical Society.

The as-prepared NiFe-LDHs exhibited significant OER activity in both half- and single cells, outperforming IrO$_x$ catalysts. It required a low overpotential of 247 mV to reach a current density of 10 mA cm^{-2}, comparable to 281 mV for IrO$_x$ at 10 mA cm^{-2}. In a full AEM cell, the MEA utilizing ultra-fine NiFe-LDHs as the anode electrocatalyst demonstrated superior performance. It achieved a conversion efficiency of 74.7% and a cell voltage of 1.59 V under working conditions of 1.0 A cm^{-2} at 80 °C. This performance was recognized as the highest among PGM-free catalyst-decorated MEAs, as illustrated in **Figure 4**.

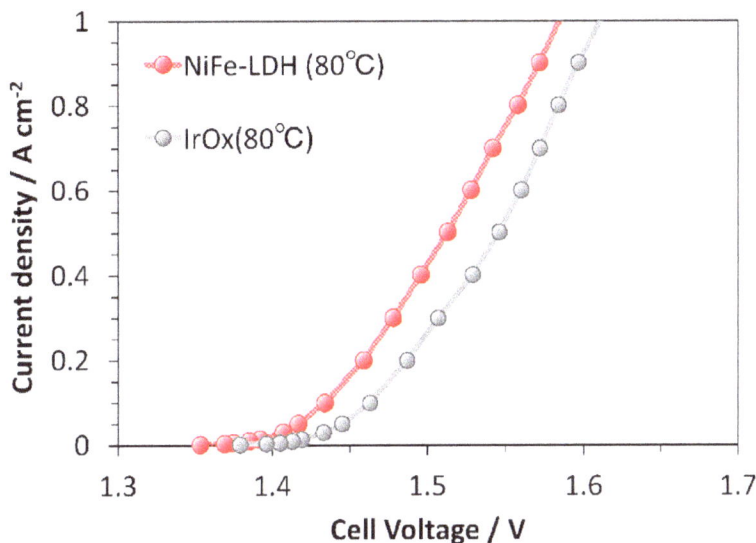

Figure 4. Current-voltage curves obtained for MEAs using (red) NiFe-LDH and (gray) IrO$_x$ at 80 °C as an anode catalyst. Reproduced with permission from ref [36]. Copyright 2020, American Chemical Society.

While Ni/Fe-based materials have been extensively explored as OER catalysts, their application in AEM electrolysis has been limited due to their poor stability. To improve their performance in practical devices, further research should focus on refining their composition, architecture, and employing facile synthesis methods.

The anion exchange membrane is a crucial component in AEM electrolysis systems as it facilitates the transport of hydroxyl ions from the cathode chamber to the anode chamber while preventing gas crossover during the electrochemical operation [37, 38]. AEMs are typically constructed using a hydrocarbon polymer backbone with side chains containing anion exchange functional groups. Polysulfone (PSF) or polystyrene (PS) linked with divinylbenzene (DVB) are commonly used as the polymeric backbone, and ion exchange groups such as ammonium (-NH_3^+, -RNH_2^+, -RN^+, =R_2N^+) or phosphonium (-R_3P^+) groups are incorporated [39, 40].

An advanced ion exchange membrane should exhibit high permselectivity, excellent ionic conductivity, robust thermal and mechanical stability, as well as superior chemical stability [41-43]. Efforts are being made to incorporate N-spirocyclic quaternary ammonium (QA) cations more efficiently into the polymers. Pham et al. [16] proposed a cyclo-polycondensation strategy to anchor the spiro-centered QA cations on the polymer chain. This approach, illustrated schematically in **Figure 5**, resulted in the formation of "spiro-ionenes" type AEMs. These AEMs exhibited high thermal and alkaline stability, as confirmed by NMR and SAXS analysis. The AEM showed no degradation for over 1800 hours in a hydrogen isotope-marked 1.0 M KOH electrolyte at 80 °C [16]. The success of the spiro-ionenes in this study suggests a straightforward method for preparing highly hydroxide ion (OH⁻) conductive and alkali-stable AEMs.

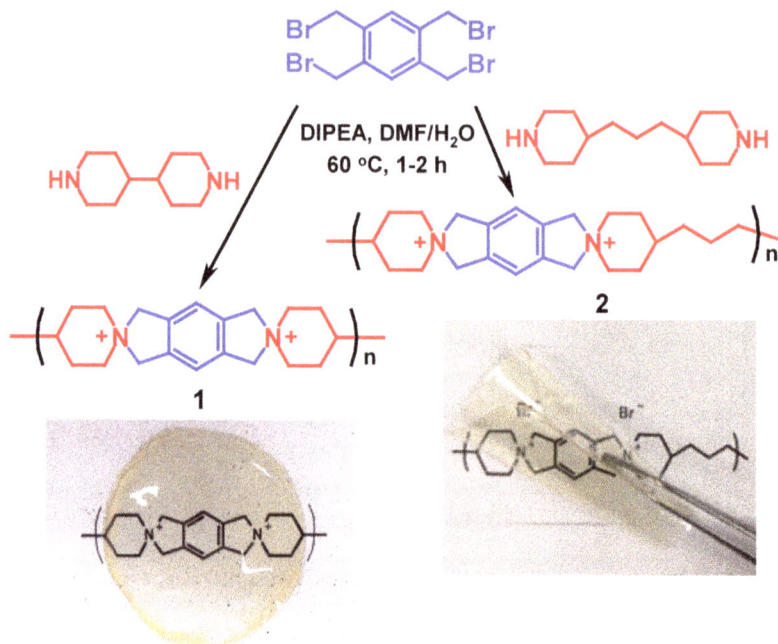

Figure 5. Synthetic pathway to spiro-ionenes 1 and 2 via cyclo-polycondensation of tetrakis(bromomethyl) benzene (4BMB) with 4,4'-piperidine (BP) and 4,4'-trimethylenedipiperidine (TMDP) respectively, with photographs of their films cast from water solutions at 120 °C. Reproduced with permission from ref [16]. Copyright 2017, American Chemical Society.

The performance of ionomers in practical applications can be influenced by various factors, including not only cation cross-linkers but also the polymer matrix itself [44]. Additionally, the type of solvent, catalyst coating method, operating temperature, and catalyst characteristics can all play a role in determining the extrinsic performance of ionomers [45, 46]. When designing ionomers, it is essential to consider the compatibility of all the components in the AEM system as well as the expected performance levels.

3.   Acidic Electrolyzers

The development of an acidic electrolyzer is of great importance for hydrogen production through water electrolysis [21]. Firstly, acidic electrolyzers efficiently catalyze the HER, thereby enhancing hydrogen production efficiency, which is crucial for sustainable and clean hydrogen energy generation. Secondly, integrating acidic electrolyzers with traditional industries, especially in metal processing, chemical engineering, and energy sectors, is more seamless. Acidic electrolysis technology can directly align with existing industrial systems, improving overall industrial production efficiency [47]. Additionally, constructing acidic electrolyzers is relatively cost-effective due to the lower cost of acidic electrolytes. While manufacturing and maintaining acidic electrolyzers may have higher economic aspects, their overall cost-effectiveness remains favorable. Technologically, acidic electrolysis has matured and demonstrated success in practical applications, providing robust support for the widespread adoption and implementation of acidic electrolyzers [48]. Therefore, acidic electrolyzers offer a viable and effective technological approach for clean hydrogen production, playing a significant role in driving the revolution of traditional industries.

In proton exchange membrane water electrolysis (PEMWE), the core reaction involves the splitting of water into oxygen and hydrogen, driven by electricity. This process can be expressed by the equation (Reaction 1):

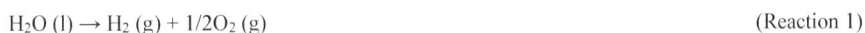

$$H_2O \ (l) \rightarrow H_2 \ (g) + 1/2O_2 \ (g) \qquad \qquad \text{(Reaction 1)}$$

The electrolysis process comprises two electrochemical half-reactions, they are OER occurring at the anode and HER taking place at the cathode. At the anode, water molecules decompose to form oxygen, protons, and electrons. The protons migrate through the membrane and are ultimately reduced to hydrogen at the cathode. The process can be represented by the equations in Reactions 2 and 3:

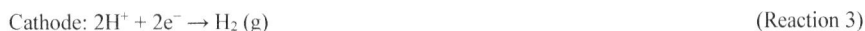

$$\text{Anode: } H_2O \ (l) \rightarrow 1/2O_2 \ (g) + 2H^+ + 2e^- \qquad \qquad \text{(Reaction 2)}$$

$$\text{Cathode: } 2H^+ + 2e^- \rightarrow H_2 \ (g) \qquad \qquad \text{(Reaction 3)}$$

Unlike aqueous electrolysis cells, PEMWE utilizes solid acidic membranes as solid electrolytes instead of liquid electrolytes. This solid acidic membrane allows the two half-reactions to occur in separate compartments, enabling the collection of highly purified hydrogen and oxygen. The thin membranes facilitate short-distance proton transport, reducing ohmic losses. Additionally, catalysts on the membrane surface efficiently transfer protons from the reaction sites to the solid electrolyte, mitigating mass transport limitations. Due to these properties, PEMWE does not require a strong electrolyte solution to enhance ionic conductivity, simplifying the system design as only pure water is supplied into the electrolyzer.

The core components of PEMWE include bipolar plates (BPs), gas diffusion layers (GDL), proton exchange membranes (PEM), and electrocatalysts for the cathode and anode [49]. **Figure 6** illustrates the schematic diagram of a single cell in a PEMWE stack. The PEM separates the two half-cells, facilitating proton transport during the reaction while blocking the passage of the product gas. Catalysts are applied directly to the membrane or porous transport layers. In the typical cell design, catalyst layers are deposited on the membrane, forming the MEA [50]. Two porous transport layers (GDL) are sandwiched on both sides of the MEA. Flow-field plates BPs encapsulate the two half-cells, facilitating charge, mass, and heat transfer while establishing contact with the external power supply [51]. Hydrogen and oxygen products pass through the catalyst surface, GDL, and BPs sequentially to be released from the cells. Sealing elements are also necessary to prevent gas and water leakage in the half-cells.

The mentioned core components significantly impact the cost, performance, and longevity of PEMWE for hydrogen production. BPs and GDL constitute the largest proportion of the stack cost, accounting for 51% and 17%, respectively [52]. The oxidative and corrosive operating environment often limits the materials used primarily to advanced titanium-based materials, and protective coatings such as Pt and Au are also necessary [53]. Developing inexpensive alternative materials is crucial for reducing the cost of the entire electrolytic cell. PEM and catalysts account for a smaller portion of the PEMWE stack, at 5% and 8%, respectively [54]. Additionally, the manufacture of MEA consumes 10% of the whole expense [30]. Achieving high-efficiency water electrolysis performance while minimizing material costs remains a significant challenge for PEMWE.

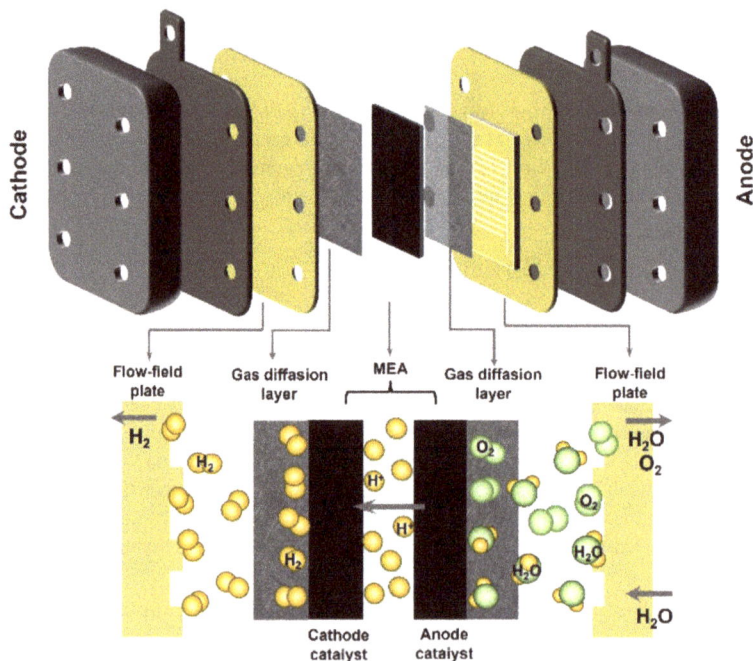

Figure 6. *Stack structure and key components of PEMWE. Reproduced with permission from ref [15]. Copyright 2023, Elsevier.*

The MEA is a critical component of PEMWE and plays a significant role in determining hydrogen production performance. The PEM within the MEA needs to meet several functional requirements, including low gas permeability, excellent proton conductivity, good water absorption, low swelling ratio, outstanding chemical and mechanical stability, low cost, and high durability [55]. Commercially available PEMs, such as the Nafion series, are commonly used in PEMWE [56]. These perfluorosulfonic acid (PFSA) membranes have hydrophobic Teflon-like backbones with hydrophilic sulfonic acid side chains. Different membranes within the Nafion series offer variations in equivalent weight (EW), side-chain chemistry, and length [57]. The choice of Nafion membrane with different EW and thickness significantly impacts the overall performance of the electrolyzer [58]. Research has shown that the thickness of the membrane affects the ionic conductivity of PEMWE, with thinner membranes resulting in lower ohmic resistances and improved performance [59].

However, using excessively thin membranes in PEMWE can lead to challenges such as increased gas permeability, reduced hydrogen purity, decreased mechanical strength and durability, and potential safety hazards. Striking a balance is crucial because crossover in PEMWE is considered detrimental [60]. If gases permeate through the membrane, hydrogen can react with oxygen, releasing a large amount of heat that can damage the membrane and the entire stack. To address these issues, researchers have proposed solutions such as a Nafion/graphene/Nafion PEM sandwich structure, which allows protons to pass through while blocking the transmission of hydrogen gas. This design results in high proton conductivity and an eightfold reduction in hydrogen crossover [61, 62].

BPs serve multiple functions in a PEM electrolytic cell. Primarily, they electrically connect adjacent cells in a stack and facilitate the supply and removal of reactants (such as water) and gaseous products (H_2 and O_2). BPs also play roles in mass transport and heat transfer within the cell. These functions need to be maintained under high pressure, oxidation (at the anode), and reduction (at the cathode) conditions, requiring the BPs to be highly conductive, corrosion-resistant, impermeable, cost-effective, and mechanically robust. Currently, materials such as graphite, titanium, and stainless steel are used as BPs, but each has its drawbacks, including operational defects and high cost. The development of low-cost and high-performance BPs is crucial for the commercial success of PEMWE [63]. Graphite has been previously used in fuel cells due to its high electrical conductivity.

4.   Solid Oxide Electrolyzers

The solid oxide electrolyzer (SOE), also known as a high-temperature electrolyzer (THE) differs from traditional acidic and alkaline electrolyzers [64]. It utilizes a solid electrolyte with high ionic conductivity, improved electrical conductivity, and reduced safety risks associated with liquid-phase operations [65]. The corrosion-resistant solid electrolyte enhances system stability, longevity, and hydrogen gas purity by minimizing gas mixing. Although it requires high temperatures, the SOE improves electrolysis efficiency and reduces overpotential, making it an efficient, stable, and secure hydrogen production technology. Additionally, the SOE can be adapted for various applications such as steam electrolysis and co-electrolysis, providing flexibility to meet diverse hydrogen production needs.

At a fundamental level, electrolysis cells, including SOE cells, consist of an electrolyte and two electrodes (anode and cathode), as depicted in **Figure 7**. Additional components such as interconnects and sealing materials are necessary for fabricating complete cells and stacks. It is crucial to have appropriate thicknesses for both the electrolyte and electrodes to minimize electric and diffusion resistance. The microstructure and thickness of these functional materials play vital roles in determining device performance. The electrolyte typically comprises a pure ceramic, while the anode and cathode are ceramic-metal composites known as cermets. A thin and dense electrolyte is required to separate oxidation gases from fuel gases within the cell. Reducing the electrolyte's thickness, especially in electrode-supported cells, significantly decreases overall ohmic resistance. However, the applicability of certain manufacturing technologies may limit the use of thinner electrolytes. The cathode and anode usually consist of a mixture of electrolyte and electrode materials, forming a composite that helps reduce polarization and expand the triple phase boundaries (TPBs), thereby enhancing the electrolyzer's performance.

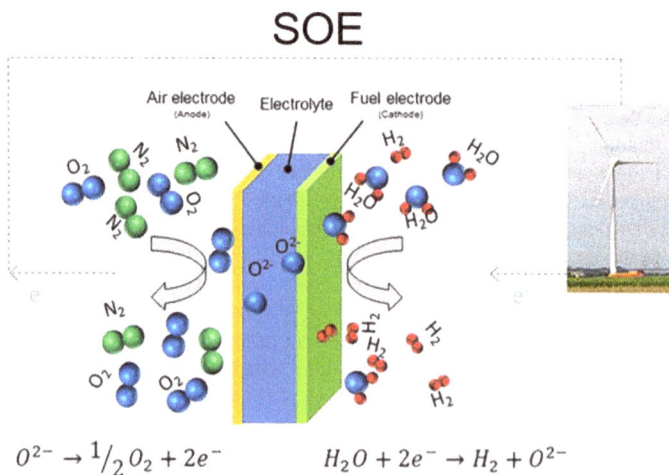

$$O^{2-} \rightarrow {}^1\!/_2\, O_2 + 2e^- \qquad H_2O + 2e^- \rightarrow H_2 + O^{2-}$$

Figure 7. The basic scheme of electrochemical reaction in the anodic and cathodic compartments of the SOE. Reproduced with permission from ref [66]. Copyright 2019, Elsevier.

The cathode of the SOE typically consists of a nickel-yttria-stabilized zirconia (Ni-YSZ) cermet. YSZ provides ionic conductivity and structural support, while nickel acts as a catalyst and electronic conductor [67]. The anode is usually made of mixed conductors such as lanthanum-strontium cobalt ferrite (LSCF) [68] or lanthanum-strontium cobaltite (LSC). In the case of LSC, it acts as a catalyst for the oxygen electrode. Another alternative material for the cathode/anode is strontium-doped LaMnO$_3$ (LSM) in a cermet with YSZ. LSM provides electronic conduction and catalytic function, while YSZ serves as the structural component and provides ionic conduction. Additionally, a thin layer of gadolinium-doped ceria (GDC) may be used as a buffer layer between the electrolyte and the LSCF cathode to prevent reactions between the oxygen electrode materials and YSZ [69]. In terms of recycling and the circular economy in SOE, materials such as nickel and lanthanum (La) have identified environmental burdens. The environmental impact of these materials can potentially be reduced by around 70% through recycling initiatives and the adoption of a circular economy approach.

The versatility of SOE makes it suitable for both hydrogen production and oxygen electrolysis, enhancing its flexibility in energy systems. The required key materials for SOE are relatively abundant, reducing concerns about cost and sustainability. The produced hydrogen is typically of high purity, and suitable for various applications. The rapid response of SOE provides an advantage in applications requiring quick adjustments to hydrogen or oxygen production levels. Additionally, SOE can be integrated with renewable energy systems, utilizing sources like solar or wind power, to achieve green hydrogen production, as schematically illustrated in **Figure 7**.

However, challenges such as material durability, thermal cycling, and cost need to be addressed to achieve the widespread commercial deployment of intermediate-temperature electrolyzers. As research and development efforts continue, solid oxide electrolyzers play a crucial role in advancing the field of hydrogen production and contributing to a sustainable future for renewable energy.

5. Intermediate-Temperature Electrolyzers

Intermediate-temperature electrolyzers (ITE) have emerged as a promising electrolysis technology, addressing the limitations of both high-temperature solid oxide electrolyzers (SOE) and low-temperature alkaline electrolyzers. ITE operates at lower temperatures, leading to reduced equipment manufacturing costs and less corrosiveness to materials and devices, enhancing system stability. It offers a compromise solution with electrolysis efficiency levels between high-temperature SOE and low-temperature alkaline electrolyzers, making it suitable for hydrogen production and energy conversion in scenarios where moderate temperatures are preferred. ITE demonstrates extensive prospects in the field of green hydrogen energy production due to its moderate temperature operation, efficiency, and relatively lower costs.

The development of intermediate-temperature SOE (IT-SOE) has gained much attention, particularly in the temperature of 500-700 °C, due to improved cell stability and reduced material issues [70]. Sm^{3+} doped ceria (SDC) is a promising electrolyte material for IT-SOE, offering high oxygen ion conductivity of up to 10^{-2} S cm^{-1} at 600 °C [71, 72]. CeO$_2$ (Fluorite-type) or BaCeO$_3$ (perovskite-type) have been investigated as the hydrogen electrode materials for the ceria-based SOE, demonstrating similar thermal expansion coefficients to the ceria electrolyte [73]. At elevated temperatures, the migration of Ba ions towards the interface of the hydrogen electrode and the SDC electrolyte can create an electron-blocking layer that impedes the reduction of Ce^{4+} to Ce^{3+}, which has a significant impact on the effectiveness of SOE [74]. The NiO-BaZr$_{0.1}$Ce$_{0.7}$Y$_{0.2}$O$_{3-\delta}$ (BZCY) cermet, with Ni serving as the catalyst supported on BZCY, is recognized as a highly promising material for hydrogen electrodes due to its exceptional combination of high ionic and electronic conductivity [75, 76]. BZCY demonstrates outstanding ionic conductivity within the intermediate-temperature range of 500-700 °C and remarkable stability when exposed to water vapor [77, 78].

To enhance the electrocatalytic performance of ITE, Luo et al. [79] reported modifying the interface between the hydrogen electrode and the electrolyte. They successfully synthesized uniformly distributed NiO and BZCY nanocomposites via a one-step combustion method and utilized them to fabricate a hydrogen electrode functional layer (HFL). Transmission electron microscopy (TEM) and energy-dispersive X-ray spectroscopy (EDS) confirmed the uniform distribution of NiO-BZCY composite materials. The dip-coating method was employed to create a gradient porous hydrogen electrode structure. The uniformly distributed Ni in the HFL provided ample active reaction sites for electrocatalytic reactions. By constructing the NiO-BZCY HFL between the NiO-BZCY hydrogen electrode and the SDC electrolyte, the activation polarization resistance of the hydrogen electrode was reduced. **Figure 8** illustrates the structure model of the NiO-BZCY HFL between the NiO-BZCY hydrogen electrode and the SDC electrolyte, highlighting the enhancement in electrocatalytic performance achieved through interface modification. The polarization resistance of the cell decreased from 2.12 to 1.17 Ω cm^2 at an operating temperature of 700 °C. Moreover, the optimized electrolyzer achieved a high current density of 0.74 A cm^{-2} at 1.3 V. The improved hydrogen electrode structure and electrode/electrolyte interface contacts effectively reduced polarization losses.

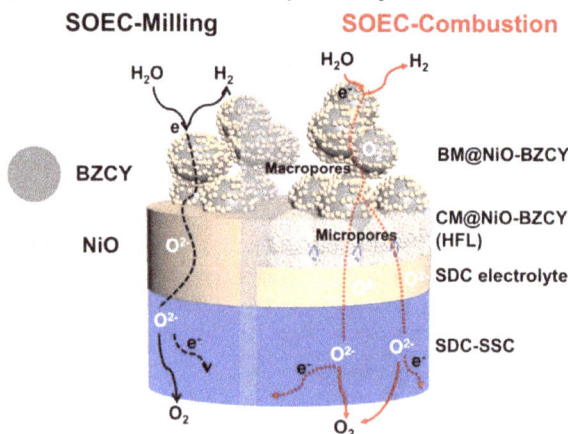

Figure 8. *Schematic diagram of ion and electron transfer in control SOEC-Milling (left) and SOEC-Combustion with modified hydrogen electrode structure (right). BM: ball-milling; CM: one-step combustion method; HFL: hydrogen electrode function layer; BZCY: BaZr$_{0.1}$Ce$_{0.7}$Y$_{0.2}$O$_{3-\delta}$; SDC: Ce$_{0.8}$Sm$_{0.2}$O$_{2-\delta}$; SSC: Sm$_{0.5}$Sr$_{0.5}$CoO$_{3-\delta}$. Reproduced with permission from ref [79]. Copyright 2023, Elsevier.*

6.   Conclusion and Prospects

In conclusion, the chapter emphasizes the critical role of electrolyzers in the electrocatalytic hydrogen production process. The chapter highlighted the different types of electrolyzers and their working mechanisms. Various types of electrolysis technologies, including alkaline Electrolyzer, anion exchange membrane electrolyzers, acidic electrolyzers, solid oxide electrolyzers, and intermediate-temperature electrolyzers, have been discussed. The chapter's insights provide a valuable foundation for researchers, engineers, and policymakers working towards the widespread adoption of electrocatalytic hydrogen production as a viable solution for a clean and renewable energy future. Additionally, it underscores the need for further research and development to overcome technical barriers, enhance efficiency, and reduce costs. The challenges and opportunities in scaling up electrocatalytic hydrogen production for industrial applications are concluded as follows.

Indeed, the forthcoming advancements in hydrogen electrolysis systems are expected to focus on innovative technologies, including pioneering electrode materials, advancements in membranes, and sophisticated intelligent control systems. These advancements aim to enhance the efficacy, durability, and overall performance of electrolytic cells for hydrogen production.

(i) Acidic and alkaline electrolytic cells commonly utilize sulfuric acid or sodium hydroxide as electrolytes due to their ability to facilitate the ionization and conduction of ions during the electrolysis process. However, it is true that these electrolytes pose certain challenges and drawbacks, including corrosion susceptibility and high energy consumption.

(ii) Solid-state electrolytic cells, which utilize solid electrolytes for stability enhancement, indeed face certain hurdles, primarily related to high costs and material complexities. The development of solid-state electrolytic cells is a high cost, which is associated with the materials used in solid electrolytes. Solid electrolytes need to possess specific properties such as high ionic conductivity, chemical stability, and compatibility with electrode materials. These requirements often lead to the use of expensive materials, which can significantly drive up the overall cost of the cell.

(iii) Concurrently, breakthroughs in membrane technology hold the promise of heightening efficiency and stability. Recent advancements in membrane technology have opened up new possibilities for improving process efficiency, enhancing selectivity, and ensuring long-term stability.

(iv) The anticipated advancements in intelligent control systems hold great promise for enabling real-time adjustments tailored to diverse operational conditions. These advancements are expected to revolutionize various industries and sectors, offering unprecedented levels of automation, efficiency, and adaptability. By leveraging predictive modelling and machine learning techniques, these systems will be able to anticipate energy demand patterns, identify energy-saving opportunities, and automatically adjust energy usage in real time. This will not only result in substantial cost savings but also contribute to sustainability efforts by reducing carbon emissions.

References

[1]   X. Kong, K. Xu, C. Zhang, J. Dai, S. Norooz Oliaee, L. Li, X. Zeng, C. Wu, Z. Peng, Free-standing two-dimensional Ru nanosheets with high activity toward water splitting, ACS Catal., 6 (2016) 1487-1492. https://doi.org/10.1021/acscatal.5b02730

[2]   J. Li, C. Hou, C. Chen, W. Ma, Q. Li, L. Hu, X. Lv, J. Dang, Collaborative Interface optimization strategy guided ultrafine RuCo and MXene heterostructure electrocatalysts for efficient overall water splitting, ACS Nano, 17 (2023) 10947-10957. https://doi.org/10.1021/acsnano.3c02956

[3]   B. Fei, Z. Chen, J. Liu, H. Xu, X. Yan, H. Qing, M. Chen, R. Wu, Ultrathinning nickel sulfide with modulated electron density for efficient water splitting, Adv. Energy Mater., 10 (2020) 2001963. https://doi.org/10.1002/aenm.202001963

[4]   K. Chang, D.T. Tran, J. Wang, S. Prabhakaran, D.H. Kim, N.H. Kim, J.H. Lee, Atomic heterointerface engineering of Ni_2P-$NiSe_2$ nanosheets coupled ZnP-based arrays for high-efficiency solar-assisted water splitting, Adv. Funct. Mater., 32 (2022) 2113224. https://doi.org/10.1002/adfm.202113224

[5]   X. Zhou, Y. Mo, F. Yu, L. Liao, X. Yong, F. Zhang, D. Li, Q. Zhou, T. Sheng, H. Zhou, Engineering active iron sites on nanoporous bimetal phosphide/nitride heterostructure array enabling robust overall water splitting, Adv. Funct. Mater., 33 (2022) 2209465. https://doi.org/10.1002/adfm.202209465

[6]   L. An, J. Feng, Y. Zhang, R. Wang, H. Liu, G.C. Wang, F. Cheng, P. Xi, Epitaxial heterogeneous interfaces on N-$NiMoO_4$/NiS_2 nanowires/nanosheets to boost hydrogen and oxygen production for overall water splitting, Adv. Funct. Mater., 29 (2019) 1805298. https://doi.org/10.1002/adfm.201805298

[7]   R. Andaveh, A. Sabour Rouhaghdam, J. Ai, M. Maleki, K. Wang, A. Seif, G. Barati Darband, J. Li, Boosting the electrocatalytic activity of NiSe by introducing MnCo as an efficient heterostructured electrocatalyst for large-current-density alkaline seawater splitting, Appl. Catal. B: Environ., 325 (2023) 122355. https://doi.org/10.1016/j.apcatb.2022.122355

[8]   R.K. Hona, S.B. Karki, F. Ramezanipour, Oxide electrocatalysts based on earth-abundant metals for both hydrogen- and oxygen-evolution reactions, ACS Sustain. Chem. Eng., 8 (2020) 11549-11557. https://doi.org/10.1021/acssuschemeng.0c02498

[9]   H. Shi, X.-Y. Sun, S.-P. Zeng, Y. Liu, G.-F. Han, T.-H. Wang, Z. Wen, Q.-R. Fang, X.-Y. Lang, Q. Jiang, Nanoporous nonprecious high-entropy alloys as multisite electrocatalysts for ampere-level current-density hydrogen evolution, Small Struct., 4 (2023) 2300042. https://doi.org/10.1002/sstr.202300042

[10] X. Liu, W. Chen, W. Wang, Y. Jiang, K. Cao, PdZn alloys decorated 3D hierarchical porous carbon networks for highly efficient and stable hydrogen production from aldehyde solution, Int. J. Hydrogen Energ., 46 (2021) 33429-33437. https://doi.org/10.1016/j.ijhydene.2021.07.193

[11]   A. Qayum, X. Peng, J. Yuan, Y. Qu, J. Zhou, Z. Huang, H. Xia, Z. Liu, D.Q. Tan, P.K. Chu, Highly durable and efficient Ni-FeO$_x$/FeNi$_3$ electrocatalysts synthesized by a facile in situ combustion-based method for overall water splitting with large current densities, ACS Appl. Mater. Interfaces, 14 (2022) 27842-27853. https://doi.org/10.1021/acsami.2c04562

[12]   X. Huang, H. Zheng, G. Lu, P. Wang, L. Xing, J. Wang, G. Wang, Enhanced water splitting electrocatalysis over MnCo$_2$O$_4$ via introduction of suitable Ce content, ACS Sustain. Chem. Eng., 7 (2018) 1169-1177. https://doi.org/10.1021/acssuschemeng.8b04814

[13]   L. Zhang, J. Xiao, H. Wang, M. Shao, Carbon-based electrocatalysts for hydrogen and oxygen evolution reactions, ACS Catal., 7 (2017) 7855-7865. https://doi.org/10.1021/acscatal.7b02718

[14]   W. Li, C. Wang, X. Lu, Integrated transition metal and compounds with carbon nanomaterials for electrochemical water splitting, J. Mater. Chem. A, 9 (2021) 3786-3827. https://doi.org/10.1039/D0TA09495A

[15]   K. Zhang, X. Liang, L. Wang, K. Sun, Y. Wang, Z. Xie, Q. Wu, X. Bai, M.S. Hamdy, H. Chen, X. Zou, Status and perspectives of key materials for PEM electrolyzer, Nano Res. Energy, 1 (2022) e9120032. https://doi.org/10.26599/NRE.2022.9120032

[16]   T.H. Pham, J.S. Olsson, P. Jannasch, N-spirocyclic quaternary ammonium ionenes for anion-exchange membranes, J. Am. Chem. Soc., 139 (2017) 2888-2891. https://doi.org/10.1021/jacs.6b12944

[17]   A. Rodríguez-Gómez, F. Dorado, A. de Lucas-Consuegra, A.R. de la Osa, Influence of Pt/Ru anodic ratio on the valorization of ethanol by PEM electrocatalytic reforming towards value-added products, J. Energy Chem., 56 (2021) 264-275. https://doi.org/10.1016/j.jechem.2020.07.061

[18]   S.A.H. Naqvi, T. Taner, M. Ozkaymak, H.M. Ali, Hydrogen production through alkaline electrolyzers: A techno-economic and enviro-economic analysis, Chem. Eng. Technol., 46 (2022) 474-481. https://doi.org/10.1002/ceat.202200234

[19]   M. Bodner, A. Hofer, V. Hacker, H$_2$ generation from alkaline electrolyzer, WIREs Energy Environ., 4 (2015) 365-381. https://doi.org/10.1002/wene.150

[20]   M. Sánchez, E. Amores, D. Abad, L. Rodríguez, C. Clemente-Jul, Aspen Plus model of an alkaline electrolysis system for hydrogen production, Int. J. Hydrogen Energ., 45 (2020) 3916-3929. https://doi.org/10.1016/j.ijhydene.2019.12.027

[21]   J. Chi, H. Yu, Water electrolysis based on renewable energy for hydrogen production, Chin. J. Catal., 39 (2018) 390-394. https://doi.org/10.1016/S1872-2067(17)62949-8

[22]   A. Buttler, H. Spliethoff, Current status of water electrolysis for energy storage, grid balancing and sector coupling via power-to-gas and power-to-liquids: A review, Renew. Sustain. Energy Rev., 82 (2018) 2440-2454. https://doi.org/10.1016/j.rser.2017.09.003

[23]   N. Tenhumberg, K. Büker, Ecological and economic evaluation of hydrogen production

by different water electrolysis technologies, Chem. Ing. Tech. , 92 (2020) 1586-1595. https://doi.org/10.1002/cite.202000090

[24] Z. Ren, J. Wang, Z. Yu, C. Zhang, S. Gao, P. Wang, Experimental studies and modeling of a 250-kW alkaline water electrolyzer for hydrogen production, J. Power Sources, 544 (2022) 231886. https://doi.org/10.1016/j.jpowsour.2022.231886

[25] N. Du, C. Roy, R. Peach, M. Turnbull, S. Thiele, C. Bock, Anion-exchange membrane water electrolyzers, Chem. Rev., 122 (2022) 11830-11895. https://doi.org/10.1021/acs.chemrev.1c00854

[26] Y. Yang, P. Li, X. Zheng, W. Sun, S.X. Dou, T. Ma, H. Pan, Anion-exchange membrane water electrolyzers and fuel cells, Chem. Soc. Rev., 51 (2022) 9620-9693. https://doi.org/10.1039/D2CS00038E

[27] I. Gottschalk, R. Knight, The development of a machine-learning approach to construct a field-scale rock-physics transform, Geophysics, 87 (2021) MR35-MR48. https://doi.org/10.1190/geo2020-0811.1

[28] K. Artyushkova, A. Serov, H. Doan, N. Danilovic, C.B. Capuano, T. Sakamoto, H. Kishi, S. Yamaguchi, S. Mukerjee, P. Atanassov, Application of X-ray photoelectron spectroscopy to studies of electrodes in fuel cells and electrolyzers, J Electron Spectrosc, 231 (2019) 127-139. https://doi.org/10.1016/j.elspec.2017.12.006

[29] G. Huang, M. Mandal, N.U. Hassan, K. Groenhout, A. Dobbs, W.E. Mustain, P.A. Kohl, Ionomer optimization for water uptake and swelling in anion exchange membrane electrolyzer: Oxygen evolution electrode, J. Electrochem. Soc., 167 (2020) 164514. https://doi.org/10.1149/1945-7111/abcde3

[30] E. Liu, J. Li, L. Jiao, H.T.T. Doan, Z. Liu, Z. Zhao, Y. Huang, K.M. Abraham, S. Mukerjee, Q. Jia, Unifying the hydrogen evolution and oxidation reactions kinetics in base by identifying the catalytic roles of hydroxyl-water-cation adducts, J. Am. Chem. Soc., 141 (2019) 3232-3239. https://doi.org/10.1021/jacs.8b13228

[31] Y.-C. Cao, X. Wu, K. Scott, A quaternary ammonium grafted poly vinyl benzyl chloride membrane for alkaline anion exchange membrane water electrolysers with no-noble-metal catalysts, Int. J. Hydrogen Energ., 37 (2012) 9524-9528. https://doi.org/10.1016/j.ijhydene.2012.03.116

[32] L. Xiao, S. Zhang, J. Pan, C. Yang, M. He, L. Zhuang, J. Lu, First implementation of alkaline polymer electrolyte water electrolysis working only with pure water, Energy Environ. Sci., 5 (2012) 7869. https://doi.org/10.1039/c2ee22146b

[33] A. Faid, A. Oyarce Barnett, F. Seland, S. Sunde, Highly active nickel-based catalyst for hydrogen evolution in anion exchange membrane electrolysis, Catalysts, 8 (2018) 614. https://doi.org/10.3390/catal8120614

[34] M. Gong, H. Dai, A mini review of NiFe-based materials as highly active oxygen evolution reaction electrocatalysts, Nano Res., 8 (2014) 23-39. https://doi.org/10.1007/s12274-014-0591-z

[35] Z. Qiu, C.-W. Tai, G.A. Niklasson, T. Edvinsson, Direct observation of active catalyst surface phases and the effect of dynamic self-optimization in NiFe-layered double hydroxides for alkaline water splitting, Energy Environ. Sci., 12 (2019) 572-581. https://doi.org/10.1039/C8EE03282C

[36] H. Koshikawa, H. Murase, T. Hayashi, K. Nakajima, H. Mashiko, S. Shiraishi, Y. Tsuji, Single nanometer-sized NiFe-layered double hydroxides as anode catalyst in anion exchange membrane water electrolysis cell with energy conversion efficiency of 74.7% at 1.0 A cm^{-2}, ACS Catal., 10 (2020) 1886-1893. https://doi.org/10.1021/acscatal.9b04505

[37] M.R. Kraglund, M. Carmo, G. Schiller, S.A. Ansar, D. Aili, E. Christensen, J.O. Jensen, Ion-solvating membranes as a new approach towards high rate alkaline electrolyzers, Energy Environ. Sci., 12 (2019) 3313-3318. https://doi.org/10.1039/C9EE00832B

[38] S. Kim, S. Yang, D. Kim, Poly(arylene ether ketone) with pendant pyridinium groups for alkaline fuel cell membranes, Int. J. Hydrogen Energ., 42 (2017) 12496-12506. https://doi.org/10.1016/j.ijhydene.2017.03.187

[39] C. Vogel, J. Meier-Haack, Preparation of ion-exchange materials and membranes, Desalination, 342 (2014) 156-174. https://doi.org/10.1016/j.desal.2013.12.039

[40] N. Lee, D.T. Duong, D. Kim, Cyclic ammonium grafted poly (arylene ether ketone) hydroxide ion exchange membranes for alkaline water electrolysis with high chemical stability and cell efficiency, Electrochim. Acta, 271 (2018) 150-157. https://doi.org/10.1016/j.electacta.2018.03.117

[41] M.M. Nasef, O. Güven, Radiation-grafted copolymers for separation and purification purposes: Status, challenges and future directions, Prog. Polym. Sci., 37 (2012) 1597-1656. https://doi.org/10.1016/j.progpolymsci.2012.07.004

[42] A. Marinkas, I. Strużyńska-Piron, Y. Lee, A. Lim, H.S. Park, J.H. Jang, H.-J. Kim, J. Kim, A. Maljusch, O. Conradi, D. Henkensmeier, Anion-conductive membranes based on 2-mesityl-benzimidazolium functionalised poly(2,6-dimethyl-1,4-phenylene oxide) and their use in alkaline water electrolysis, Polymer, 145 (2018) 242-251. https://doi.org/10.1016/j.polymer.2018.05.008

[43] S. Miyanishi, T. Yamaguchi, Highly conductive mechanically robust high Mw polyfluorene anion exchange membrane for alkaline fuel cell and water electrolysis application, Polym. Chem., 11 (2020) 3812-3820. https://doi.org/10.1039/D0PY00334D

[44] A.Y. Faid, L. Xie, A.O. Barnett, F. Seland, D. Kirk, S. Sunde, Effect of anion exchange ionomer content on electrode performance in AEM water electrolysis, Int. J. Hydrogen Energ., 45 (2020) 28272-28284. https://doi.org/10.1016/j.ijhydene.2020.07.202

[45] A. Lim, H.-j. Kim, D. Henkensmeier, S. Jong Yoo, J. Young Kim, S. Young Lee, Y.-E. Sung, J.H. Jang, H.S. Park, A study on electrode fabrication and operation variables affecting the performance of anion exchange membrane water electrolysis, J. Ind. Eng. Chem., 76 (2019) 410-418. https://doi.org/10.1016/j.jiec.2019.04.007

[46] A. Amel, N. Gavish, L. Zhu, D.R. Dekel, M.A. Hickner, Y. Ein-Eli, Bicarbonate and

chloride anion transport in anion exchange membranes, J. Membr. Sci., 514 (2016) 125-134. https://doi.org/10.1016/j.memsci.2016.04.027

[47]   S. Shiva Kumar, V. Himabindu, Hydrogen production by PEM water electrolysis-A review, Mater. Sci. Energy Technol., 2 (2019) 442-454. https://doi.org/10.1016/j.mset.2019.03.002

[48]   M. Rodríguez-Peña, J.A. Barrios Pérez, J. Llanos, C. Saez, C.E. Barrera-Díaz, M.A. Rodrigo, Electrochemical generation of ozone using a PEM electrolyzer at acidic pHs, Sep. Purif. Technol., 267 (2021) 118672. https://doi.org/10.1016/j.seppur.2021.118672

[49]   K.W. Ahmed, M.J. Jang, M.G. Park, Z. Chen, M. Fowler, Effect of components and operating conditions on the performance of PEM electrolyzers: A review, Electrochem, 3 (2022) 581-612. https://doi.org/10.3390/electrochem3040040

[50]   M. Chen, C. Zhao, F. Sun, J. Fan, H. Li, H. Wang, Research progress of catalyst layer and interlayer interface structures in membrane electrode assembly (MEA) for proton exchange membrane fuel cell (PEMFC) system, eTransportation, 5 (2020) 100075. https://doi.org/10.1016/j.etran.2020.100075

[51]   S.A. Atyabi, E. Afshari, S. Wongwises, W.-M. Yan, A. Hadjadj, M.S. Shadloo, Effects of assembly pressure on PEM fuel cell performance by taking into accounts electrical and thermal contact resistances, Energy, 179 (2019) 490-501. https://doi.org/10.1016/j.energy.2019.05.031

[52]   K. Bareiß, C. de la Rua, M. Möckl, T. Hamacher, Life cycle assessment of hydrogen from proton exchange membrane water electrolysis in future energy systems, Appl. Energy, 237 (2019) 862-872. https://doi.org/10.1016/j.apenergy.2019.01.001

[53]   R. Haider, Y. Wen, Z.-F. Ma, D.P. Wilkinson, L. Zhang, X. Yuan, S. Song, J. Zhang, High temperature proton exchange membrane fuel cells: Progress in advanced materials and key technologies, Chem. Soc. Rev., 50 (2021) 1138-1187. https://doi.org/10.1039/D0CS00296H

[54]   M. Maier, K. Smith, J. Dodwell, G. Hinds, P.R. Shearing, D.J.L. Brett, Mass transport in PEM water electrolysers: A review, Int. J. Hydrogen Energ., 47 (2022) 30-56. https://doi.org/10.1016/j.ijhydene.2021.10.013

[55]   E. Middelman, Improved PEM fuel cell electrodes by controlled self-assembly, Fuel Cells Bull., 2002 (2002) 9-12. https://doi.org/10.1016/S1464-2859(02)11028-5

[56]   P. Shirvanian, F. van Berkel, Novel components in Proton Exchange Membrane (PEM) Water Electrolyzers (PEMWE): Status, challenges and future needs. A mini review, Electrochem. Commun., 114 (2020) 106704. https://doi.org/10.1016/j.elecom.2020.106704

[57]   G. Palanisamy, T. Sadhasivam, W.-S. Park, S.T. Bae, S.-H. Roh, H.-Y. Jung, Tuning the ion selectivity and chemical stability of a biocellulose membrane by PFSA ionomer reinforcement for vanadium redox flow battery applications, ACS Sustain. Chem. Eng., 8 (2020) 2040-2051. https://doi.org/10.1021/acssuschemeng.9b06631

[58]   M.E. Günay, N.A. Tapan, G. Akkoç, Analysis and modeling of high-performance polymer electrolyte membrane electrolyzers by machine learning, Int. J. Hydrogen Energ., 47 (2022) 2134-2151. https://doi.org/10.1016/j.ijhydene.2021.10.191

[59]   B. Han, S.M. Steen, J. Mo, F.-Y. Zhang, Electrochemical performance modeling of a proton exchange membrane electrolyzer cell for hydrogen energy, Int. J. Hydrogen Energ., 40 (2015) 7006-7016. https://doi.org/10.1016/j.ijhydene.2015.03.164

[60]   A. Martin, D. Abbas, P. Trinke, T. Böhm, M. Bierling, B. Bensmann, S. Thiele, R. Hanke-Rauschenbach, Communication-Proving the importance of Pt-interlayer position in PEMWE membranes for the effective reduction of the anodic hydrogen content, J. Electrochem. Soc., 168 (2021) 094509. https://doi.org/10.1149/1945-7111/ac275b

[61]   D. Ion-Ebrasu, B.G. Pollet, A. Spinu-Zaulet, A. Soare, E. Carcadea, M. Varlam, S. Caprarescu, Graphene modified fluorinated cation-exchange membranes for proton exchange membrane water electrolysis, Int. J. Hydrogen Energ., 44 (2019) 10190-10196. https://doi.org/10.1016/j.ijhydene.2019.02.148

[62]   M.E. Scofield, H. Liu, S.S. Wong, A concise guide to sustainable PEMFCs: Recent advances in improving both oxygen reduction catalysts and proton exchange membranes, Chem. Soc. Rev., 44 (2015) 5836-5860. https://doi.org/10.1039/C5CS00302D

[63]   T. Husaini, I. Alshami, J. Goh, M. Masdar, K.S. Loh, Review on bipolar plates for low-temperature polymer electrolyte membrane water electrolyzer, Int. J. Energy Res. , 45 (2021) 20583-20600. https://doi.org/10.1002/er.7182

[64]   S. Sebbahi, N. Nabil, A. Alaoui-Belghiti, S. Laasri, S. Rachidi, A. Hajjaji, Assessment of the three most developed water electrolysis technologies: Alkaline water electrolysis, proton exchange membrane and solid-oxide electrolysis, Mater. Today: Proc., 66 (2022) 140-145. https://doi.org/10.1016/j.matpr.2022.04.264

[65]   T. Ye, L. Li, Y. Zhang, Recent progress in solid electrolytes for energy storage devices, Adv. Funct. Mater., 30 (2020) 2000077. https://doi.org/10.1002/adfm.202000077

[66]   J. Kupecki, K. Motylinski, S. Jagielski, M. Wierzbicki, J. Brouwer, Y. Naumovich, M. Skrzypkiewicz, Energy analysis of a 10 kW-class power-to-gas system based on a solid oxide electrolyzer (SOE), Energy Convers. Manag., 199 (2019) 111934. https://doi.org/10.1016/j.enconman.2019.111934

[67]   P.J. Gasper, Y. Lu, A.Y. Nikiforov, S.N. Basu, S. Gopalan, U.B. Pal, Detailed electrochemical performance and microstructural characterization of nickel-Yttria stabilized zirconia cermet anodes infiltrated with nickel, gadolinium doped ceria, and nickel-Gadolinium doped ceria nanoparticles, J Power Sources., 447 (2020) 227357. https://doi.org/10.1016/j.jpowsour.2019.227357

[68]   D. Kim, J.W. Park, M.S. Chae, I. Jeong, J.H. Park, K.J. Kim, J.J. Lee, C. Jung, C.-W. Lee, S.-T. Hong, K.T. Lee, An efficient and robust lanthanum strontium cobalt ferrite catalyst as a bifunctional oxygen electrode for reversible solid oxide cells, J. Mater. Chem. A, 9 (2021) 5507-5521. https://doi.org/10.1039/D0TA11233J

[69]   M. Riegraf, F. Han, N. Sata, R. Costa, Intercalation of thin-film Gd-doped ceria barrier layers in electrolyte-supported solid oxide cells: Physicochemical aspects, ACS Appl. Mater. Interfaces, 13 (2021) 37239-37251. https://doi.org/10.1021/acsami.1c11175

[70]   T. Ishihara, N. Jirathiwathanakul, H. Zhong, Intermediate temperature solid oxide electrolysis cell using $LaGaO_3$ based perovskite electrolyte, Energy Environ. Sci., 3 (2010) 665. https://doi.org/10.1039/b915927d

[71]   M.A. Laguna-Bercero, Recent advances in high temperature electrolysis using solid oxide fuel cells: A review, J. Power Sources, 203 (2012) 4-16. https://doi.org/10.1016/j.jpowsour.2011.12.019

[72]   S. Zhu, Y. Wang, Y. Rao, Z. Zhan, C. Xia, Chemically-induced mechanical unstability of samaria-doped ceria electrolyte for solid oxide electrolysis cells, Int. J. Hydrogen Energ., 39 (2014) 12440-12447. https://doi.org/10.1016/j.ijhydene.2014.06.051

[73]   D. Luo, Q. Xu, J. Qian, X. Li, Interface-engineered intermediate temperature solid oxide electrolysis cell, Energy Technol., 7 (2019) 1900704. https://doi.org/10.1002/ente.201900704

[74]   D. Medvedev, V. Maragou, E. Pikalova, A. Demin, P. Tsiakaras, Novel composite solid state electrolytes on the base of $BaCeO_3$ and CeO_2 for intermediate temperature electrochemical devices, J. Power Sources, 221 (2013) 217-227. https://doi.org/10.1016/j.jpowsour.2012.07.120

[75]   B. Wang, L. Bi, X.S. Zhao, Exploring the role of NiO as a sintering aid in $BaZr_{0.1}Ce_{0.7}Y_{0.2}O_{3-\delta}$ electrolyte for proton-conducting solid oxide fuel cells, J. Power Sources, 399 (2018) 207-214. https://doi.org/10.1016/j.jpowsour.2018.07.087

[76]   S. Rajendran, N.K. Thangavel, H. Ding, Y. Ding, D. Ding, L.M. Reddy Arava, Tri-doped $BaCeO_3$-$BaZrO_3$ as a Chemically stable electrolyte with high proton-conductivity for intermediate temperature solid oxide electrolysis cells (SOECs), ACS Appl. Mater. Interfaces, 12 (2020) 38275-38284. https://doi.org/10.1021/acsami.0c12532

[77]   Z. Zhu, J. Hou, W. He, W. Liu, High-performance $Ba(Zr_{0.1}Ce_{0.7}Y_{0.2})O_{3-\delta}$ asymmetrical ceramic membrane with external short circuit for hydrogen separation, J. Alloys Compd., 660 (2016) 231-234. https://doi.org/10.1016/j.jallcom.2015.11.065

[78]   C. Zuo, S. Zha, M. Liu, M. Hatano, M. Uchiyama, $Ba(Zr_{0.1}Ce_{0.7}Y_{0.2})O_{3-\delta}$ as an electrolyte for low-temperature solid-oxide fuel cells, Adv. Mater., 18 (2006) 3318-3320. https://doi.org/10.1002/adma.200601366

[79]   D. Luo, J. Qian, H. Zhang, C. Lin, Q. Xiao, X. Li, Nanoarchitecturing hydrogen electrode with intimately coupled ion-conducting and catalytic phases for enhanced intermediate-temperature steam electrolysis, J. Alloys Compd., 969 (2023) 172362. https://doi.org/10.1016/j.jallcom.2023.172362

About the Author

Xiang Peng received his PhD in Physics and Materials Science from the City University of Hong Kong in 2017. After completing his doctoral studies, he continued to expand his expertise as a postdoctoral fellow at the City University of Hong Kong from 2017 to 2018. Presently, he is a professor and PhD supervisor of Materials Science and Engineering at the Wuhan Institute of Technology (WIT). He took the initiative to establish the Department of New Energy Materials and Devices at WIT, serving as its inaugural head. His research focuses on the design of functional nanomaterials and their application in energy storage and conversion. He has authored over 80 peer-reviewed papers, which have garnered significant recognition within the scientific community, accumulating more than 6200 citations with an H-index of 45 (data from Google Scholar). He has also secured a total of 17 patents for his groundbreaking inventions. He was honored as one of the World's Top 2% Scientists in 2023, conferred by Elsevier and Stanford University. He currently serves as the guest editor-in-chief, associate editor, and editorial board member of approximately 10 journals.

www.ingramcontent.com/pod-product-compliance
Lightning Source LLC
Chambersburg PA
CBHW061204220326
41597CB00015BA/1345